CANADIAN COPPER & BRASS DEVELOPMENT ASSOCIATION

65 Overlea Boulevard, Suite 210, Toronto, Ontario, Canada M4H 1P1

Telephone: (416) 391-5599 / 1-877-640-0946 Fax: (416) 391-3823

ALUMINUM RECYCLING AND PROCESSING FOR ENERGY CONSERVATION AND SUSTAINABILITY

JOHN A.S. GREEN
EDITOR

The Materials Information Society

ASM International®
Materials Park, Ohio 44073-0002
www.asminternational.org

First printing, December 2007

Comments, criticisms, and suggestions are invited, and should be forwarded to ASM International.

Prepared under the direction of the Aluminum Advisory Group (2006–2007), John Green, Chair.

ASM International staff who worked on this project include Scott Henry, Senior Manager of Product and Service Development; Steven R. Lampman , Editor; Ann Britton, Editorial Assistant; Bonnie Sanders, Manager of Production; Madrid Tramble, Senior Production Coordinator; Patty Conti, Production Coordinator; and Kathryn Muldoon, Production Assistant.

Library of Congress Control Number: 2007932444
ISBN-13: 978-0-87170-859-5
ISBN-10: 0-87170-859-0
SAN: 204-7586

ASM International®
Materials Park, OH 44073-0002
www.asminternational.org

Printed in the United States of America

Contents

Preface

ALUMINUM AND ENERGY—energy and aluminum—the two have been intimately linked since the industry started in 1886. That was when both Charles Martin Hall, in the United States, and Paul Heroult, in France, working independently, almost simultaneously discovered an economical process to produce aluminum from a fused salt using electrolysis. From this relatively recent discovery, the use of aluminum has grown rapidly and overtaken other older metals, such as copper, tin, and lead. It is now the second most widely used metal after steel.

Although the actual chemistry of the winning of aluminum from its oxide, alumina, has not changed greatly since 1886, the growth of the industry has brought about huge changes in production scale and sophistication. Also, there have been considerable reductions in the amount of energy used per unit of production. However, the production of aluminum is still energy-intensive, and the smelting process requires approximately 15 MWh per metric ton of aluminum production. For the United States, aluminum production consumes approximately 2% of the total industrial energy used.

For most of the 120 years of aluminum production, the growth of aluminum and energy production from hydroelectric sources were essentially symbiotic in nature. Aluminum production requires large quantities of stable and low-cost power, while hydroelectric projects need steady baseline users to ensure the viability of a hydroelectric project. Nowhere was this linkage between the aluminum industry and hydroelectric power producers better demonstrated than in the Pacific Northwest of the United States. There is now a concentration of both smelters and hydroelectric dams in the Columbia River basin. Although this mutually beneficial relationship was tested on occasion by market recession, drought, or lack of sufficient snowpack, the linkage persisted for several decades. It was not until 2000 and 2001 that the severe economic recession, coupled with the extreme energy crisis in California, caused the linkage between the industry and hydroelectric power producers to finally rupture. At this time, several aluminum smelters "mothballed" their operations, and power producers discontinued their supply arrangements with the aluminum smelters. About this same time, the importance of recycled secondary aluminum grew, and, in fact, in 2002, the percentage of recycled metal exceeded the primary smelted metal in the total U.S. metal supply for the first time.

Recycling of aluminum is vitally important to the sustainability of the aluminum industry. When the metal has been separated from its oxide in the smelting process, it can be remelted and recycled into new products numerous times, with only minimal metal losses each time. In fact, as the life-cycle and sustainability studies discussed in Chapters 3, 4, and 5 indicate, the recycling of aluminum saves ~95% of the energy used as compared to making the metal from the original bauxite ore. This enormous energy savings has accounted for the continuing growth of the secondary industry and has led to the concept that aluminum products can be considered as a sort of "energy bank." The energy embedded in aluminum at the time of smelting remains in an aluminum product at the end of its useful life and effectively can be recovered through the recycling process. Probably the best example of this is the ubiquitous aluminum beverage can that, on average, is recovered, recycled, and fabricated into new cans that are put back on the supermarket shelves in approximately 60 days!

With an increasing awareness of environmental and climate-change issues in the public arena, it is considered that the publication of this sourcebook will be most timely. The purpose of this book is to provide a comprehensive source for all aspects of the sustainability of the aluminum industry. It is anticipated that issues of sustainability will become increasingly important in the next couple of decades as individuals, companies, various agencies, governments, and societies in general strive to seek a responsible balance between using materials to maintain and improve living standards while not despoiling the planet of its limited mineral and material resources for future generations.

This publication is a collection of basic factual information on the modeling of material flow in the aluminum industry, the life-cycle materials and energy inputs, and the products, emissions, and wastes. The energy savings involved with recycling, various scrap-sorting technologies, and future energy-saving opportunities in aluminum processing are outlined. Finally, the positive impact of the growing use of lightweight aluminum in several segments of the transportation infrastructure and its benefit on greenhouse gas production is also highlighted. This book should provide much-needed basic information and data to reduce speculation and enable fundamental analysis of complex sustainability issues associated with the aluminum industry.

Regarding the specific contents of this book, Chapter 1 is a brief introduction to the concept of life-cycle analysis by Hans Portisch and coworkers. Portisch has pioneered in the field of life-cycle studies and has helped to establish many of the life-cycle protocols developed by the European Union and International Standardization Organization (ISO) for working groups. Chapter 1, entitled "Life-Cycle Engineering and Design," is an opportunity for the reader to become familiar with the concept of life-cycle analysis and its terminology that will be important in appreciating several of the subsequent chapters.

Chapter 2, entitled "Sustainability—The Materials Role," by Lyle Schwartz, is probably the real introduction to the complex subject of sustainability. This chapter was first presented by the author as the Distinguished Lecture in Materials and Society in 1998. The chapter sets out the case for sustainability and life-cycle analysis and is introductory in nature. The huge worldwide growth of the automobile is used to illustrate the enormity of the materials and sustainability issues facing the technical community and society in general. The chapter traces some recent history and proposes several paths for future direction, such as:

- Cleaner processing
- The development of alternative materials
- Dematerialization, or the use of less material per capita to accomplish the necessary material requirements
- Reuse and recycling

The latter, recycling, is of course one of the key attributes of aluminum. This chapter also contrasts other materials, such as magnesium, advanced steels, and polymer composites, with aluminum in the context of reducing the weight of automobiles to enhance fuel efficiency. The chapter ends with a call to action by the professional societies and the individual materials scientists. It is indeed a rallying call for materials responsibility!

The *Life-Cycle Inventory for the North American Aluminum Industry*, discussed in Chapter 3, represents the original (year of 1995) study of the industry and is probably still the most comprehensive. It has since become the basis for future studies by the International Aluminum Institute (IAI). The study was conducted in response to a request from Chrysler, Ford Motor Company, and General Motors under the United States Automotive Materials Partnership. This automotive materials partnership was enabled by the PNGV program established by the U.S. Government. Recently, the PNGV activities have transitioned to FreedomCAR and its emphasis has been expanded to include other light materials, e.g. Mg, Ti and composites, as well as aluminum. The purpose of this study was to provide the participating companies with detailed life-cycle inventories of the various processes within the aluminum product life cycle. This information provides a benchmark for improvements in the management of energy, raw material use, waste elimination, and the reduction of air and water emissions. Although this is titled a North American study, it was in fact global in reach due to the international operations of the 13 companies taking part. The study incorporated data from 15 separate unit processes located in 213 plants throughout North and South America, Africa, Australia, Europe, and the Caribbean. The results were tabulated by an independent contractor (Roy F. Weston, Inc.) and were peer reviewed by a distinguished panel of experts prior to publication in accord with ISO methodologies. One excellent feature of this chapter is the graphical presentation of the results. For example, for any particular process, such as aluminum extrusion or cold rolling, it is possible to see at a glance what the materials and energy inputs are and what are the products, air emissions, and wastes generated to that stage in the fabrication process.

Following the publication of the comprehensive life-cycle study discussed in Chapter 3, the leaders of the aluminum industry vested in the IAI, based in London, the responsibility of maintaining and extending the database to include significant areas of aluminum production that were not included in the initial study, namely Russia and China. Also, the IAI was requested to develop several global performance indicators and to track these indicators toward key sustainability goals agreed upon by the international industry. The global performance indicators chosen include such items as primary production; electrical energy used for production; emissions of greenhouse gases during electrolysis; specific emissions of perfluorocarbon gases, which are potent global-warming gases; consumption of fluoride materials; as well as injury rates and loss time severity rates. The considerable progress that the industry has made toward achieving many of these voluntary objectives is described in Chapter 5. The addendum report, updated to the end of 2005, illustrates quantitatively the industry's progress toward the 12 voluntary objectives. Significant progress has been achieved and documented.

The sixth chapter, entitled "Material Flow Modeling of Aluminum for Sustainability," by Kenneth Martchek of Alcoa, describes the development of a global materials flow model. Annual statistical data since 1950 from all the significant market segments have been combined with the most recent life-cycle information from the IAI to develop this global model. The model has demonstrated good agreement between estimated and reported worldwide primary production over the past three decades. Probably one of the most interesting features of the chapter is the table citing the worldwide collection rates and recycle rates for each market segment. The model also demonstrates that approximately 73% of all aluminum that has ever been produced is contained in products that are currently in service—surely a good testament to the recyclability and versatility of the metal!

Chapter 7 is devoted to a detailed discussion of the recycling of aluminum. As noted previously, recycling is a critical component of the sustainability of aluminum because of the considerable energy savings and the equivalent reduction in emissions from both energy and metal production. The chapter starts with a discussion of the recycling process and reviews the steps to remelt, purify the molten metal, and fabricate new products. The chapter also contains a discussion of the life-cycle trends in each major market area and how these factors impact recyclability. For example, one significant development that is discussed is the growing importance of automotive scrap. It is now estimated from modeling approaches that automotive scrap became more dominant than the traditional recycling of beverage containers at some stage during the 2005 to 2006 time period. This transition has occurred because of the marked increases of aluminum being used in automotives to enhance fuel efficiency, the fact that auto shredders are now commonly used, and shredder scrap can be economically sorted on an industrial scale. The transition has also occurred because the rates for the collection of can scrap have recently declined from the peak values of 1997, when ~67% of all cans were bought back by the industry, to the time of writing, when the recycling rates are hovering around 50%.

One dominant issue in the recycling of automotive scrap is the control of impurities, especially iron and silicon, that inevitably build up during the recycling process. Cast aluminum alloys, with their higher silicon content, are better able to tolerate this increase of impurity content than wrought alloys. Future trends, potential solutions, and research directions to resolve this issue of impurity control are outlined in this chapter. Also, the chapter mentions the potential impact of government regulations in European Union countries that now mandate that vehicles be 95% recyclable by the year 2015. Chapter 7 concludes with a brief discussion of the safety issues related to melting and casting aluminum.

Aluminum products are formed from an extremely wide array of alloys. These range from the soft alloys used in foil and packaging material, to the intermediate alloys used in the construction of boats and trains, to the hard alloys used in aircraft and aerospace applications. It is inevitable that some amount of all these alloys will end up in the products from the industrial shredder. Accordingly, to achieve the optimum recycling, it is most economical to identify and separate scrap and to reuse the specific alloying elements in the most advantageous manner. This is why the recent advances in scrap sorting by Adam Gesing and his coworkers at Huron Valley Steel Corporation are so significant. These developments are detailed in Chapter 8. This chapter is a comprehensive discussion of the complexities of automotive alloys and recycling issues. The chapter provides a state-of-the-art description of sorting technologies and demonstrates alloy sorting by color, x-ray absorption, and laser-induced breakdown spectroscopy technology. Color sorting of cast material from wrought alloys is now fully established on a commercial scale, and LIBS sorting has become commercially viable in the past couple of years.

The next chapter, Chapter 9, explores some of the emerging trends in municipal recycling from the perspective of the operation of a municipal recycling facility. More importantly, the chapter discusses at length the issue of impurities and alloy content and how best to assimilate the recycled material stream into the existing suite of aluminum alloys. At present, sorted material can contain a wide range of elemental content, and this can modify and impact the physical, chemical, and mechanical properties of recycled alloys. This chapter, contributed by Secat and the University of Kentucky with partial support of the Sloan Foundation, suggests several routes to optimize the economical and property benefits achieved through recycling. It also explores the development of aluminum alloys that are more tolerant of recycling content, that is, recycling friendly alloys.

The final chapter of the sourcebook, Chapter 10, was originally prepared by BCS, Inc. for the U.S. Department of Energy in Washington, D.C., in February 2003 but has since been updated with the latest available data as of early 2007. This chapter looks at the whole production system for aluminum, from the original bauxite ore, through refining of alumina and smelting of aluminum, to various rolling, extrusion, and casting technologies. From an historical perspective, the chapter explores the energy requirements for aluminum production. The theoretical energy limits for each process step are compared to the actual current industry practice, and new opportunities for saving energy are highlighted. The original report was commissioned as the baseline study of the industry by the Department of Energy and contains extensive discussions of potential advances in aluminum processing and fabrication. For example, the potential of wettable cathodes, inert anode technology, carbothermic reduction, and various melting and fabrication technologies are discussed at length. Finally, the chapter is most valuable because it is supported by numerous appendixes with almost 50 years of industry data and statistics. Much of the energy data used for the energy calculations evaluating competing technologies is drawn from the industry life-cycle outlined in Chapter 3.

It is hoped that this compilation of published material can be a contribution to the sustainability debate and, specifically, can help to increase the understanding about the sustainability and recyclability of aluminum. The availability of credible information can only help sustain rational debate and the development of optimal actions and policies for the future.

Much progress has been made in recent years, although a lot still remains to be achieved. At the time of writing, the *Baltimore Sun* newspaper (dated January 24, 2007), in an article entitled "Plane Trash," refers to a report by the National Resources Defense Council that says that the aviation industry is pitching enough aluminum cans each year to build 58 Boeing 747s! This is blamed on a lack of understanding and on a mishmash of conflicting regulations and procedures at various airports around the country. While many airports are in fact recycling much of their trash and thereby reducing operating costs and landfill fees, many airlines and airports are not doing so. Under the present conditions and with the potential gains of energy and environmental emissions that are available through recycling of beverage cans, this situation seems remarkably shortsighted, especially when all cans are collected before the termination of a flight! On the other hand, enormous progress has been made in recycling and sustainability. Especially, it is noteworthy that computer models now indicate that the aluminum industry will become "greenhouse gas neutral" by the year 2020. This is indicated by the fact that the potential savings in emissions of greenhouse gases from the transportation use of aluminum for lightweighting of vehicles and increased fuel efficiency is growing at a faster rate than the emissions from the production of the aluminum itself. For all of us with children and grandchildren, this is indeed a hopeful sign.

John Green, Ellicott City, MD
January 2007

CHAPTER 1

Life-Cycle Engineering and Design*

ENVIRONMENTAL CONSIDERATIONS play an increasingly important role in design and development efforts of many industries. "Cradle-to-grave" assessments are being used not only by product designers and manufacturers but also by product users (and environmentalists) to consider the relative merits of various available products and to improve the environmental acceptability of products.

Life-cycle engineering is a part-, system-, or process-related tool for the investigation of environmental parameters based on technical and economic measures. This chapter focuses on life-cycle engineering as a method for evaluating impacts, but it should be noted that other techniques also can be used to analyze the life-cycle costs of products (e.g., see the article "Techno-Economic Issues in Materials Selection" in *Materials Selection and Design,* Volume 20, *ASM Handbook, 1997.*

Products and services cause different environmental problems during the different stages of their life cycle. Improving the environmental performance of products may require that industry implement engineering, process, and material changes. However, a positive change in one environmental aspect of a product (such as recyclability) can influence other aspects negatively (such as energy usage). Therefore, a methodology is required to assess trade-offs incurred in making changes. This method is called life-cycle analysis or assessment (LCA).

Life-cycle analysis aims at identifying improvement possibilities of the environmental behavior of systems under consideration by designers and manufacturers. The whole life cycle of a system has to be considered. Therefore, it is necessary to systematically collect and interpret material and energy flows for all relevant main and auxiliary processes (Fig. 1.1).

Life-cycle analysis methods have been developed by governmental, industrial, academic, and environmental professionals in both North America and Europe. Technical documents on conducting LCA have been published by the Society of Environmental Toxicology and Chemistry (SETAC), the U.S. Environmental Protection Agency (EPA), the Canadian Standards Association (CSA), the Society for the Promotion of LCA Development (SPOLD), and various practitioners.

For meaningful comparisons of the life-cycle performance of competing and/or evolving product systems, it is important that associated LCAs be conducted consistently, using the same standards. Although the common methodologies developed by SETAC, EPA, CSA, and SPOLD are a step in that direction, a broad-based international standard is needed. Such an effort is being undertaken by ISO 14000 series (TC207).

Life-cycle thinking and techniques can be applied to products, processes, or systems in various ways: it can help assess life-cycle economic costs (LCA_{econ}), social costs (LCA_{soc}), or environmental costs (LCA_{env}).

A primary objective of LCA is to provide a total life-cycle "big-picture" view of the interactions of a human activity (manufacturing of a product) with the environment. Other major goals are to provide greater insight into the overall environmental consequences of industrial

---

*Adapted from an article by Hans H. Portisch, Krupp VDM Austria GmbH (Committee Chair), with contributions from Steven B. Young, Trent University; John L. Sullivan, Ford Motor Company; Matthias Harsch, Manfred Schuckert, and Peter Eyerer, IKP, University of Stuttgart; and Konrad Saur, PE Product Engineering, which was published in *Materials Selection and Design*, Volume 20, *ASM Handbook*, ASM International, 1997, p 96–104.

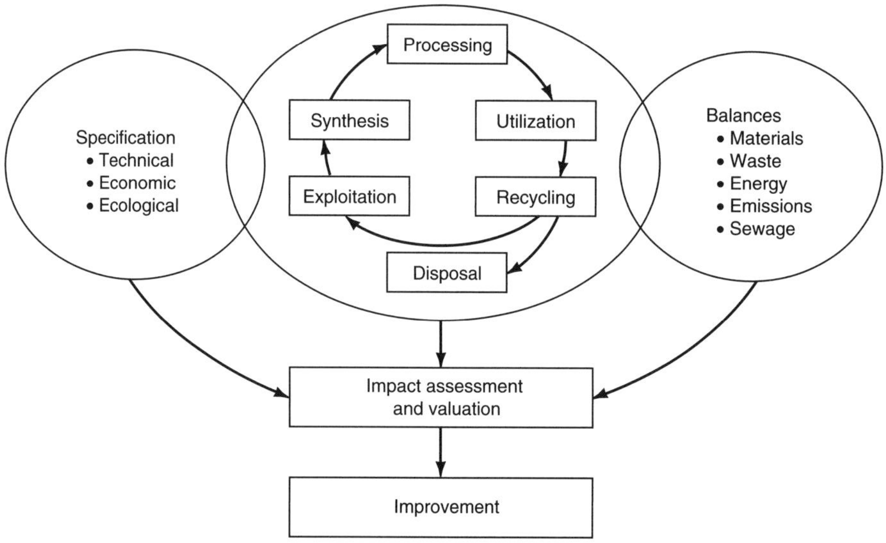

Fig. 1.1 Factors considered in the life-cycle engineering approach. Source: Ref 1.1

activities and to provide decision makers with a quantitative assessment of the environmental consequences of an activity. Such an assessment permits the identification of opportunities for environmental improvement.

Life-Cycle Analysis Process Steps

Life-cycle analysis is a four-step process; each of these steps is described in detail as follows. The process starts with a definition of the goal and scope of the project; because LCAs usually require extensive resources and time, this first step limits the study to a manageable and practical scope. In the following steps of the study, the environmental burdens (including both consumed energy and resources, as well as generated wastes) associated with a particular product or process are quantitatively inventoried, the environmental impacts of those burdens are assessed, and opportunities to reduce the impacts are identified.

All aspects of the life cycle of the product are considered, including raw-material extraction from the earth, product manufacture, use, recycling, and disposal. In practice, the four steps of an LCA are usually iterative (Fig. 1.2).

Step 1: Goal Definition and Scoping. In the goal definition and scoping stage, the purposes of a study are clearly defined. Subsequently, the scope of the study is developed, which defines the system and its boundaries, the assumptions, and the data requirements needed to satisfy the study purpose. For reasons of economy and brevity, the depth and breadth of the study is adjusted, as required, to address issues regarding the study purpose. Goal definition and project scope may need to be adjusted periodically throughout the course of a study, particularly as the model is refined and data are collected.

Also during this stage, the functional unit is defined. This is an important concept because it defines the performance of a product in measured practical units and acts as a basis for product system analysis and comparison to competing products. For example, the carrying

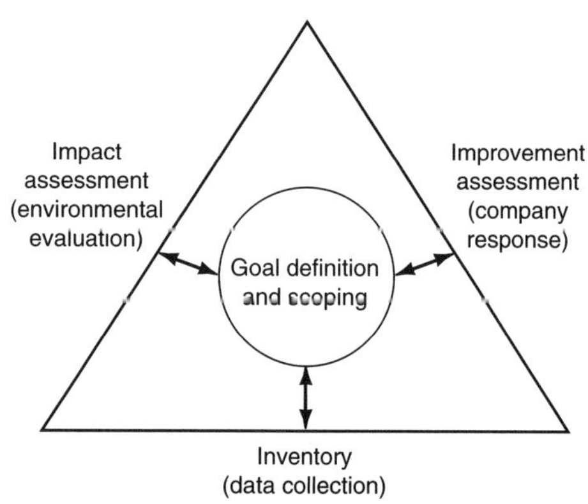

Fig. 1.2 The life-cycle assessment triangle. Source: Ref 1.2

capacity of a grocery bag may be a sensible functional unit.

Finally, the quality of the life-cycle data must be assessed in order to establish their accuracy and reliability. Typically, factors such as data age, content, accuracy, and variation need to be determined. Clearly, data quality affects the level of confidence in decisions that are based on study results.

Step 2: Inventory Analysis. The second stage of LCA is a life-cycle inventory (LCI). It is in this stage that the various inputs and outputs (energy, wastes, resources) are quantified for each phase of the life cycle. As depicted in Fig. 1.3, systems boundaries are defined in such a way that the various stages of the life cycle of a product can be identified. The separation of burdens (inputs and outputs) for each stage facilitates improvement analysis.

For the purposes of LCI, a "product" should be more correctly designated as a "product system." First, the system is represented by a flowchart that includes all required processes: extracting raw materials, forming them into the product, using the resulting product, and disposing of and/or recycling it. The flowchart is particularly helpful in identifying primary and ancillary materials (such as pallets and glues) that are required for the system. Also identified are the sources of energy, such as coal, oil, gas, or electricity. Feedstock energies, which are defined as carbonaceous materials not used as fuel, are also reported.

After system definition and materials and energy identification, data are collected and model calculations performed. The output of an LCI is typically presented in the form of an inventory table (an example is shown in Table 1.1), accompanied by statements regarding the effects of data variability, uncertainty, and gaps. Allocation procedures pertaining to co-product generation, re-cycling, and waste treatment processes are clearly explained.

Step 3: Impact Assessment and Interpretation. Impact assessment is a process by which the environmental burdens identified in the inventory stage of an LCA are quantitatively or qualitatively characterized as to their effects on local and global environments. More specifically, the magnitude of the effects on ecological and human health and on resource reserves is determined.

Life-cycle impact assessment is, at this time, still in an early phase of development. Although some impact assessment methods have been advanced as either complete or partial approaches, none has been agreed upon. Nevertheless, an approach to impact analysis, known as "less is better," is typically practiced. With this approach, process and product changes are sought that reduce most, if not all, generated wastes and emissions and consumed resources. However, situations in which such reductions are realized are not yet typical. Usually, a change in product systems is accompanied by trade-offs between burdens, such as more greenhouse gases for fewer toxins. A fully developed impact analysis methodology would help in the environmental impact assessment of such cases.

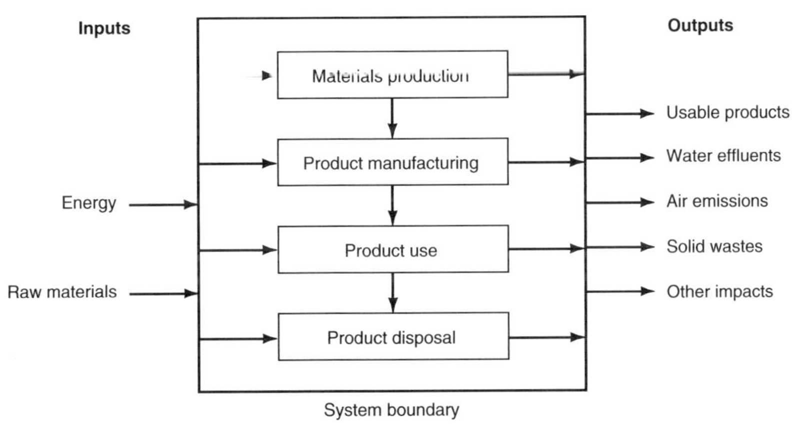

Fig. 1.3 Generalized system boundaries for a life-cycle inventory of a generic product. Source: Ref 1.2

Table 1.1 Example of a life-cycle inventory for an unspecified product

Substance	Amount	Substance	Amount
Inputs		**Outputs**	
Energy from		Air emissions, mg	
fuels, MJ		Dust	2000
Coal	2.75	Carbon	
Oil	3.07	monoxide	800
Gas	11.53	Carbon dioxide	11×10^5
Hydro	0.46	Sulfur oxides	7000
Nuclear	1.53	Nitrogen oxides	11,000
Other	0.14	Hydrogen	
Total	**19.48**	chloride	60
		Hydrogen	
Energy from		fluoride	1
feedstocks, MJ		Hydrocarbons	21,000
Coal	<0.01	Aldehydes	5
Oil	32.75	Other organics	5
Gas	33.59	Metals	1
Other	<0.01	Hydrogen	1
Total feedstock	**66.35**	Solid wastes, mg	
Total energy		Mineral waste	3100
input, MJ	**85.83**	Industrial waste	22,000
		Slags and ash	7000
Raw materials, mg		Toxic chemicals	70
Iron ore	200	Nontoxic	
Limestone	150	chemicals	2000
Water	18×10^6	Water effluents, mg	
Bauxite	300	COD	1000
Sodium chloride	7000	BOD	150
Clay	20	Acid, as H^+	75
Ferromanganese	<1	Nitrates	5
		Metals	300
		Ammonium ions	5
		Chloride ions	120
		Dissolved	
		organics	20
		Suspended solids	400
		Oil	100
		Hydrocarbons	100
		Phenol	1
		Dissolved solids	400
		Phosphate	5
		Other nitrogen	10
		Sulfate ions	10

COD, chemical oxygen demand; BOD, bacteriological oxygen demand. Source: Ref 1.2

As advanced by SETAC, impact analysis comprises three stages:

- *Classification:* In this stage, LCI burdens are placed into the categories of ecological health, human health, and resource depletion. Within each of these categories, the burdens are further partitioned into subcategories, for example, greenhouse gases, acid rain precursors, and toxins of various kinds. Some burdens may fall into several categories, such as sulfur dioxide, which contributes to acid rain, eutrophication, and respiratory-system effects. Environmental burdens are sometimes called stressors, which are defined as any biological, chemical, or physical entity that causes an impact.
- *Characterization:* In the characterization step of impact assessment, the potential impacts within each subcategory are estimated. Approaches to assessing impacts include relating loadings to environmental standards, modeling exposures and effects of the burdens on a site-specific basis, and developing equivalency factors for burdens within an impact subcategory. For example, all gases within the global-warming category can be equated to carbon dioxide, so that a total aggregate "global-warming potential" can be computed.
- *Valuation:* In the valuation step of impact assessment, impacts are weighted and compared to one another. It should be noted that valuation is a highly subjective process with no scientific basis. Further, attaching weighting factors to various potential impacts for comparison purposes is intrinsically difficult. For example, what is more important: the risk of cancer or the depletion of oil reserves? Who would decide this? Because a consensus on the relative importance of different impacts is anticipated to be contentious, a widely accepted valuation methodology is not expected to be adopted in the foreseeable future, if ever.

It is important to recognize that an LCA impact assessment does not measure actual impacts. Rather, an impact in LCA is generally considered to be "a reasonable anticipation of an effect," or an impact potential. The reason for using impact potentials is that it is typically difficult to measure directly an effect resulting from the burdens of a particular product. For example, are the carbon dioxide emissions of any individual's vehicle specifically causing the world to get warmer? It is unlikely that this could ever be shown, although it is reasonable to assume that any individual vehicle contributes its share to the possible effect of global warming caused by human-generated carbon dioxide in proportion to the amount of emissions.

Inventory Interpretation. It is argued by some that, due to the difficulties cited previously, the notion of impact assessment should be dropped and replaced by *inventory interpretation*. Classification and characterization could still be used, but all suggestion that environmental effects are assessed is avoided. In comparative assessments, "less is better" is the principle in identifying the environmentally preferable alternative.

Step 4: Improvement Analysis. This step involves identifying chances for environmental improvement and preparing recommendations. Life-cycle assessment improvement analysis is an activity of product-focused pollution prevention and resource conservation. Opportunities for improvement arise throughout an LCA study. Improvement analysis is often associated with design for the environment or total quality management. With both of these methodologies, improvement proposals are combined with environmental cost and other performance factors in an appropriate decision framework.

Application of Life-Cycle Analysis Results

The results of an LCA can be used by a company internally, to identify improvements in the environmental performance of a product system; and externally, to communicate with regulators, legislators, and the public regarding the environmental performance of a product. For external communications, a rigorous peer-review process is usually required. Virtually all of the peer-reviewed studies conducted to date represent analyses of simple product systems. However, studies for systems as complicated as automobiles are being conducted.

Whether used qualitatively or quantitatively, LCAs often lead to products with improved environmental performance. In fact, an often-over-looked, important qualitative aspect of LCA is that it engenders a sense of environmental responsibility. Beyond this development within manufacturers, LCA has the potential to become a tool to regulate products, or perhaps even for "eco-labeling." However, such uses are contentious and are expected to remain so.

The bulk of LCA efforts to date have been focused on preparing LCIs, with the impact assessment stage currently seen as the weakest link in the process. Indeed, some companies have even decided to skip this phase of the process altogether, opting to carry out a brief life-cycle review before moving straight on to the improvement stage.

Large or small companies and other users will find LCA of value at a number of different levels. Indeed, groups such as SETAC and SPOLD now see LCA playing a key role in three main areas:

- *Conceptually:* As a framework for thinking about the options for the design, operation, and improvement of products and systems
- *Methodologically:* As a set of standards and procedures for the assembly of quantitative inventories of environmental releases or burdens—and for assessing their impacts
- *Managerially:* With inventories and—where available—impact assessments serving as a platform on which priorities for improvement can be set

Not surprisingly, perhaps, the bulk of current LCA efforts is devoted to the second of these areas, particularly initiatives such as the 1993 Code of Practice by SETAC (Ref 1.3). However, the scope of LCA is rapidly spreading to embrace the other two application areas. The "supplier challenges" developed by companies such as Scott Paper, which has incorporated environmental performance standards in its supplier selection process, underscore the very real implications of the managerial phase for suppliers with poor environmental performances. Also, the "integrated substance chain management" approach developed by McKinsey & Company Inc. (Denmark) for VNCI (Association of the Dutch Chemical Industry), covering three chlorine-base products, shows that LCA can produce some fairly pragmatic tools for decision making.

Longer term, the prospects for LCA are exciting. Within a few years, product designers worldwide may be working with "laptop LCAs"—small, powerful systems networked with larger databases and able to steer users rapidly around the issues related to particular materials, products, or systems. This process would be greatly aided by a widely accepted, commonly understood environmental accounting language.

In the meantime, however, LCA is still quite far from being simple or user-friendly, as is illustrated in the following example.

Example: Life-Cycle Analysis of a Pencil. Anyone who has had even a brief encounter with an LCA project will have seen flow charts rather similar to the one in Fig. 1.4, which shows the key life-cycle stages for one of the simplest industrial products, a pencil. Most such diagrams are much more complicated, but, as is evident in the figure, even the humble pencil throws an extraordinarily complex environmental shadow.

For example, imagine the flow chart in Fig.1.4 is on the pencil maker's PC screen as the computer menu for an electronic information

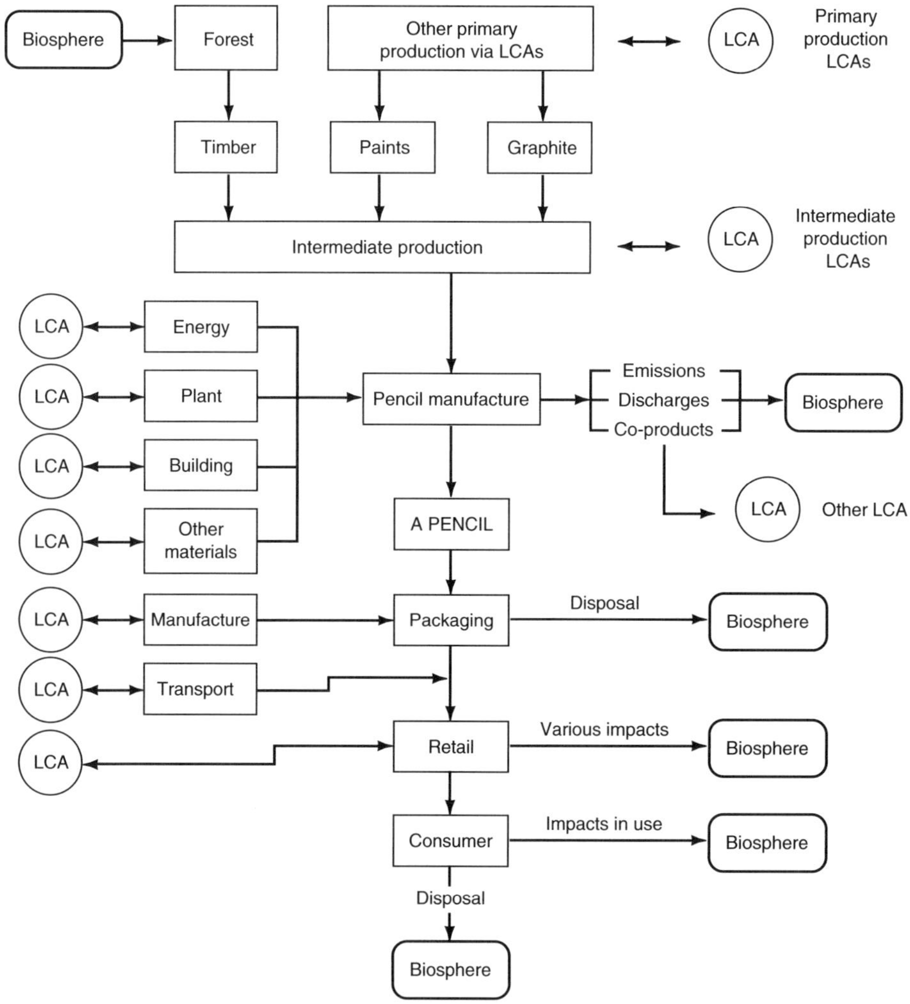

Fig. 1.4 Simplified life-cycle analysis (LCA) process for a pencil. Source: Ref 1.4

system. When the pencil maker clicks on "Timber," a wealth of data begins to emerge that makes one realize things are not as simple as may have been imagined. Not only is there a potential problem with tropical timber because of the rain forest issue, but the pencil maker now notes that suppliers in the U.S. Pacific Northwest have a problem with the conflict between logging operations and the habitat of the Northern Spotted Owl.

At this point, a pencil maker recognizes the need to examine the LCAs produced by the companies supplying timber, paints, and graphite. Working down the flowchart, the pencil maker sees a total of ten points at which other LCA data should be accessed. This is where complex business life gets seriously complicated. At the same time, however, LCA projects can also be fascinating, fun, and a potential gold mine of new business ideas.

Different Approaches to LCA. As Fig. 1.5 indicates, the LCA practitioner can look at the life cycle of a product through a number of lenses, focusing down of life-cycle costs or focusing out to the broader sociocultural effects. One example is the Eco-Labeling Scheme (Fig. 1.6 administered by the European Commission Directorate General XI (Environment, Nuclear Safety, and Civil Protection). This scheme is committed to assessing environmental impacts from cradle to grave.

The sheer variety of data needs, and of data sources, makes it very important for LCA producers and users to keep up to date with the

	Raw material acquisition	Bulk processing	Engineered materials processing	Assembly and manufacture	Use and service	Retirement	Treatment and disposal
1. Cost							
2. Value/ performance							
3. Environment: compliance							
4. Environment: beyond compliance							
5. Socio-economic							
6. Socio-cultural							

Fig. 1.5 Matrix showing some possible different approaches to LCA. Source: Ref 1.4

	Product life cycle				
ENVIRONMENTAL FIELDS	Preproduction	Production	Distribution	Utilization	Disposal
Waste relevance					
Soil pollution and degradation					
Water contamination					
Air contamination					
Noise					
Consumption of energy					
Consumption of natural resources					
Effects on ecosystems					

Fig. 1.6 The European Community eco-labeling scheme "indicative assessment matrix." Source: Ref. 1.4

debate and build contacts with other practitioners. Among the biggest problems facing the LCA community today are those associated with the availability of up-to-date data and the transparency of the processes used to generate such data.

Most LCA applications, however, focus—and will continue to focus—on single products and on the continuous improvement of their environmental performance. Often, too, significant improvements will be made after a relatively simple cradle-to-grave, or perhaps cradle-to-gate, analysis.

A detergent company, for example, may find that most of the energy consumption associated with a detergent relates to its use, not its manufacture. So, instead of just investing in a search for ingredients that require less energy to make, the company may decide to develop a detergent product that gives the same performance at lower wash temperatures.

In short, LCA is not simply a method for calculation but, potentially, a completely new framework for business thinking.

Case History: LCA of an Automobile Fender

A detailed LCA for an automotive fender as performed by IKP (University of Stuttgart, Germany) and PE Product Engineering

(Dettingen/Teck, Germany) is included to illustrate the present status and limitations of this methodology (Ref 1.1).

Goal and Scope. The specific goal of this investigation was to compare four different fender designs for an average compact class automobile in Germany. The comparison should result in the identification of the best material in terms of resource use, impact on global climate, and recyclability.

The four options were steel sheet; primary aluminum sheet; an injection-molded polymer blend of polyphenylene oxide and nylon (PPO/PA); and sheet molding compound (SMC), a glass-fiber-reinforced unsaturated polyester resin. The mechanical requirements for the four fenders were identical; this ensures that the functional unit is well defined and that they are equivalent. Table 1.2 shows the materials and weights of the four different fender designs.

Data Origin and Collection. *Data* in this context means all pieces of information that may be relevant for the calculation of processes and materials. Such information includes material and energy flows of processes, process descriptions, materials and tools, suppliers, local energy supply, local energy production, production and use of secondary energy carriers (e.g., pressurized air, steam), and location of plants. Which processes are the most relevant and must be considered in more detail depends on the goal and scope of the study. Within this study, the following information (supplier specific, if possible) had to be identified, collected, and examined:

- Production processes, with all links in the process chain
- Primary data concerning energy and material flow with respect to use of energy carriers (renewable and nonrenewable), use of mineral resources (renewable and nonrenewable), emissions into the air, waterborne emissions, and waste and production residues
- Coupled and by-products as well as entries from other process steps (internal loops)
- Transportation needs with respect to distance, mode, and average utilization rate
- Primary energy carriers and their means of production and distribution
- Secondary energy carriers and their means of production and distribution
- Air and water treatment measures and disposal of residues

Table 1.2 Material and weight of the different fender designs

Material	Thickness mm	Thickness in.	Weight kg	Weight lb
Steel	0.7	0.0275	5.60	12.35
Sheet molding compound	2.5	0.10	4.97	11.00
Polyphenylene oxide/polyamide	3.2	0.125	3.35	7.40
Aluminum	1.1	0.043	2.80	6.20

Source: Ref 1.1

Data collection is not a linear process. Good data collection and evaluation requires iteration steps for identifying relevant flows or additional information, and experience is needed to interpret the collected data. Calculation of modules should be carried out with special regard to the method of data collection (e.g., measured, calculated, or estimated) and the complexity of the system.

Materials production is an important factor. The consideration of aluminum shows that not only the main production chain has to be considered but also the process steps for alumina production (Fig. 1.7). The steps in electrolysis must be calculated, and the energy use connected with caustic soda and the anode coke has to be

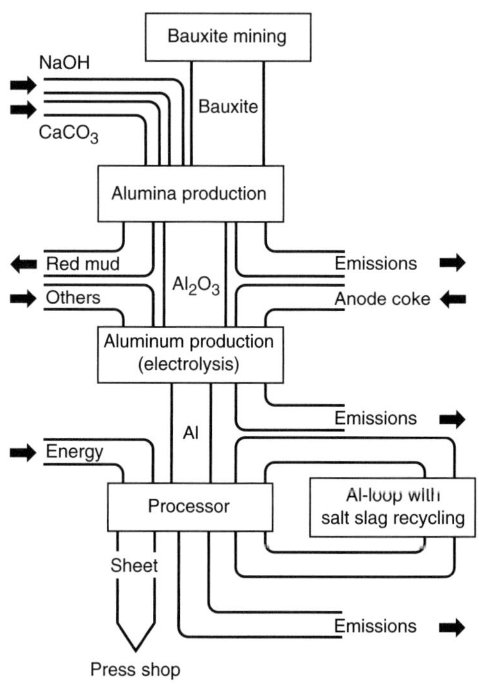

Fig. 1.7 Main material flow for the production of aluminum sheet parts. Source: Ref 1.1

examined. The four steps shown in Fig. 1.7, which must be considered along with a long list of others, demonstrate the difficulty of balancing costs and environmental impacts.

In electrolysis, the source of electric power is important because of the differences in carbon dioxide emissions between plants that are water-power driven and those that burn fossil fuels. Another significant factor is how electrolysis is controlled. Modern plants use technologies that prevent most of the anode effects responsible for the production of fluorocarbon gases, but many older plants emit four or five times as much. This shows the importance of calculating on a site-specific or at least on a country-specific basis.

Because aluminum is globally merchandised, the user frequently does not know the exact source of the metal. The solution to this problem is to calculate the average aluminum import mix. However, this calculation requires detailed information about the different ways aluminum is produced all over the world.

Material weight must also be considered. In selection of automotive parts, the usage phase is of great interest. The main environmental factor during this phase is weight difference. Each part contributes to the energy demand for operating an automobile. The share a fender contributes depends only on its mass. However, no data are available for the same car carrying different fenders. Therefore, this study calculated the fuel consumption assuming a steel fender, because average fuel consumption is known for the complete car with the traditional fender. In the same way, possible weight savings are known. Measurements and judgments from all automobile producers show that the assumptions for fuel reduction from weight savings vary within a range of 2.5 to 6% fuel reduction per 10% weight savings. For this study, 4.5% was assumed to be an average value for the kind of cars considered.

Recycling of the SMC fender shows another weight-related issue. After the useful life of the product, a decision has to be made about whether the part should be dismantled for recycling or otherwise disposed of. Within this study, the recycling solution was considered because the SMC part can be dismantled easily and ground into granules. Furthermore, SMC can replace virgin material as reinforcement, and granules can be used as filler up to 30%. In addition to the possibility of using recycled material in new parts, the SMC recycling process offers another advantage because the reformulated material has a lower density than the primary material. This means that the use of recycled SMC leads to further weight savings of approximately 8%, while fulfilling the same technical requirements. This example shows that recycling is not only useful for the purpose of resource conservation but many provide other benefits as well. However, successful recycling requires more than technical feasibility—it is highly dependent on viable economics.

Inventory Results. The discussion of the whole inventory process is not possible here, because it includes up to 30 resource parameters, approximately 80 different emissions into the air, more than 60 water effluents, and many different types of waste. Therefore, this example concentrates on energy demand, selected airborne emissions, and resource use (recyclability).

Energy use is one of the main parameters to consider when selecting automotive parts. It is a reliable basis for judgment because energy use generates waste and emissions, and it requires depletion of resources. Figure 1.8 shows the energy demand for the different fender materials over two complete usage phases, including production out of raw material and recycling for the second application.

The values at the zero kilometer line represent the energy needed for both material and part production. It is easy to see that aluminum has the highest energy demand of all four materials. This comes mainly from the electrolysis process and the alumina production process.

SMC has the lowest energy demand, needing approximately one-third of the energy required for the aluminum fender. This is due to the fact that SMC is a highly filled material in which the extender is a heavy, relatively inexpensive material. Second best is steel, which requires only a little more energy than SMC. Somewhere in the middle is the PPO/PA blend; the reason for the relatively high energy demand is the feedstock energy of the materials used in polymer production.

The ascending gradients represent the differences arising from the weights of the fenders. The larger the gradient, the higher the weight. It is easy to see that steel, as the heaviest material, loses a lot of its advantage from the production phase. This points out the importance of lightweight designs. The energy demand for the usage phase is approximately four times higher than that required for part production. As a result, the most significant improvements can be made in the usage phase.

Nevertheless, SMC still has the lowest energy demand after the first usage phase, and aluminum is still the worst.

After the first life cycle of the fender, it is recycled into a new part. The energy needed for recycling of SMC, steel, and aluminum is relatively low; PPO/PA requires much more energy for recycling. The disadvantage of PPO/PA is that although recycling is possible and very energy efficient, the production of the 70% virgin material required in the part is very energy intensive.

The second utilization phase shows the same results as the first. In the final analysis, steel turns out to be the most energy-intensive material, followed by the PPO/PA blend. While steel has the disadvantage of its weight, the polymer blend has disadvantages concerning recyclability for external body parts. The situation would be totally different if more material could be recycled, or if the polymer blend could be used more extensively in heavier cars with a longer usage phase. The weight advantage is especially high for aluminum. However, SMC turns out to be the most energy-efficient material over-all.

Emissions of carbon dioxide, nitrogen oxides, sulfur dioxide, and fluorocarbons were estimated for each material because of their effects

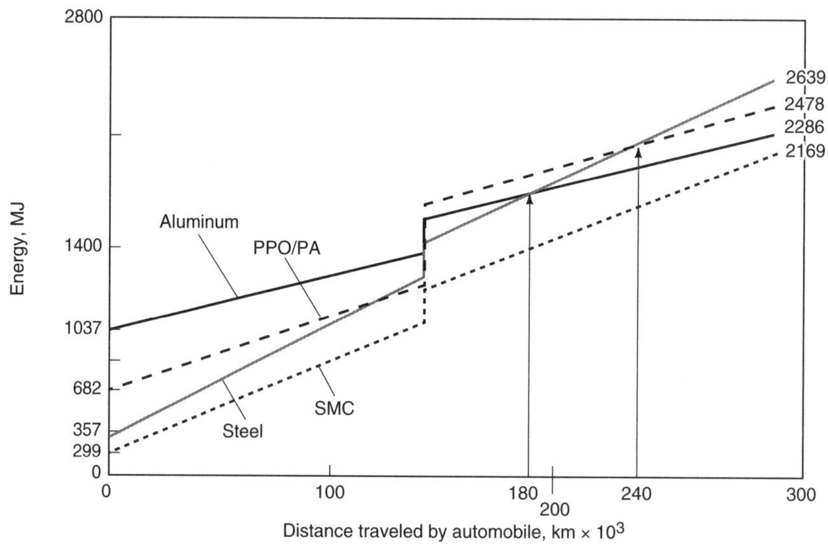

Fig. 1.8 Energy consumption for the production, use, recycling, and reuse of different fender materials considering the distance traveled by the automobile. PPO/PA, polyphenylene oxide and nylon; SMC, sheet molding compound. Source: Ref 1.1

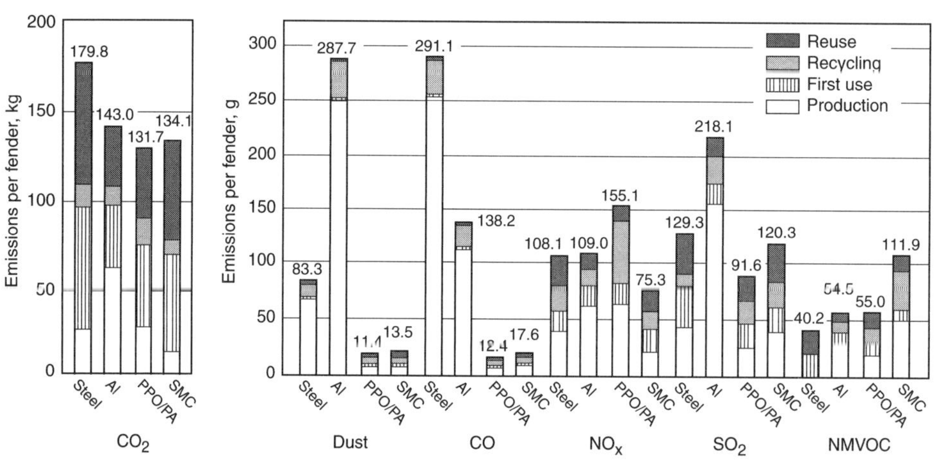

Fig. 1.9 Selected airborne emissions for the production, use, recycling, and reuse of different fender materials. NMVOC, nonmethane volatile organic compound; PPO/PA, polyphenylene oxide and nylon; SMC, sheet molding compound. Source: Ref 1.1

on ozone depletion and global warming (Fig. 1.9). These pollutants were also chosen because they are generated by nearly every manufacturing process, all over the world.

As mentioned before, a high percentage of atmospheric emissions is caused by energy generation. In the case of polymers, emissions are lower than expected because so much energy is stored as material feedstock. Aluminum is the material with the highest energy demand, but emissions are comparatively low because water power is used for a high percentage of aluminum electrolysis. The highest levels of carbon dioxide are emitted during steel production, mainly from the ore reduction process. Carbon dioxide emissions for the production of both polymers are dominated by hydrocarbon processing and refining.

For aluminum, most emissions come from earlier process steps. Alumina is produced mainly in bauxite mining countries, where the least expensive locally available energy is typically generated by burning heavy fuel and coal. Carbon dioxide emissions from aluminum production are dominated by this source, plus the electric power demand of those electrolysis processes that are not based on water power.

Carbon dioxide emissions during usage are directly related to fuel consumption: heavier fenders result in the generation of more carbon dioxide. This is also true for all other emissions considered here. One important approach for a possible improvement is certainly to reduce this main impact on global warming.

Impact assessment is a special step within the framework of LCA. Based on the results of the inventory, conclusions can be drawn, and judgments and valuations are possible. The impact assessment supplies additional information that enables the practitioner to interpret the results from the inventory.

Impact assessment also should allow the practitioner to draw the right conclusions concerning improvement approaches. However, it should be noted that consideration of environmental effects as a consequence of environmental releases is additional information that is not covered by the inventory step. This case history provides only a brief overview.

Impact assessment involves three steps. First is the definition of "environmental problems" or "themes." The problems to be addressed are defined in the scope of the project. Second, emissions are grouped to show their specific contribution to the environmental themes. Third, their shares are calculated. A standard list covers the following themes, which are more or less identical with most of the approaches taken in LCA literature:

- *Global criteria*: Resource use (energy carriers and mineral resources, both renewable and nonrenewable, and water and land use), global warming, ozone depletion, and release of persistent toxic substances
- *Regional criteria*: Acidification and landfill demand
- *Local criteria:* Spread of toxic substances, eutrophication, and formation of photochemicals
- Others: Noise, odor, vibration, and so on

In most of the studies conducted by IKP and PE Engineering, resource use and the global climate problems are considered. The methodology for their consideration is broadly accepted. Sometimes, acidification or eutrophication is considered as well. All others are more difficult to handle, and appropriate methods are still under discussion.

For the fender example, the contribution to the global-warming problem is calculated by taking into account production, use, recycling, and second use of each material (Fig. 1.10). The results are mainly influenced by carbon dioxide emissions and energy use and show that lightweight materials have advantages during utilization. However, aluminum is far worse than the others during production because electrolysis is accompanied by fluorocarbon emissions (CF_4 and C_2F_6), which have a very high global-warming potential.

Valuation. The second step in the judgment of the environmental impacts is the valuation step. This step may be divided into the normalization process and the final weighing.

Normalization involves scaling absolute contributions to single environmental themes on the same level, because absolute numbers may vary within six to ten decades. The effect scores are normalized with the amount of the annual global effect score or the contribution of one process to the theme per year, and so on.

Final weighing involves a personal judgment about the importance of each environmental theme, and the effect of each score on overall impact. This final step is part of the decision-making process. Scientists create tools for this process and help decision makers use and

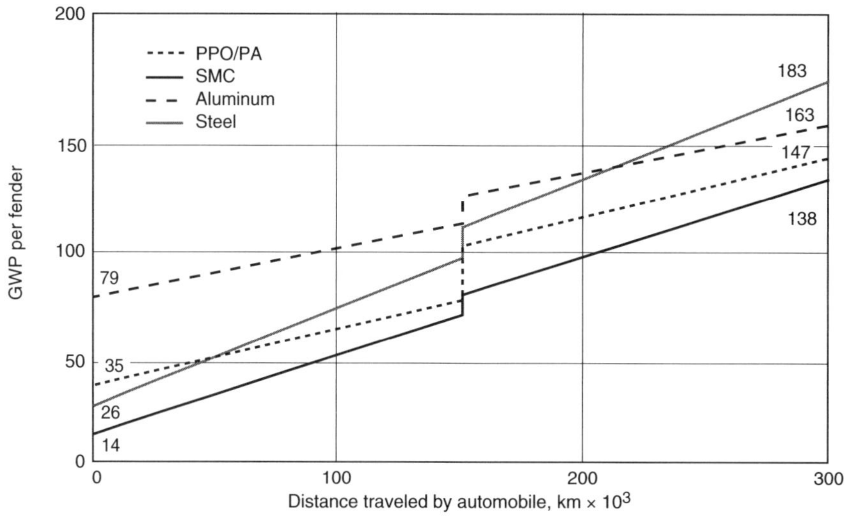

Fig. 1.10 Calculated contribution to global warming for the production, use, recycling, and reuse of different fender materials considering the distance traveled by the automobile. GWP, greenhouse warming potential (CO_2 equivalents); PPO/PA, polyphenylene oxide and nylon; SMC, sheet molding compound. Source: Ref 1.1

understand them, but the final decisions depend on company policies, not scientific or consultancy work.

Improvement Options. From this study, the following conclusions for improvement can be drawn:

- The usage phase is dominated by fuel consumption and the resulting carbon dioxide emissions. For other emissions, the production phase and recycling is also of great importance.
- Reducing part weight may improve energy use and reduce the contribution to global warming. However, reducing part weight may require higher environmental investments during production or recycling. In some cases, these investments are very useful.
- Recycling is more important for expensive and energy-intensive materials.

Experience gained from the evaluation of fender materials shows that the following general conclusions can be made:

- The fuel production has great impact and is not well known today.
- The best basis for decision making is a supplier-specific LCA.
- Close cooperation between producers and suppliers is necessary to find processes that will reduce environmental impacts.

Conclusions

Life-cycle engineering—in particular, LCA—is gaining importance for design and materials engineers because environmental considerations are increasingly important factors in design and materials selection. The creation and development of environmental management systems, including extended producer responsibility and product stewardship responsibility, pollution prevention strategies, "green" procurement guidelines, and eco-labeling programs, are evidence of the growing importance of life-cycle concerns.

To make a proper assessment, the total life cycle of a material, all forms of energy use, waste production, reuse, and recycling have to be considered. Many of these factors are site specific, which complicates calculations and comparisons. While LCAs for simple products have been performed, more complicated systems are only now being tackled.

Many industry trade organizations have developed or are in the process of developing LCI databases for their products. The Association of Plastics Manufacturers in Europe (APME), the European Aluminum Association (EAA), Finnboard, and the International Iron and Steel Institute are just a few examples. This publication addresses efforts with respect to aluminum and energy conservation.

A wide variety of reports and software packages containing inventory data are available.

Table 1.3 Life-cycle assessment and life-cycle inventory software tools

Tool	Vendor	URL
BEES 3.0	NIST Building and Fire Research Laboratory	http://www.bfrl.nist.gov/oae/software/bees.html
Boustead Model 5.0	Boustead Consulting	http://www.boustead-consulting.co.uk/products.htm
CMLCA 4.2	Centre of Environmental Science	http://www.leidenuniv.nl/cml/ssp/software/cmlca/index.html
Dubo-Calc	Netherlands Ministry of Transport, Public Works and Water Management	http://www.rws.nl/rws/bwd/home/www/cgi-bin/index.cgi?site=1&doc=1785
Ecoinvent 1.2	Swiss Centre for Life Cycle Inventories	http://www.ecoinvent.ch
Eco-Quantum	IVAM	http://www.ivam.uva.nl/uk/producten/product7.htm
EDIP PC-Tool	Danish LCA Center	http://www.lca-center.dk
eiolca.net	Carnegie Mellon University	http://www.eiolca.net
Environmental Impact Indicator	ATHENA Sustainable Materials Institute	http://www.athenaSMI.ca
EPS 2000 Design System	Assess Ecostrategy Scandinavia AB	http://www.assess.se/
GaBi 4	PE Europe GmbH and IKP University of Stuttgart	http://www.gabi-software.com/software.html
GEMIS	Öko-Institut	http://www.oeko.de/service/gemis/en/index.htm
GREET 1.7	DoE's Office of Transportation	http://www.transportation.anl.gov/software/GREET/index.html
IDEMAT 2005	Delft University of Technology	http://www.io.tudelft.nl/research/dfs/idemat/index.htm
KCL-ECO 4.0	KCL	http://wwwl.kcl.fi/eco/softw.html
LCAIT 4.1	CIT Ekologik	http://www.lcait.com/01_1.html
LCAPIX vl.1	KM Limited	http://www.kmlmtd.com/pas/index.html
MIET 3.0	Centre of Environmental Science	http://www.leidenuniv.nl/cml/ssp/software/miet/index.html
REGIS	Sinum AG	http://www.sinum.com/htdocs/e_software_regis.shtml
SimaPro 6.0	PRé Consultants	http://www.pre.nl/simapro.html
SPINE@CPM	Chalmers	http://www.globalspine.com
SPOLD	The Society for Promotion of Life-Cycle Assessment	http://lca-net.com/spold/
TEAM 4.0	Ecobalance	http://www.ecobalance.com/uk_lcatool.php
Umberto	ifu Hamburg GmbH	http://www.ifu.com/en/products/umberto
US LCI Data	National Renewable Energy Lab	http://www.nrel.gov/lci

Source: Ref 1.5

Examples are given in Table 1.3 (Ref 1.5). In addition, a large number of national and international database projects exist. A comprehensive listing can be found in Ref 1.6. Also, Ref 1.4, while concentrating on Europe, gives an excellent overview and many useful examples and addresses.

Steps of a complete LCA are being standardized with respect to methods and data. Simplification and standardization will lead to more reliable, timely, and cost-effective LCIs. An operational guide for ISO standards is given in Ref 1.7. When consensus about an acceptable impact assessment methodology is reached, LCA for simple and then more complex units and systems will be possible.

ACKNOWLEDGMENTS

Portions of this article were adapted from Ref 1.1 and 1.2. The authors wish to thank Sustainability Ltd. (United Kingdom) and the Secretariat of SPOLD (Belgium) for allowing the use of some of their information.

REFERENCES

1.1. M. Harsch et al., Life-Cycle Assessment, *Adv. Mater. Proc.*, June 1996, p 43–46
1.2. J.L. Sullivan and S.B. Young, Life Cycle Analysis/Assessment, *Adv. Mater. Proc.*, Feb 1995, p 37–40
1.3. "Guidelines for Life Cycle Assessment: A Code of Practice," Society of Environmental Toxicology and Chemistry (SETAC), Europe (Brussels), 1993
1.4. *The LCA Sourcebook*, Sustainability Ltd., London, 1993
1.5. "Life Cycle Assessment: Principles and Practice," EPA/600/R-06/060, Scientific Applications International Corporation (SAIC), May 2006
1.6. "Directory of Life Cycle Inventory Data Sources," Society for the Promotion of LCA Development (SPOLD), Brussels, Nov 1995
1.7. J.B. Guinée, *Handbook on Life Cycle Assessment: Operational Guide to the ISO Standards,* Kluwer Academic Publishers, 2002

SELECTED REFERENCES

- "Life Cycle Assessment: Inventory Guidelines and Principles," U.S. Environmental Protection Agency (EPA), Office of Research and Development, Cincinnati, OH, 1993
- "Life Cycle Design Manual: Environmental Requirements and the Product System," University of Michigan, 1993
- "A Technical Framework for Life Cycle Assessment," SETAC USA, Washington, D.C., 1991

CHAPTER 2

Sustainability—The Materials Role*

HARVEY BROOKS (Ref 2.1) focused on the links between materials, energy, and the environment. At that time, the issue of sustainability had emerged but appeared first as an exploration of the possibility of materials depletion in the face of predicted population growth. This theme of sustainability is made more relevant by the current debate about greenhouse gases and the global climate, but one which is central to the materials industries and the users of their products. One cannot do justice to such a broad topic in this brief overview, but several themes do establish the basis for several recommendations. There are roles to be played by industry, by academia, by government, and by each of us individually and collectively through our societies. With some passing remarks about the other arenas, the aim of the recommendations is for government and professional societies, based on current knowledge.

Our economy and that of the other nations in the world depends on materials to an extent that most nontechnical persons do not realize. Materials are, after all, "the stuff that things are made of." At issue today, is can the current level of use of this "stuff" maintain? If, as expected, both economic activity and population will grow significantly worldwide in the next half-century, the current per capita use of materials will certainly be unsustainable. Indeed, the Organization for Economic Cooperation and Development (OECD) recently adopted a long-range goal that industrial countries should decrease their materials intensities by a factor of 10 over the next four decades. That would be equivalent to using only 66 lb of materials per $100 gross domestic product (GDP) compared to the present value of approximately 660 lb per $100 GDP. If achievable at all, much of these savings would come from the construction and mining industries, the largest users of materials in tonnage, but opportunities for more efficient use of materials would need to be found in every aspect of our industrial and service economy. Are such efficiencies achievable, and, if so, can they be obtained through the application of existing technologies, or must new technology be developed? There is no comprehensive answer to such questions, yet there is hope that governments would turn to members of the materials community for answers as they make public policy on issues relating to sustainable development.

Today, there is a sensitivity to the desire to achieve a sustainable world economic and social system, which both satisfies human needs and does not despoil the earth. Brown air in the capitals of many of the world's countries, holes in the ozone layer, and increases in greenhouse gases, with their attendant climactic consequences, have made some of the negative impacts of current technology apparent to all. The centrality of materials usage to this subject should make the achievement of sustainability our issue, but environmental issues have not always been visible on our lists as we identified our priorities for future attention. Indeed, it is more common today to hear the mechanical and electrical engineering communities discussing dematerialization and alternate materials technologies with ecologists and economists. What has been our role in this increasingly important arena of human concern, and what should it be?

Think back to the times in which this series of lectures originated. Those were the days of

* Adapted from the 1998 Distinguished Lecture in Materials and Society of ASM International, by Lyle H. Schwartz, *Metallurgical and Materials Transactions A*, Vol 30, April 1999, p 895.

awakening consciousness: the first Earth Day on April 22, 1970; the development of "green" political parties and action organizations; the beginning of a period of rapidly escalating regulation; and the inception of a heightened awareness of the links between materials, energy, the environment, and rising population. Two landmark studies captured the thinking of that era about the role of materials. In 1973, the report of the Congressionally mandated National Commission on Materials Policy appeared. Entitled "Materials Needs and the Environment Today and Tomorrow" (Ref 2.2), this document contained a detailed discussion of the materials cycle and many recommendations for government action. Shortly thereafter, a major Nuclear Regulatory Commission (NRC) study was published. This monumental effort was a product of the ad hoc committee on the study of materials (COSMAT), chaired by Morris Cohen. Titled "Materials and Man's Needs," this text (Ref 2.3) set the stage for serious consideration about materials science and engineering for the next 20 years, clarified concerns with structure-property relationships, and graphically focused attention on the whole materials cycle. Indeed, it was from the COSMAT study that the representation of the materials cycle (Fig. 2.1) was derived, which still defines the scope of our field. These two documents, the first emphasizing policy and the second exploring technical issues, education, and research and development (R&D), represented high-water marks for focus of attention by the materials community on the links between what they do and the consequences to the environment.

While many of the specific recommendations in these volumes are a bit dated, the general principles still apply and are worth repeating here. The three summary directives for policy makers were as follows (Ref 2.4):

> "Strike a balance between the 'need to produce goods' and the 'need to protect the environment' by modifying the materials system so that all resources, including environmental, are all paid for by users."
>
> "Strive for an equilibrium between the supply of materials and the demand for their use by increasing primary production and by conserving materials through accelerated waste recycling and greater efficiency-of-use of materials."
>
> "Manage materials policy more effectively by recognizing the complex interrelationships of the materials-energy-environment system so that laws, executive orders, and administrative practices reinforce policy and not counteract it."

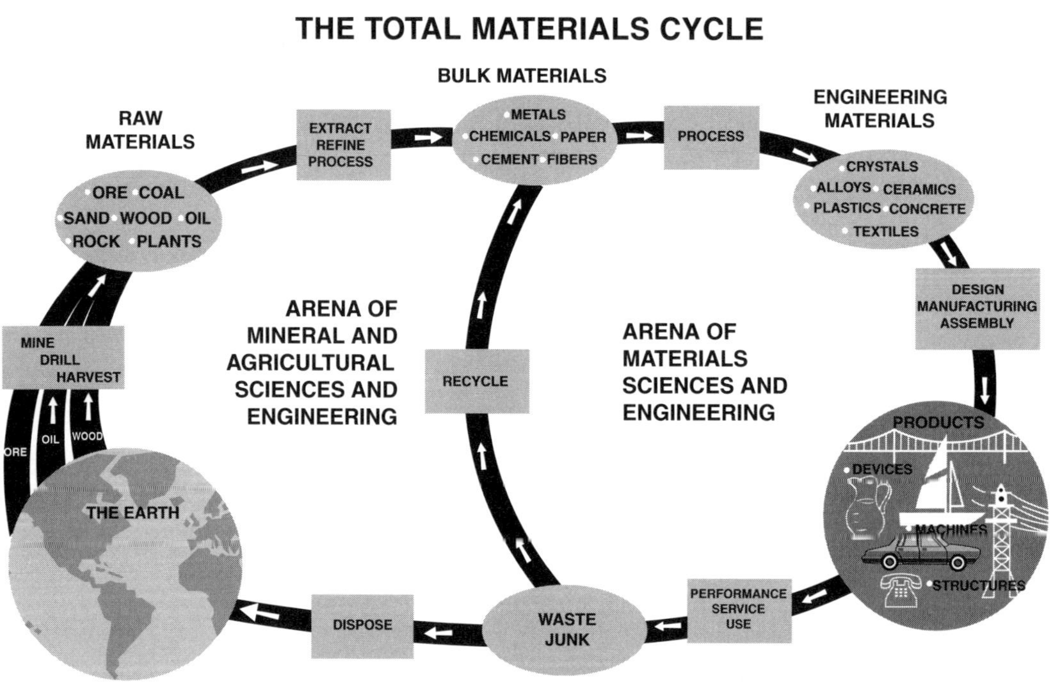

Fig 2.1 The total materials cycle

While significant progress since the early 1970s may be found on all three fronts, these principles could well be taken today as guidance for our continued efforts in both public and private arenas.

As the decade of the 1970s passed, other lecturers in this series returned to environmental issues. James Boyd devoted his 1973 lecture (Ref 2.5) to resource limitation in the context of what he termed the resource trichotomy: materials, energy, and the environment. Michael Tennebaum (Ref 2.6) included references to the environment in his 1975 lecture, describing what he saw as lack of balance in regulation efforts, with too much focus on the near term. Herbert Kellogg (Ref 2.7) in 1978 returned to the subject of conservation and anticipated the current term *dematerialization* to describe the more efficient use of materials. However, with the exception of several passing allusions to the need to be environmentally sensitive in everything we do, later speakers in this series have not devoted any serious consideration to the subject.

The visible evidence of our apparent disinterest in the subject was most clearly presented to the world in our comprehensive self-study organized by the National Academy of Sciences-National Academy of Engineering-Nuclear Regulatory Commission in the late 1980s. It is difficult to even find the word *environment* in that volume entitled *Materials Science and Engineering for the 1990's; Maintaining Competitiveness in the Age of Materials* (Ref 2.8). What were the reasons for the disappearance of environmental concern from the center of our focus?

Some History

There have certainly been many issues of concern in the past quarter-century in terms of economic challenges, dramatic restructuring of industrial sectors, shifts of employment on massive scales, and globalization of manufacturing and R&D. Governmental focus also shifted from energy and the environment in the 1970s to strategic defense concentration in the 1980s, and onto industrial competitiveness and technology transfer in the 1990s. Most dramatically, this era brought the end of the Cold War and a world increasingly focused on economic competition and the desires of all people to achieve the standards of living exemplified by that small fraction living in the so-called developed lands. It is no wonder that in the midst of this major restructuring in the United States and the apparent invincibility of alternate technology policy and strategies in several Asian nations, the last major study on materials by the NRC focused its attention on maintaining industrial competitiveness.

The past quarter of a century has been one to reckon with, and our colleagues have explored the ramifications of many of these societal changes in the series of lectures that preceded this one. Many of these issues have drawn our attention away from the necessity of organizing our activity to achieve a sustainable economic system.

Certainly, the issue has not gone away. We've made progress and expanded our knowledge and environmental sensitivities, and our attention has moved beyond cleaning up the water and the air and controlling pollution. We now explore a systems approach to the development of a world in which human society and commerce can be sustained throughout the coming generations. Emphasis on pollution and environmental impact caused us to focus our attention on effluents and process modification. By contrast, emphasis on sustainability will direct our attention increasingly toward materials usage and the flow of materials through what has been called the materials cycle.

This transition in the way environmental goals are viewed was made concrete in the 1992 United Nations Conference on Environment and Development (UNCED) held in Rio de Janeiro, Brazil. The strategies emphasized to achieve a sustainable state of development were as follows: improving efficiency and productivity through frugal use of energy and materials, substituting environmentally detrimental materials with ones that are less so, as well as recycling and reusing products at the end of their lives. These are remarkably similar to the agenda espoused by COSMAT almost two decades earlier and remain the technical framework today. It was particularly striking that UNCED defined waste as "material out of place." In that simple phrase, the challenge for the materials community is clearly etched. While we may be increasingly in agreement around the world about desirable outcomes, we are not always in such close accord about the path to achieve those goals, and the record of actions is more variable yet.

A comprehensive picture of where we stand on the road toward achieving sustainability could fill a book, and indeed many articles and books have been written on the subject. When we move from generalities to specifics, there is no consensus on even the definition of sustainability. We cannot precisely define it, but we will know it when we see it. In lieu of such a precise definition, the following discussion extracts a few facts from one of those many articles: that written by Ausubel et al. (Ref 2.9), who, in 1995, explored "The Environment Since 1970."

During this period, population grew from 3.7 to 5.7 billion and is expected to continue to grow to a steady state of double or triple current levels. Global per capita commercial energy consumption has stayed level, but because of population growth, 8 billion tons of oil were used in 1995 compared to 5 billion in 1970. The oil "crisis" of the 1970s has not returned, as proven oil reserves have increased from 600 to 1000 billion barrels during the period, while using more than 500 billion barrels, which were pumped from the ground. Meanwhile, technology has been shifting our source of energy toward the less polluting natural gas, a form of decarbonification, while proven reserves of natural gas have tripled. In short, while we may see some improvement in reduction of greenhouse gases per unit of energy extracted, we cannot count on imminent shortages of fuels to drive us in the direction of seeking alternate, higher-priced sources of energy, even if they may represent more "sustainable" options.

Automobiles are, of course, a major aspect for structural and functional materials of all sorts. Their manufacture, repair, and recycle occupy a large fraction of the working populace. They use substantial amounts of energy, contributing greatly to greenhouse gas production. Since 1970, as the world became more affluent and the population grew, the number of motor vehicles more than doubled to the staggering figure of approximately 600 million, more than offsetting any gains in fuel efficiency that may have been achieved. As the developing nations strive to emulate our affluent style of living, we may expect to see the number of vehicles double again, and once again see the fruits of more efficient fuel utilization per unit negated by the sheer numbers of units.

Ausubel et al. (Ref 2.9) summarize the situation by noting that "production, consumption, and population have grown tremendously since 1970. . . . Globally, and on average, economic and human development appears to have outpaced population growth."

There can be no doubt of the continuing presence of major global consequences of humankind's expanding numbers and current lifestyle. Among these, we may number increasing emission of greenhouse gases with projected climate change, depletion of stratospheric ozone layer by chlorofluorocarbons, tropical forest depletion, and, with it, decreases in biological diversity, air quality in densely populated cities, and waste disposal in increasingly limited landfill space.

Public concern for the environment has taken many forms. In the United States, the number of federal laws for environmental protection has more than doubled since 1970, and government involvement in industrial operations has been correspondingly increased. As one indicator of that impact, spending on pollution abatement has also doubled and exceeded $90 billion annually by 1995. Nongovernmental environmental organizations in the United States have roughly tripled since 1970, and, with their presence, more information is available in the popular press, not all of sound scientific basis.

Again quoting Ausubel et al. (Ref 2.9), "People are demanding higher environmental quality. The lengthening list of issues and policy responses reflects not only changing conditions and the discovery of new problems, but also changes in what human societies define as problems and needs" (Ref 2.9). This observation is nowhere more appropriate than when applied to the automobile. In the 25-year period we are looking at, cars were lightened by 30%; catalytic converters dramatically cleaned up the noxious effluents of combustion; gas mileage increased by a factor of 2; corrosion protection and damage-tolerant materials lengthened the use of vehicles, contributing to dematerialization; and, at the same time, the vehicle became safer and more attractive to owners. Materials substitution has reduced the weight of the average family sedan from approximately 4000 to approximately 3000 lb and dramatically changed the mix of materials. At the same time, however, we have seen the development of changing consumer buying practice with rapid growth in the more materials-intensive small trucks, sport utility vehicles, and vans. In fact, since 1973, all of the increase in U.S. highway fuel consumption has been due to these other vehicles.

To offset these effects and to make a dramatic change in materials usage and energy use will require a total redesign of the vehicles and their power trains. In the United States, the Big Three auto manufacturers have joined together in R&D partnerships to explore many of the requisite new technologies under the banner of the United States Council for Automotive Research (USCAR) and are intensively engaged with the federal government in a Partnership for New-Generation Vehicles (PNGV) to achieve such radically new vehicles. The situation in transportation has become even more complex as we confirm the consequences of increases of greenhouse gases on the global environment. In this context, the basis for power in most vehicles, the gasoline engine itself, is viewed as a culprit, and we must look to other power sources or at least to much greater efficiency via burning carbon in power plants and then using electricity to power the vehicle.

In summary, there has certainly been much accomplished in the last 25 years, but instead of the issues of materials and the environment disappearing, they have become more demanding, raising questions in many arenas about whether our advanced technologically-based prosperity is in fact sustainable.

So, is it that others have taken this issue and made it their agenda while we have remained in the background? Many other communities have embraced the issue of sustainability with open arms. Although the general initial reaction of industry has been to resist regulations, many have now recognized that environmentally friendly manufacturing can be a plus to the bottom line. This practical observation and the desirable strategic responses have now been codified in the term *industrial ecology*.

Selected with obvious reference to the complex interactive natural world around us, this term has as many definitions as there are commentators, like that advanced by Frosch and Uenohara (Ref 2.10):

> "Industrial ecology provides an integrated systems approach to managing the environmental effects of using energy, materials, and capital in industrial ecosystems. To optimize resource use (and to minimize waste flows back to the environment), managers need a better understanding of the metabolism (use and transformation) of materials and energy in industrial ecosystems, better information about potential waste sources and uses, and improved mechanisms (markets, incentives, and regulatory structures) that encourage systems optimization of materials and energy use."

The systems approach leads to thinking about both the productive output and the waste from one industrial arena as the input for others; it leads to a demand for extensive databases on materials flow throughout the total materials cycle, and it leads to the need for sophisticated decision-making tools to enable enlightened technical and business decisions. The materials community has been engaged in such activities to a greater or lesser extent throughout the last decades, but we have always given too little priority to such activities in favor of more glamorous, and usually more "scientific," efforts in new materials development and characterization. Increasingly, as industrial rather than government priorities dictate, we will have to change our focus as well.

The Materials Role in Industrial Ecology

The U.S. government has been a major factor in driving environmental fixes for some time now. The quarter-century since the first Earth Day has frequently been characterized as one of regulatory command and control. Certainly, this philosophy has led to the development of quite an array of new technology, but much of this has been aimed at "end-of-pipe" and cleanup. These old regulatory strategies, enacted as a quick dose of strong medicine, may have played themselves out. Many now question whether the cost/benefit of further regulatory reform is supportable. Increasingly over the last 10 years or so, many in the government have recognized the opportunities for new technology development as the next phase in the achievement of sustainable development. This new philosophy was expressed in the first 2 years of the Clinton-Gore Administration as an Environmental Technology Initiative (ETI) and described in glossy brochures as the National Environmental Technology Strategy (Ref 2.11, 2.12). Products of the National Science and Technology Council, these referenced documents take a broad-brush view of the societal requirements to achieve sustainability and touch lightly on the technical specifics. For example, the following generalizations about materials are made (Ref 2.11):

> "Many materials will need to be discontinued and new materials employed in order to achieve environmental technologies that conform with the principles of industrial ecology. Advances in the materials used in the

manufacturing process and the development of lighter materials for transportation have the potential to reduce the production of wastes, minimize the extraction and use of virgin organic resources, mitigate pollution, and improve energy efficiency. These new materials would have a predictable lifespan and, when they are retired from their original use would be designed to be used for other purposes."

To translate such lofty goals into reality would take planning and execution over decades. And it would take dedicated resources over a long time horizon. And, of course, it would take public-private cooperation on a scale not yet achieved nor perhaps even imagined. This environmental strategy envisioned a broad multiagency effort, which was to include those agencies in which much of the materials research is funded; however, most of the "new" money appeared in the Environmental Protection Agency (EPA), which was designated lead agency. Not surprisingly, given the regulatory mission and historical strategy of the EPA, these funds were targeted at expansions of pre-existing projects to implement best existing practices, and little found its way into the development of new materials technologies. The ETI was launched with an initial appropriation of $36 million in fiscal year (FY) 1994, grew to $68 million in FY 1995, but was then slashed to $10 million for FY 1996 by the 104th Congress, which directed that the remaining appropriation be directed toward environmental verification. This start-stop record is consistent with the general history of technology investment by the federal government, but it is further complicated by the lack of clarity within the Congress regarding their regulatory role. In no statute has Congress explicitly given the regulatory agencies any mission to encourage technological change as a means toward environmental improvement (Ref 2.13).

We may properly ask here about the desired technical agenda. If new funds were to be made available, toward what ends may they be applied? The technical agenda has not really changed much since it was outlined in the COS MAT report in the early 1970s. In that overview, the subject was divided into effluent abatement, materials substitution, functional substitution (or redesign), waste disposal, and increased use of recycling. This list, expanded to the next level of detail, is included as Table 2.1. The specific issues vary in importance for different industries, and the degree of progress made since 1973 is

Table 2.1 COSMAT list of materials tasks for environmental issues

Effluent abatement
- Process restructuring
- Containment
- Recycling

Materials substitution
- Through alteration of existing devices
- Through substitute devices

Functional substitution

Waste disposal
- Increased degradability
- Reduction in noxiousness

Increased recyclability
- Through design
- Through suitable materials choice

similarly quite varied; however, when the issue was addressed in detail in a workshop held in 1995 (Ref 2.14), the same general list of topics emerged. Using a slightly more up-to-date terminology, we may say that the opportunities may be found in cleaner (greener) processing, alternative materials, dematerialization (lighter and less to do "same" function), and reuse or recycle. Since the 1970s, substantial progress has been made in each of these areas, but much more will clearly have to be done if we are to achieve a sustainable level of materials usage.

Some general comments about each area follow.

Cleaner Processing. Industry has clearly been the leader here; driven largely by regulation, dramatic improvements have been made in reducing effluents, cleaning up scrap, and minimizing energy uses. Linked strongly to the recycling issue, new technologies have been introduced, which have radically transformed the materials-producing industries. Recall that before the 1970s, the economic prowess of a nation was schematized by a picture of tall smokestacks, belching smoke and other noxious fumes into the atmosphere. How differently we regard that image today, and how much more common it is to see photos of grass-covered campuses on which green factories ply their work. Manufacturing moves ever closer to a scrap-free environment, as we introduce one after another of the desirable net-shape technologies. Continued R&D in this arena is certain to produce further benefits, but a delicate balance must be struck here. In many of the industries where process improvement is most likely to be of greatest impact, such as casting, coatings, and specialty alloys, the disaggregated nature of the business and the small size of most companies make research difficult to do and unlikely to pay off in the near term. Regulations intended to force cleaner processing in these industries may have the undesirable alternative of forcing the companies to locate off-shore, with a consequent loss of jobs to the United States and no net improvement in the world's environmental position. For this, if for no other reason, close cooperation must be sought within government between the regulators and the technology developers.

Alternative Materials. This topic appears similar to dematerialization, and some definitions are appropriate. It is useful to distinguish between those technologies intended to eliminate "bad" things from the waste stream (referred to here as alternative materials) and those dematerialization technologies intended to reduce material usage, either through reduction in weight in the product or in scrap. When considering alternative materials, we must focus our attention on the materials producers and be sensitive to the source of R&D funding. In the case of polymers, it has clearly been industry that has footed the bill, with some rare exceptions. It is quite interesting that these companies have strongly linked their business goals to their perceptions of an environmentally conscious public. Refer, for example, to the Dupont ads that appear in every issue of *Scientific American* and describe the newest biodegradeable polymers as alternatives for manufacturing. Dupont clearly believes that developing new materials from renewable sources is in their best corporate interests as well as those of the planet.

On the other hand, in the case of metals, we see a mixed picture, due in part to the source or lack of R&D funding. At one time, extensive alloy development could be expected from the private sector. This went in two directions. On the commercial side, autos and beer cans, with their significant environmental implications, have produced fierce competitions for market share, which have driven both aluminum and steel. On the other hand, in the aerospace arena, it is government funding that was dominant as the driver for materials development. We have seen the dramatic improvement in materials properties over the last 25 years, largely driven by performance needs and often financed by the federal government. Higher-temperature operation to achieve higher performance has been the motivation behind programs focused on engine materials and largely funded by the Department of Energy (DoE) and Department of Defense (DoD). While higher performance may translate into higher fuel efficiency with positive environmental implications, this was clearly not the driver. As DoD procurement needs decrease, we are finding them less ready to invest in such costly and time-consuming efforts as new materials development, and few of our industrial concerns can capture enough economic benefit to justify private sector investment. This subject was addressed in some detail in the 1997 Distinguished Lecture by Jim Williams (Ref 2.15). As DoD disappears from the picture, will DoE fill the void, and can we expect the motivation for energy reduction, coupled with the desire for sustainability, to achieve a substantial new investment by that agency?

In any discussion of alternative materials, we must deal with the materials selection in the context of a system, and that the system must be explored over its full life cycle. Life-cycle analysis (LCA) is the generic terminology to describe such thought processes and accounting exercises, but current versions of this technology have not yet reached the levels of user-friendliness and economy, which would make LCA common practice for all designers. The weaknesses noted by many authors include: full analysis is too costly to be justified, and, instead, limited environmental impacts usually suffice; results are too sensitive to input data and critical boundary conditions, but sensitivity analysis is not readily accomplished; and no standards exist, limiting intercomparisons of alternate methodologies. These latter failings often lead to conflicting conclusions, which decrease technical and public acceptance of the results. If we cannot resolve the paper cup versus plastic cup choice, how can we be expected to use this methodology to choose between aluminum and steel in auto bodies?

We have a long way to go to resolve these issues, but progress continues on the technical issues of data and sensitivity analysis. Data limitations include proprietary ownership and lack of accurate, materials-specific information, but the most significant uncertainties in LCA arise from options in the attribution of environmental burden between the original materials producers and the ultimate users. This attribution, in turn, depends on assumptions regarding anticipated recycling prospects. In one article (Ref 2.16), Clark et al. describe a technique they dub "product stream life-cycle inventory" for quantifying this allocation. Most importantly, this approach allows the analyst to consider the sensitivity of results to assumptions. Extending this sensitivity analysis to the full LCA was done by Newell in his Ph.D. thesis (Ref 2.17), creating the new technique he calls Explicit LCA (XLCA).

It remains to be seen whether XLCA will be accepted by those engaged in LCA, but strategies like this and other steps, most notably the introduction of standards, will be required to transform this technology from a subjective one to a truly objective one. Only then can we expect to see LCA take its place as a requisite tool on the palette of every design engineer, and only then will we see materials selection done in a manner most consistent with sustainability.

Dematerialization. Some writers suggest that advanced technologies have already led to dematerialization, a reduction in the per capita consumption of materials needed to sustain our societal economic hunger. Others are more cautious, because the data are hard to acquire and even more difficult to interpret. For example, in a study of the change over time of per capita lead usage in the United States, carried out by the United States Geological Survey (USGS) (Ref 2.18), an erroneous observation of dematerialization could be obtained if consumption were estimated by tracking the mining production to manufacturer materials flow, rather than doing the much more difficult, bottoms-up analysis of actual consumption. What has actually been happening is complicated by the movement of lead acid battery manufacture offshore. In fact, U.S. per capita consumption has actually increased during the period. Similarly, while we have seen more efficient use of materials in home construction in the United States, lifestyle demands for leisure and comfort have led to expanded space per person and consequent increases in per capita materials usage. The obvious dematerialization in the weight of some automobiles is easily demonstrated in the reduction in body weight of a standard passenger vehicle by one-quarter during this period. However, during the same period, the number of vehicles per capita has increased, and the product mix on the road has dramatically changed. Clearly, we cannot begin to evaluate real progress toward dematerialization without significant improvement in our worldwide materials flow database.

Reuse/Recycle. One of the great successes of the last quarter-century is the growth of recycling as a natural way of life for consumers. More than 80% of the states in the United States have comprehensive recycling laws, and curbside recycling programs have grown from a few hundred to more than 4000. There can be no doubt that younger generations will be ever more desirous of eliminating the path from use to landfill. There are television programs and children's cartoons dedicated to the subject of recycling. These television programs, and many others like it, are sponsored in part by funds from the National Science Foundation (NSF) and the DoE, and represents an aspect of the federal role in promoting recycling.

Increasingly in other nations, the governmental role in recycling is moving from promotion to mandating, and with such a transition comes opportunity and challenge. Germany has gone farther than most with its "take-back" legislation. The Closed Substance Cycle and Waste Management Act of 1996, requires manufacturers to recover, recycle, or dispose of assembled products such as automobiles, electronics, and household appliances when consumers retire them. Japan has introduced similar requirements in its Law Promoting the Utilization of Recycled Resources. In the global economy, which now governs manufacturing, products manufactured in the United States will not be sold in such lands if they do not conform to local standards, so we are already feeling the impact of such take-back philosophy without any legislative mandate in this country. Shared ownership, manufacturing in the lands of sale, and now international mergers such as that of Chrysler and Daimler-Benz will accelerate this trend. The technical, business, and political issues underlying this trend will demand the continued intense involvement of the materials science and engineering community.

Ultimately, the success of any recycling program depends on creating markets for the recycled material, and therein we find much of the accomplishments of materials engineering in the last quarter-century. Recycle/reuse of the automobile is one of the most visible and most complex of these stories. A discarded car, sent to a junkyard, is first denuded of reusable components, then crushed, shredded, and separated. Approximately 75% by weight is recycled. The remainder, known as "fluff," is treated as waste and buried in landfills. This process has spawned and depends for its success on an industrial infrastructure composed of disassemblers, distributors, and reusers; an infrastructure which is driven by profit and sensitive to change. This infrastructure is continually at risk as the product changes; the cost of labor, landfill, and transportation varies; and the nature and market for the recovered components is modified.

Evolution of the automobile in response to desires to improve fuel economy and reduce pollution has not only lightened the vehicle but has also changed the materials mix. More complex alloy steels, while "reusable" in the sense that they appear once again as useful products, are not truly "recycled" in the sense of being used again and again for the same product. While this is certainly better than no reuse at all, when viewed in the context of a sustainable economy, it must be viewed as falling short of the desired goal. The current mix of materials will soon be disrupted significantly as a new generation of highly fuel-efficient vehicles emerges. Whatever the specific design and power source of a given car, such vehicles taken as a fleet must be lighter yet than today's (2007) vehicles, ensuring the use of a still larger fraction of high-strength steels or substitution on a massive scale by aluminum alloys and organic-matrix composites. One of the major bogeys to be met in any such redesign should be an improvement in the reuse of components and true recycling of a larger fraction of materials.

Research on materials substitution in automobiles has been a continuing subject for original equipment manufacturers (OEMs) and their suppliers for many years, but in recent years, prompted in large part by the generic, precompetitive nature of such research from the perspective of the OEMs, cooperative joint research ventures have become more common. As part of their more general cooperation under the banner of USCAR, Chrysler, Ford, and General Motors have formed a vehicle recycling partnership for R&D to recover and recycle materials from scrap autos and to develop tools to evaluate the recyclability of new designs. It is certain that this area of R&D will grow and spread to other manufactured products. The challenge is to maintain profitability and utility in the product while sustaining the recycle/reuse infrastructure so that the consequences of government mandates for socially desirable goals are not merely passed on to the consumer as higher prices.

The U.S. Government Role—Organizational

The U.S. federal government has a role in support of the technology base for sustainability. The organizational structure includes several agencies, each with their own missions and each with their complex arrays of governing Congressional committees and various private sector constituencies and customers. It represents a system that may certainly better be characterized as a collection rather than an organization. And yet, as we approach such issues as complex as sustainability, defining the proper role for the federal government demands an organized approach. In the late 1990s, such a cross-agency involvement was displayed in the committee efforts that led to the policy positions on enviromental issues cited earlier and to more detailed efforts focused on research and on information about materials flows.

For some time, the efforts to achieve organizational approach within government have been applied to R&D through the Frederal Coordinating Council on Science and Technology, established in the mid-1970s, and the National Science and Technology Council (NSTC), which replaced it in 1993. Under the NSTC, the materials R&D was coordinated by the interagency committee called Materials Technology (MatTec). MatTec organized around the areas of focus of the civilian technologies identified by the NSTC and developed working groups to consider materials needs in automotive, building and construction, electronics, and aeronautics. While environmental issues came up in each of these areas, it was felt that a more comprehensive view of sustainability demanded a working group devoted to defining the issues for "materials and the environmental." Formed in 1996, this group began to work with others to identify the cross-agency and government/private sector issues that must be addressed as the government role in sustainability evolves. During 1997 to 1998, this committee collaborated with the Federation of Materials Societies in two workshops intended to intensify interest in the subject, identify activities that societies may fruitfully pursue individually and collectively, and search for closer society/public sector interaction.

One of the roles of interagency groups such as MatTech and Environmental Management and Technology (EMAT) is to examine the portfolio of federal R&D programs in the relevant areas of science and technology; identify opportunities for synergism through cooperative ventures; and, where appropriate, reveal gaps in the portfolio. No such inventory has yet been made by EMAT, so we may only make rough statements about the magnitude of federal funding in this area. Two useful sources were used for this "analysis."

Teich (Ref 2.19) examined the data for fiscal 1995 and estimated a total of approximately $5 billion on "environmental research" as so characterized in nondefense agency budget justifications. This substantial sum is concentrated primarily in the National Aeronautics and Space Administration (NASA), DoE, NSF, Department of Interior (DoI) and Department of Agriculture (DoA). However no matter how hard we look, we are not going to find much materials research included in this total, because this survey focused on programs relating to pollution control and abatement, conservation, and management of natural resources. To avoid double counting in such inventories, agencies went out of their way to put all materials research in the "advanced materials and processing" category, so what we are searching for must be found among the approximately $2 billion included in the survey published by MatTech (Ref 2.20).

When this materials survey was put together, no funding breakdown identifying environmental issues was made, so no quantitative information can be derived. The report does call out many examples of such activity, especially in the DoE and DoA, with some other examples in other agencies. What is most apparent, however, is that few of these examples are being justified primarily because of their environmental impact, and words such as *sustainability* have not even penetrated into the vocabulary used to describe this work. It is generally believed by members of EMAT that most of the current materials R&D funding, which is primarily justified as environmental in nature, is focused on issues associated with cleanup, and little is aimed at prevention. As the issues of global change become more significant in a political sense, we may expect to see a relabeling of projects, currently justified by their energy savings, as leading to sustainable use of materials. It would be very interesting to follow this relabeling over the next several years to see what is really new in the government's research portfolio. In any case, it would be desirable to assess the current portfolio to clarify what the federal government is now doing to support new technology development for sustainable use of materials.

The U.S. Government Role—Technical

Federal Support of Civilian R&D. As we explore the role of the federal government in supporting research that may influence sustainability, we must confront head-on the continuing debate regarding the place of federal investments in the private sector product arena. We all know too well the use of the term *corporate welfare* to denigrate any direct taxpayer funding of research that may have an impact on corporate profit. Certainly, there are enough examples of situations in which we already do this, so that one must recognize that the federal role is specific to the situation, not easily generalized. Arguments favoring federal investment in technology development are usually based on "market failures," idiosyncracies of the technology/regulation/capital environment that impede the development of economically and/or socially desirable technology. New materials and new materials processing may be an example of market failure, triggered particularly by the mismatch between the 10- to 20-year development times of these technologies vis-à-vis the increasingly short development times for the products that may take advantage of such new materials technologies. New materials technologies suitable for improving our goal of sustainable development would then appear to be ideal candidates for federal investment in technology development, if there is to be any at all.

By way of example, the federal role in providing national security justifies not only R&D but also test and evaluation of actual products, which are then sold to the government. The collateral benefit to the aircraft industry and its commercial return to its stockholders is a pleasant side benefit that seems to bother none of the free-marketeer critics of corporate welfare. Similarly, in the field of agriculture, a federal role has long been recognized, supported by the extensive research establishment of the Agriculture Department and augmented by the many states through the land-grant college system. What then should be the federal role in supporting the development of environmental technology?

Certainly, basic research and even applied research with a focus on more environmentally friendly technologies should be encouraged. When we get to product, the answer lies in assessing whether the desired social good—in this case, a more environmentally friendly product—can be expected without the involvement of the government. If the answer to that question is no, a government mission exists and may be addressed by one of three alternatives: taxes (financial incentive to consumer or disincentive to producer to follow desirable paths), regulations (product guidelines mandated to achieve the desired end with technology choice left to the producer), or assistance in technology development (recognizing that the free market will not invest if no near-term profit is believed to be forthcoming).

On the automotive scene, we see the mix of these three policy choices vying for ascendency in our contentious political arena. Regulation has been the predominant strategy in the United States for the last quarter-century, with significant visible results in both cleaner emission and lower fuel consumption. This path is by no means exhausted because new clean air regulations continue to be debated in the Congress, and many individual states place ever greater restrictions on the allowable emissions from vehicles. In most of the rest of the world, taxation has also been a significant factor in governmental attempts to minimize the ecological impact and energy usage for transportation. With higher resultant energy costs to the consumer as a key driver, smaller, fuel-efficient vehicles are more readily marketable. The political obstacles to achieving even a 5-cent increase on gasoline make such an approach unlikely to play a significant role in the United States.

Finally, in recent years, the U.S. federal government has expanded its research agenda to include the commercial automotive arena. First, through several unconnected efforts in the DoE, National Institute of Standards and Technology (NIST), and elsewhere, and then through the highly visible PNGV, the government and the automotive Big Three joined in a historic partnership to develop technology that would satisfy the individual customer's desire for transportation and the collective societal desire for a "greener" vehicle. The success of this partnership in both a technical and a political sense will have profound implications for the continued participation in such research in other industrial sectors. It is worth looking in a bit more detail at the technical opportunities and challenges in the PNGV.

The challenge of the PNGV can be summarized in three goal statements: advancement of manufacturing practices, implementation of innovations on current vehicles, and development of a vehicle with up to 3 times current fuel efficiency. It is this third goal and its restrictions that make up the grand challenge of the PNGV. It is possible today (2007) to build a passenger vehicle with 3 times the fuel efficiency of today's passenger sedan, but the formidable challenge set by the PNGV is to do this while maintaining the passenger and storage capacities; satisfying the driver's desires for speed, acceleration, and driving range; meeting all existing emission regulations; achieving recyclability of 80%; maintaining manufacturability; and perhaps, most daunting, maintaining affordability. To achieve this result will require dramatic mass reduction of the vehicle and simultaneous significant increases in power source efficiency. The design space allows for many solutions to this problem, and we expect that each of the Big Three will find its own unique combination of parameters. Nevertheless, it seems apparent that vehicular weight reductions of the order of 40% will be required. One scenario for distribution of the weight reductions among vehicle components is displayed in Table 2.2. Such dramatic reductions in weight can only be achieved through major materials substitution and with significant design and manufacturing changes. Some of the research topics under study are summarized in Table 2.3. This list leaves no doubt that our agenda remains formidable. One of the most complex tasks facing the materials and manufacturing communities is to achieve the requirement of recyclability while further increasing the mix of materials. The competition now underway among the principal contestants for such alternative technologies has added new vitality to the metals and polymer composites industries and will certainly lead to significant technological change in automobile manufacture for years to come.

Energy and the Environment. Our attention was diverted from environmental issues in the mid-1970s by the development of the so-called "energy crisis," a supply/demand trauma attributable to the price escalation by the Organization of the Petroleum Exporting Countries (OPEC), a near-monopolistic trade organization. As we explored alternative energy sources, new and improved materials came to the forefront. We found that no new energy technology was going to become a serious price competitor to then-dominant fuels without dramatic reductions in

Table 2.2 Vehicle mass reduction targets for Partnership for New-Generation Vehicles (PNGV) goal 3

	Current vehicle		PNGV vehicle target		
System	kg	lb	kg	lb	Mass reduction,%
Body	514	1134	257	566	50
Chassis	499	1101	249	550	50
Powertrain	394	868	354	781	10
Fuel/other	62	137	29	63	55
Curb weight	1470	3240	889	1960	40

Table 2.3 Partnership for New-Generation Vehicles widely applicable materials—Challenges

Priority of challenges	Aluminum	Magnesium	Polymer composite components	Polymer composite body	Steel
High	Feedstock cost	Feedstock cost	Low-cost carbon fiber	Low-cost carbon fiber	
				High-volume manufacturing	Weight-reduction concepts
			High-volume manufacturing	Joining	
		High-temperature alloy	Analytical design	Analytical design	
			In-service inspection	In-service inspection	
				Recycling	
Medium	Casting		Recycling		Lightweight technology (incremental)
	Forming	Improved design and manufacturing			
	Joining		Properties in service	Properties in service	
	Recycling	Machining			
		Recycling			
Low		Extrusion			

materials costs or improvements in conversion efficiencies. As the "crisis" passed, and with the arrival of the Reagan administration, budgets fell at the DoE, and the focus on alternative energy sources was reduced. Has anything changed in the subsequent 15 years? There is certainly no doubt that petroleum prices will someday rise as supplies decrease—only the timeframe is debated. Published studies by Campbell and Laherrère (Ref 2.21) and MacKenzie (Ref 2.22) conclude that the peak of global production of conventional oil is only one or at most two decades away. With the subsequent decline in production and increase in price, the pressure for alternative sources of energy will accelerate. Among those alternatives, we continue to list natural gas and coal-derived fuels, but are these still realistic alternatives?

In the last several years, we have come to recognize that the accumulation of greenhouse gases in the atmosphere will very likely cause substantial climactic change. While much more needs to be done to clarify the consequences of such change, public policy is already moving toward control of greenhouse gas emissions. With this new reality in mind, our attention should increasingly focus on alternatives to fossil fuels as the only long-term sustainable strategy for expanded energy needs of a growing population with growing per capita economic progress. Consequently, in an even more intense way than earlier, the issues of energy and environment have become intertwined. Not surprisingly, the DoE is the largest source of funds for environmental programs in the federal government. Nor is it surprising that the strong dependency of energy technology on materials technology has made the DoE by far the largest federal funder of materials science and engineering. The range of technical programs in the DoE portfolio is far too broad to be covered in such a brief overview as this, because it includes some level of effort in all fossil and nonfossil alternatives to petroleum as an energy source. The DoE is determined to take a lead role in the development of new technology for sustainability and has brought the efforts of several of its national laboratories to bear on the subject of R&D strategies. The volume entitled "Scenarios of U.S. Carbon Reductions: Potential Impacts of Energy Technologies by 2010 and Beyond" (Ref 2.23) is a gold mine of ideas and R&D needs, with many focused on materials issues. A second excellent summary of R&D opportunities was produced by a joint effort of the DoE and NSF. This report on "Basic Research Needs for Environmentally Responsive Technologies of the Future" (Ref 2.14) links the needs to the various industrial sectors and gives significant attention to the materials research agenda. Rather than attempt to summarize these technical options, the focus here is instead on the public-private interaction needed to bring such technology to bear on the issue of sustainability.

During the 1980s, the general antipathy toward "demonstration" projects left over from the years of the energy crisis was followed by a bipartisan recognition that new methods needed to be developed to garner the fruits of federally funded research. A series of experiments are now underway in an effort to link the various government programs more closely with industry. Many of these programs remain controversial, and the demise of the DoD's Technology

Reinvestment Program and reduced funding for the DoE's Cooperative Research and Development Agreements and the Department of Commerce's Advanced Technology Program are indications that these programs are still viewed as experimental by many in Congress. Whatever may come of the funding of these programs in the future, they have had significant impacts on many of the materials producers and users and will likely continue to do so in the future, if not through direct funding, then through changes in business practice. This is particularly true for many of the materials-producing and -processing industries with their disaggregated organizational structures.

One particular activity of note in the DoE was in the Office of Industrial Technology (OIT), which had an exciting new program called the Industries of the Future Program. This effort involved federal-private partnerships with seven industrial communities in the development of visions for their future development and technology roadmaps required to get there. We are most familiar with the roadmap concept in the electronics industry, where, guided by the rate-determining predictions of Moore's law, that industry can look 10 years ahead, define where the industry's technology will be, and then develop a map of development work necessary to get to that desired endpoint.

The DoE-OIT selected for its industrial partners the seven most energy-intensive industrial sectors, collectively using more than 60% of the energy consumed in the United States. These industries are also among the most intensive contributors to other waste streams besides greenhouse gases. Partnerships have been developed with the aluminum, chemical, forest products, glass, metal casting, steel, and agriculture industries, with supplementary agreements with petroleum refining, heat treating, and forging. The heart of the materials-producing community is displayed in this list. Thus, as these industries develop their roadmaps, identifying their technology needs for the future, they are also laying out a rough outline of the materials research needed to achieve sustainability in the next decades.

These roadmaps, unguided by any Moore's law for the metals and chemical industries, are less well defined than that of the electronics industry, and a great deal of work lies ahead before they will yield the required detail of research agenda. These roadmaps are now public documents, available through the internet and from the industries and the DoE. Among other common features, they all share a strong commitment to sustainable development. As one example, consider statements made by the Steel roadmap in the chapter entitled "Environment" (Ref 2.24).

Over the past 25 years, the investments of $6 billion on capital investments on environmental projects have led to reduced discharges of air and water pollutants by 90% and a reduction of solid waste production by more than 80%. Nevertheless, "further improvements to pollution prevention technologies are needed to reduce costs, improve profitability, and facilitate compliance with changing Federal regulations. The steel industry's goal is 'to achieve further reductions in air and water emissions and generation of hazardous wastes,' and the development of processes 'designed to avoid pollution rather than control and treat it.'" The report then goes on to list desired technical developments in cokemaking, ironmaking, steelmaking, refining and casting, forming, and finishing in this 34-page chapter on environmental technologies.

The Industries of the Future Technology Roadmaps represent a fertile area for planning, not only for the industry itself but also for the university and government organizations that would interact with these industries. Materials scientists and engineers must "mine" these plans for the concepts and then fill in the details to develop a materials research agenda. Professional societies also are deeply involved in developing roadmaps. The TMS and ASM International have both been playing a role with the metals industries, and part of our societal agenda should be to assure that we will hear more about these efforts in the meeting sessions and hallways in years to come. We must make our meetings the technical home for the results of R&D in the Industries of the Futures Program.

Materials Flow Data. The collective action of the federal government has been of benefit to industry and the citizenry in yet another technical arena of relevance to this discussion: the area of information. Many organizations within government assemble, organize, and disperse information of a technical nature. Most visible to us as scientists and engineers has been evaluated technical data such as thermodynamic, crystallographic, or materials properties, which allows the scientific enterprise to proceed with common basis for quantification. One concern in this presentation is with another broad arena of data: data on materials flow. When we think of the materials cycle, it is usually in a qualitative sense, but a small, growing number of our colleagues are beginning to look at the full life cycle of particular materials in a quantitative sense. Understanding how materials are derived, used, and disposed of may often focus our attention on alternative strategies and technologies to achieve an environmentally friendly and sustainable manufacturing enterprise. There are some economists who believe that in the future, materials flows will develop as much importance in economic analysis as have energy flows in the last several decades.

The focus on materials flow within the government is centered now in the EPA and in the USGS, conducted there by a few folks who are among the remnants of the gone-but-not-forgotten Bureau of Mines. However, there is interest in this subject in many other agencies. An interagency working group on industrial ecology, material, and energy flows has constructed a report on materials flow (Ref 2.25). This document significantly influences the visibility and significance this topic will receive in future government effort.

We need to be "mining" these data in two senses: on the one hand, by looking for opportunities to identify research needs and, on the other, by literally mining in the sense of identifying materials flows that will be sources of "raw" materials for processing. Consider the example of silver in water, sediment, and tissue of fish and marine mammals in the San Francisco Bay (Ref 2.25). The source, located after a detailed materials flow study by the University of California at Los Angeles, was traceable to photographic and radiographic materials used by the service sector, including dentists, x-ray labs, hospitals, photo shops, and so on, not to heavy industry. The solution was found, in part, through changes in the regulatory system and, in part, through the development of cost-effective processing and recovery systems. One such example is the establishment by Kaiser Permanente of a centralized silver recovery and fixer reprocessing plant in Northern California. This was a profit-making system that reduced silver loadings to waste treatment plants and waterways, and paid for itself in less than 1 year.

This is only one example of what could be a myriad of profit-making efforts to recover valuable minerals from what are now considered to be waste streams. Allen and Behmanesh (Ref 2.26) encourage the view that these waste streams are raw materials that are often significantly underused. They emphasize that one of the research challenges of industrial ecology will be to identify productive uses for such materials currently considered wastes. In their analysis, they focus on materials flow data generated by the EPA from the National Hazardous Waste Survey but emphasize that existing data represent only 5 to 10% of the total flow of industrial wastes. Detailed information about concentrations in the waste streams compared with prices for raw materials forces the conclusion that the concentrations of metal resources in many waste streams currently undergoing disposal are higher than for typical virgin resources. Among these materials are lead, antimony, mercury, and selenium. With appropriate R&D, we may expect to find more symbiotic developments in which small recovery plants are sited at the sources of these waste streams. We may also expect that, as in the case of silver in the San Francisco Bay, R&D will likely have to be supplemented by changes in the regulatory policies to enable efficient handling of these materials as potential resources, not wastes.

The Role of Professional Societies

Each of the individual professional societies with an interest in materials carries out activities focused on their particular constituency to a degree consistent with their resources and their perception of their member needs. In the arena of materials and sustainability, as in so many other areas of broad common interest, the individual efforts are significant but seem to be far less than what may be accomplished if we could find ways to pool our resources and talents. The Federation of Materials Societies (FMS), formed for just such collective action, has been just as strong and effective as its member societies are willing to make it. This has meant that it is not nearly as strong and effective as it needs to be. This area of materials and sustainability may be one in which our common interest will override our competitive natures to that degree required for some serious collective action.

The FMS has begun to set the stage for such action in its strategic planning and through three meetings held in the late 1990s. In the first two of these meetings, workshops were organized jointly with the interagency government group EMAT. At the December 1996 meeting, society and government agency representatives laid out a map of their current efforts on environmental issues in the materials arena. It was a sort of "get-to-know you" meeting, one which was useful in identifying the "best practices" that could be found in each of the professional societies. The report of that meeting, available to all participants and to other societies that request it, can act as a template against which to examine one's own efforts and a challenge to achieve the level of best practice in each society.

The second FMS-EMAT workshop, held in December 1997, targeted the technical arena and identified action items for government and societies. The third meeting was held in May 1998. At this, the 15th Biennial on Materials Policy, the general theme was "Maximizing Return on Investment in R&D: Case Studies in Materials," but materials and the environment were significant in the discussion. Here too, calls for action resulted, but no plan for action emerged.

This has too often been our history, as we usually tend to put too many items on the agenda for action and are then unable to begin any one of them. Furthermore, while it is all well and good to suggest things for others to do, it is time for us to take action ourselves. We must search for one good area where cooperation among the societies can be carried through and just begin together and do it. A specific suggestion for action is included in the summary, but this section lists some areas of current society activity and challenges the reader to determine whether ASM International and TMS may be doing more than they currently do.

Professional societies all engage in promoting the R&D agenda, in education and in publicity, but not all use the same mechanisms and not all have environmental activity in each area. The strongest efforts include some central committee(s) and staff with well-defined responsibilities for environmental activity. Promoting the R&D agenda includes the following: helping to define that agenda, for example, by participating in roadmapping exercises; creating the forum for discussing results, for example, meetings, workshops, and trade shows; making information widely available, for example, through publications of technical proceedings and organized data; and, in several instances, directly facilitating needed R&D, for example, by acting as manager for federally- or industry-funded projects. Some societies have workshops or symposia focused on environmental issues at every annual meeting; others do so rarely, if at all.

Educational activities include kindergarten to high school and beyond, including specific vocational training, and include formal curricular involvement as well as preparation of material that may supplement the core curriculum. Some societies now incorporate environmental activity in such educational efforts, while many do little more than pay lip service to the link between materials issues and the environment.

By publicity, that means telling our story to the world. Besides being an element of education, there is a special need for communication with individuals who know little about the science and technology that have played such a significant part in providing the standard of living to which we have now become accustomed. Such individuals, be they voters or heads of powerful Congressional appropriations committees, must understand the positive role materials has played and can continue to play, or we will fail in any attempt to achieve sustainable development. Scientists and engineers have not had a very good track record in this publicity arena, but many societies have turned their attention to this effort of public education. Specifically, the preparation of "success stories" and the direct contact with Congressional representatives and staff can help tell them what we have done for them lately and what we may yet do if given the opportunity. Many of the materials societies are still behind the curve in this activity, and in only a few cases have the successes of our efforts been focused on the positives to the environment. There is much yet to be done here.

Summary and Recommendations

In summary of efforts over the last quarter of the twentieth century, how are we doing? In general, we have made considerable progress. We have increased lifetimes for many materials, especially when we include recycling as an extension of total "life." We have cleaned up the processing lines and reduced scrap considerably in many instances, and we have lightened transportation vehicles through some extensive materials substitution. However, it should be clear from earlier remarks that we have mostly been gathering low-hanging fruit. If we are to really achieve a worldwide sustainable manufacturing paradigm, the hardest work is ahead. Some of that work will be technical, but much will be organizational. We have a pretty good track record on the technical agendas that do not need any further discussion here. The organizational agenda is a major concern and should be addressed in two arenas—government and professional societies.

On the government side, it is time to revitalize the environmental technology initiative focused on addressing the technical opportunities and needs associated with sustainability. Heaton and Banks (Ref 2.13) argue persuasively for such an effort and suggest three reasons why we cannot depend solely on the private sector to produce such technologies. First, the current regulatory system acts as an impediment both to the innovation and the long-term predictability that R&D investments demand; second, much of the environmental R&D that does occur is concentrated in the large firms, beyond the reach of many small companies and disaggregated industries most in need of improvement; and third, the kinds of research that may help most over the long term often produce results of such generic applicability that no single company can justify the investment. The federal investment seems justified in such instances as one element of a strategy to achieve the desired social goal; however, this time, it should not be lodged within the EPA, but rather in the DoE or NIST Advanced Technology Program. This would eliminate the conflict between regulation and the best technical solutions, which always plagues any agency. In anticipation of such an initiative and to clarify the present baseline, the MatTec/EMAT committee should carry out an inventory of materials R&D in support of environmental sustainability. Even with all of the uncertainties inherent in such an analysis of existing programs, most of them currently justified on some other, mission-focused basis, a beginning must be made to clarify the likely impacts of the government's investment.

What should we demand of our professional societies? We have seen in the FMS-sponsored discussions over the last several years that there are an enormous number of environmental-related activities underway when one examines all of the societies collectively, but few are doing all things, and many things are less than adequately covered even by the separate, disaggregated efforts of the individual societies. We need more technical activity and we need to do more to tell the story of what we have already accomplished. On the technical side, we need to give a home to the R&D activity on sustainability, through special symposia and in our regular meetings. In many instances, this may require collaboration among several societies to broaden the coverage of science through engineering or range of materials, and, in all cases, it will require more visibility and overt statement of the agenda of sustainability. The technical agenda also extends to the arena of helping in the process of identifying the needed research agenda. Here, the most significant needed steps will require closer cooperation between the professional societies and trade associations. We have seen significant steps here in the American Ceramic Society, the Heat Treat and Thermal Spray Societies within ASM International, the American Welding Society, and the American Foundry Society. Whether it be through such formal arrangements or through informal cooperation, as happened with the chemical industry associations and the American Chemical Society and the American Institute of Chemical Engineers in the development of the chemical industry roadmap, such future vision exercises demand an industrial perspective to give broad outlines, followed by detailed technical analysis to translate those guidelines into useful bases for action by researchers.

Finally, and to close with a specific call for action, we must tell the world about our successes. For several years, stimulated largely by threats against continuing funding of science, several large societies, most notably the American Chemical Society and American Physical Society have been working to develop carefully crafted, well-documented, graphically attractive examples of how basic research in their fields has led to technological advances that the general public would recognize as positively affecting their lives. In general, while those efforts have included some materials examples, the materials story has yet to be told. Several materials societies have made starts on such an effort, but the high cost and too diffuse an agenda have both contributed to a weak showing. On the other hand, a focused effort on success stories about materials and sustainability carried out jointly by several of our societies and coordinated by the FMS can succeed and would be of significant value on several fronts. It would increase the armamenta that we need to make the case on the importance of materials research to all levels of government; it would help us clarify to the public at large and to young people in particular that we wear the "white hats" in the environmental arena; it would provide a resource for educational use in early school years; and it would provide a solid, well-defined project for the diverse members of the FMS to engage in to test, once and for all, if we are capable of working together for our common good. Such a project would require resources that no single society is likely to advance but that would be feasible through collective action, particularly if the trade associations join the action. This idea has already been brought up at the FMS Biennial meeting with generally positive reactions from those assembled, but others, representing the FMS and its member societies, will have to come together to give flesh to this skeleton of an idea.

The targets are becoming clear; our role is central; it is time for the materials community to accept the burden and step forward as leaders in the quest for sustainable development.

REFERENCES

2.1. H. Brooks, *Metall. Trans.*, Vol 3, 1972, p 759–768

2.2. "Final Report of The National Commission on Materials Policy," Library of Congress Card No. 73-600202, June 1973

2.3. "Report of the Committee on the Survey of Materials Science and Engineering," National Academy of Sciences, Washington, D.C., 1974, followed by four appendixes published in 1975

2.4. "Final Report of The National Commission on Materials Policy," June 1973, p 1–4

2.5. J. Boyd, *Metall. Trans.*, Vol 5, 1974, p 5–10

2.6. M. Tenenbaum, *Metall. Trans. A*, Vol 7, 1976, p 339–350

2.7. H.H. Kellogg, *Metall. Trans. A*, Vol 9, 1978, p 1695–1704

2.8. "Report of the Committee on Materials Science and Engineering," National Academy Press, Washington, D.C., 1989

2.9. J.H. Ausubel, D.G. Victor, and I.K. Wernick, *Consequen. Nature Implic. Environ. Change*, Vol 1 (No. 3), 1995, p 2–15

2.10. R.A. Frosch and M. Uenohara, *Industrial Ecology: US/Japan Perspectives*, National Academy of Engineering, Washington, D.C., 1994, p 2

2.11. "Technology for a Sustainable Future," Report of the National Science and Technology Council, Office of Science and Technology Policy, U.S. Govt. Printing Office, July 1994

2.12. "Bridge to a Sustainable Future," Report of the National Science and Technology Council, Office of Science and Technology Policy, U.S. Govt. Printing Office, April 1995

2.13 G.R. Heaton, Jr. and R.D. Banks, in *Investing in Innovation*, L.M. Branscomb and J.H. Keller, Ed., MIT Press, Cambridge, MA, 1998, p 276, 292

2.14 P.M. Eisenberger, Ed., *Basic Research Needs for Environmentally Responsive Technologies of the Future,* Materials Institute, Princeton, NJ, 1996

2.15. J. Williams, *Metall. Mater. Trans. A*, Vol 29, 1998, p 0000–00

2.16. J. Clark, S. Newell, and F. Field, *Life Cycle Engineering of Passenger Cars*, VDI Verlag GmbH, Duesseldorf, 1996, p 1–19

2.17. S.A. Newell, Ph.D. thesis, MIT, June 1998

2.18. M. Sullivan and L. Wagner, "Materials Flow Studies, Total Materials Consumption—An Estimation Methodology and Example Using Lead. U.S. Geological Survey," presented at the Eighth Int. Symp. on Raw Materials, Federal Institute for Geosciences and Natural Resources, Hannover, Germany, 1997

2.19. A.H. Teich, *Linking Science and Technology to Society's Environmental Goals,* National Academy Press, Washington, D.C., 1966, p 345

2.20. "The Federal R&D Program in Materials Science and Engineering, Data for Fiscal 1995," www.msel.nist.gov

2.21. C.J. Campbell and J.H. Laherrère, *Sci. Am.*, March 1998, p 78–83

2.22. J.J. MacKenzie, *Iss. Sci. Technol.,* Summer 1996, p 48–54

2.23. *Scenarios of U.S. Carbon Reductions,* Office of Energy Efficiency and Renewable Energy, DoE, Washington, D.C., 1998

2.24. "The Steel Industry Technology Roadmap," The American Iron and Steel Institute

2.25. "Materials," Report of the Interagency Workgroup on Industrial Ecology, Materials and Energy Flows, Fall 1998. For copies, contact 1-800-363-3732

2.26. D.T. Allen and N. Behmanesh, *The Greening of Industrial Ecosystems,* National Academy Press, Washington, D.C., 1994, p 69–89

CHAPTER 3

Life-Cycle Inventory Analysis of the North American Aluminum Industry*

LIFE-CYCLE ASSESSMENT (LCA) is a methodology that uses a systems approach to understand the environmental consequences of a product, process, or activity from initial extraction of raw materials from the Earth until the point at which all residuals are returned to the Earth. The goal of LCA is to quantify, evaluate, and then identify opportunities to reduce the overall environmental effects of the system under study.

The LCA methodology, as defined by the International Standards Organization (ISO), is typically divided into four separate and interrelated components:

- *Life-cycle scoping and goal definition*: Includes the clear statement of the purpose of the study, the system to be studied, the intended use of the results, limitations on its use for other purposes, data quality goals, reporting requirements, and the relevant type of review process. The scope also defines a description of the geographical and temporal boundaries, system boundaries, data requirements, decision rules, and other assumptions.
- *Life-cycle inventory (LCI) analysis*: The phase of LCA involving the compilation and quantification of inputs and outputs through the life cycle of a product or service, including the stages of resource extraction, manufacturing, distribution, use, recycling, and ultimate disposal.
- *Life-cycle impact assessment*: The phase of the LCA aimed at understanding and evaluating the magnitude and significance of the potential environmental impacts of a product system.
- *Life-cycle interpretation*: The phase of the LCA technique in which the findings of the inventory analysis and impact assessment are combined in line with the defined goals and scope. The findings may take the form of conclusions and recommendations to decision makers, consistent with the goal and scope of the study.

This LCI study of the North American aluminum industry has the objective of generating a resource consumption and environmental profile of the aluminum-producing industry. The purposes are to:

- Provide the participating companies with detailed LCIs of the various processes and activities within the aluminum product life cycle. This information provides a baseline for improvements in the management of energy, raw material use, waste elimination, and the reduction of air and water emissions.
- Develop a comprehensive and up-to-date LCI of primary and secondary aluminum products produced in North America for the United States Automotive Materials Partnership (USAMP) initiative.

The domain of this study on behalf of the North American automotive industry aluminum system includes operations in the chain of plants that supply metal to the Big Three automobile manufacturers. This study contains data collected from 213 reporting locations representing 15 unit operations. The specific unit operations for the primary industry are shown in Fig. 3.1 and for the secondary industry in Fig. 3.2 Unit

*Adapted with permission from The Aluminum Association Report AT-2: "Life Cycle Inventory Report for The North American Aluminum Industry," November 1998

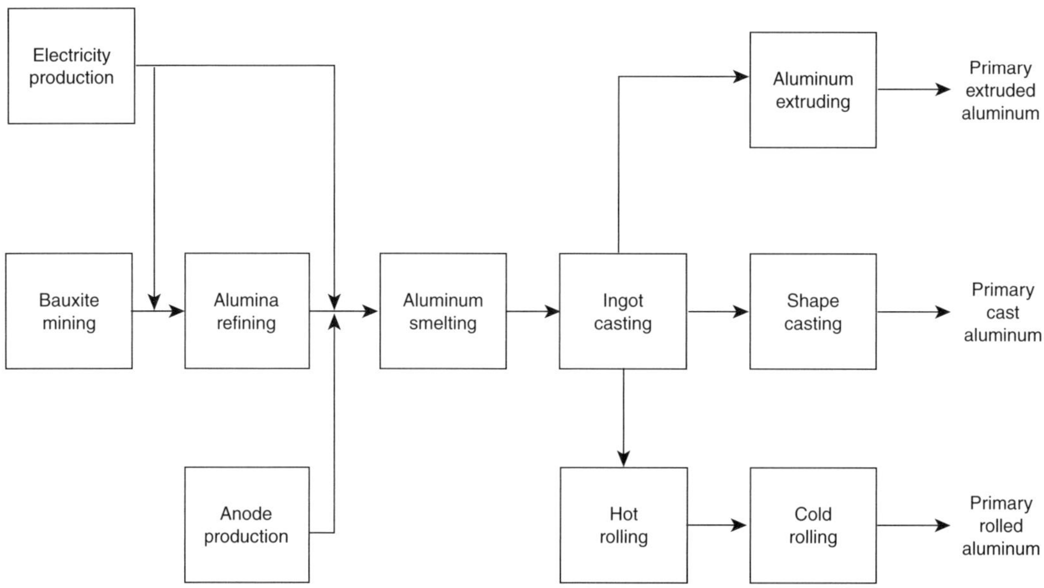

Fig. 3.1 Simplified process flow diagram for primary aluminum products

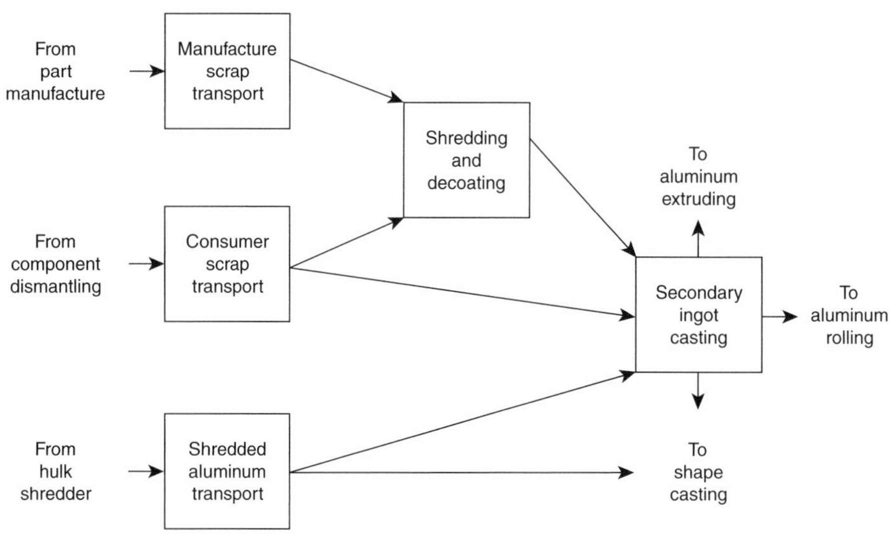

Fig. 3.2 Simplified process flow diagram for secondary aluminum products

operations ranged from bauxite mining in Australia, Africa, Brazil, and Jamaica to primary and secondary processes and manufacturing for North America. For each reporting location, information on energy and materials used, water consumption, and releases to air, water, and land was gathered. The study follows the USAMP LCI methodology (Ref 3.1) developed by Roy F. Weston, Inc. and ISO 14000 series (Ref 3.2) LCA guidelines.

A key element of this study is the scope of the data-gathering efforts. The results of this LCI report are based on input and output data of environmental relevance collected at production facilities. This information was gathered by the Aluminum Association, Inc. and Association de l'Industrie de l'Aluminium of Canada from 13 primary and secondary aluminum-producing companies in North America. The data were assessed in detail by both Weston Inc. and the

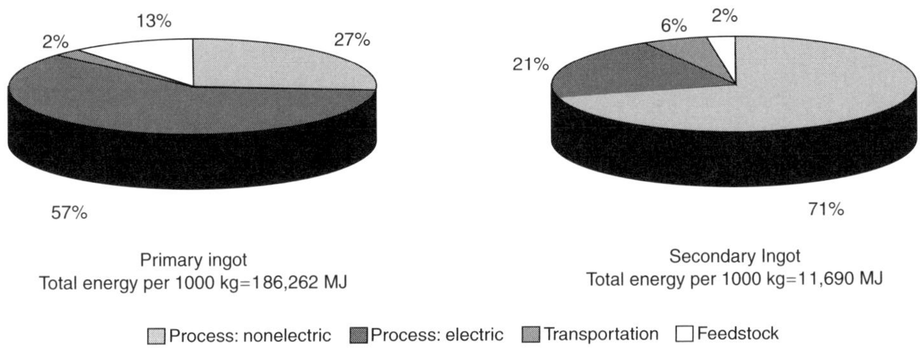

Fig. 3.3 Energy consumption for primary and secondary ingots

Aluminum Association LCA Task Force. A panel of external recognized life-cycle experts selected by the USCAR steering committee also reviewed the report.

The results of the LCI have confirmed and reinforced the progress that the aluminum industry has been making, especially in regard to energy efficiency, recycling, and the reduction of CO_2 emissions. Improvements in smelter cell operation and practice have continued to improve overall energy efficiency and have significantly lowered the releases of gases with global-warming potential. The enormous significance of recycling, which now has grown to represent approximately 37% (1995 figure) of the metal supply, has again been emphasized, because the LCI demonstrates the great energy benefits associated with recycling aluminum as compared to producing aluminum from the bauxite ore. This comparison is illustrated in Fig. 3.3 for the integrated performance of the North American industry with regard to energy-use components for both primary and secondary aluminum ingots.

The results of this LCI study are intended to:

- Establish a database describing the resource consumption and environmental releases of materials and processes
- Improve understanding of the environmental implications of aluminum product manufacture
- Facilitate assessment of the life-cycle environmental inventory of alternative product design options (for example, alternative fuels, process design, etc.), compare corresponding data sets, and guide the evaluation of modifications for improvement
- Provide environmental information for use in strategic planning

In terms of limitations, there is potential for misuse of the database and results. Specifically, the inventory database and results shall not be:

- Used as the sole criterion in raw materials selection decisions
- Quoted or published without prior consent of The Aluminum Association, Inc., except in the case where a participant* or designate wished to quote or publish their own submission of data
- Used as a basis for impact assessment

Life-Cycle Inventory Methodology

Life-cycle inventory methodology is reasonably well established. The techniques that have been used to model other systems are directly applicable to almost all aspects of the product system. However, LCI is assumption based, and many decisions are involved in the development of an inventory. Therefore, for the credibility of the results and conclusions drawn from the study, it is essential to have clearly defined assumptions, decision rules, and data quality assessment.

Aluminum product systems include:

- Primary ingot
- Primary rolled aluminum
- Primary extruded aluminum

*The Aluminum Association member companies that participated in this study include: Alcan Aluminum, Ltd.; Alumax, Inc.; Aluminum Company of America (ALCOA); Commonwealth Aluminum; Kaiser Aluminum & Chemical Corporation; Century Aluminum Corporation; Reynolds Metals Company; Southwire Company; Wabash Alloys; and Werner Company. The members of Association de l'Industrie de l'Aluminium of Canada that participated include: Aluminerie Alouette Inc.; Aluminerie de Bécancour, Inc.; Aluminerie Lauralco Inc.; and Canadian Reynolds Metals, Inc.

- Primary cast aluminum
- Secondary ingot
- Secondary rolled aluminum
- Secondary extruded aluminum
- Secondary cast aluminum

The simplified process flow diagrams for each of these processes are shown in Fig. 3.1 and 3.2. The results of the rolled-up data for each of these products (modeled for 1000 kg of the product) are presented in the section "Results by Product System" in this chapter.

In a comprehensive cradle-to-gate analysis, the system's "envelope" includes the extraction of raw materials from the Earth (bauxite mining), materials processing (anode production, alumina refining, aluminum smelting, primary ingot casting), manufacture (hot and cold rolling, extruding, and aluminum casting), and recovery/recycling (consumer aluminum transport, manufactured aluminum transport, shredded aluminum transport, shredding and decoating, and secondary ingot casting).

All of the primary data provided, with one exception, were from the calendar year 1995. One location in the bauxite mining unit process was from 1992 to 1993 from a previous aluminum industry study that met the data quality requirements for this study. Data for ancillary materials flows that were included in the study were taken from the Weston Inc. database representing data that were, for the most part, from 1995 or 1996. All ancillary data satisfied the requirements defined by the methodology.

Geographic Coverage

The goal of the USAMP study was to develop a comprehensive LCI of the North American generic vehicle. As such, the initial step in the aluminum analysis of the domain for this study included an evaluation of the sources of supply to the automotive industry for semifabricated and fabricated products (i.e., rolled, extruded, and cast aluminum). With respect to automotive rolled aluminum, the aluminum industry task force evaluated the North American rolling plants to determine those that were capable of supplying the auto industry. Only these plants were designated as part of the domain. When it came to the extruded and cast aluminum supply, there were numerous plants that could qualify for inclusion in the domain. For reasons of practicality, and recognizing that there is only minor variation in the extruding and casting processes, only a sample of these operations was selected for inclusion in the domain.

The next step in the analysis was to define the source of aluminum ingot supply to the fabrication processes; these included both primary and secondary sources. This process of defining the supply chain within the North American domain was extended to the bauxite mines that ultimately feed the automotive subsystems. The domain for the subsystems and the geographic coverage is summarized in Table 3.1.

Technology Coverage

The domain identified for the study represented the full range of technological differences in the various unit processes. These differences were defined in an analysis of the available technologies, and the review of actual reporting locations did not indicate any biases or distortions for one type of technology versus another. A summary of the unit process descriptions may be found in Table 3.2 and the following text.

Table 3.1 Domain and geographical coverage

Source	Unit process	Geographical regions	Total by unit process
Primary	Bauxite mining	Australia, Africa, Brazil, Jamaica	10
	Electricity generation	North America	8
	Alumina refining	Australia, Brazil, Germany, Jamaica, United States	13
	Anode production	North America	27
	Aluminum smelting	North America	29
	Primary ingot casting	North America	23
	Subtotal		**110**
Manufacturing	Hot rolling	North America	14
	Aluminum extruding	North America	21
	Cold rolling	North America	17
	Aluminum casting	North America	6
	Subtotal		**58**
Secondary	Manufactured scrap transport	North America	9
	Consumer scrap transport	North America	12
	Shredded aluminum transport	North America	12
	Shredding and decoating	North America	13
	Secondary ingot casting and melting	North America	39
	Subtotal		**85**
	Total		**253**

Bauxite mining activities mainly take place in tropical and subtropical areas of the Earth. Most all bauxite is mined in an open-pit mine. The known reserves of aluminous ore will sustain the present rate of mining for 300 to 400 years.

Commercial bauxite can be separated into bauxite composed of mostly alumina trihydrates and those composed of alumina monohydrates. The trihydrate aluminas contain approximately 50% alumina by weight, while monohydrates are approximately 30%. Monohydrates are normally found close to the surface in grassy plains (e.g., Australia), while trihydrates tend to be at deeper levels and in heavily forested areas (e.g., Brazil).

Table 3.2 Unit process description in aluminum life-cycle inventory analysis

Unit process	Operations of the unit process	Output
Bauxite mining	• Extraction of bauxite-rich minerals for the site • Beneficiation activities, such as washing, screening, or drying • Treatment of mining site residues and waste • Site restoration activities, such as grading, dressing, and planting • Excludes material, energy, and releases associated with personnel infrastructure	Bauxite that is transported to an alumina refinery
Electricity production	• Management of reservoirs • Generation of electrical energy • Transformation and transmission of high-voltage energy • Distribution, transformation, and rectification of low-voltage energy • Maintenance and repair of plant and equipment • Treatment of process air, liquids, and solids	Electrical energy for uses at production facility
Alumina refining	• Bauxite grinding, digestion, and processing of liquors • Alumina precipitation and calcination • Maintenance and repair of plant and equipment • Treatment of process air, liquids, and solids	Smelter-grade alumina transported to primary aluminum smelter
Anode production	• Recovery of spent anode materials • Anode mix preparation; block or briquette forming and baking • Rodding of baked anodes • Maintenance and repair of plant and equipment • Treatment of process air, liquids, and solids	Rodded anodes or briquettes for transport to aluminum smelter
Aluminum smelting	• Recovery, preparation, and handling of process materials • Manufacture of major process equipment (e.g., cathodes) • Process control activities (metal, bath, heat) • Maintenance and repair of plant and equipment • Treatment of process air, liquids, and solids	Hot metal transported to ingot casting facility
Ingot casting	• Pretreatment of hot metal (cleaning and auxiliary heating) • Recovery and handling of internal process scrap • Batching, metal treatment, and casting operations • Homogenizing, sawing, and packaging activities • Maintenance and repair of plant and equipment • Treatment of process air, liquids, and solids	Packaged aluminum ingots or hot metal transported to an aluminum fabricating facility
Aluminum extruding	• Preheating and cutting or shearing of billet lengths • Extruding of shapes, cooling, stretching, and cutting • Heat treating, aging, anodizing, or painting • Finishing and packaging activities • Maintenance and repair of plant and equipment • Treatment of process air, liquids, and solids • Excludes material, energy, or releases associated with secondary ingot casting	Extruded profiles shipped to component fabricators
Rolling hot mill	• Preheating and scalping of ingot • Hot mill rolling of coils • Maintenance and repair of plant and equipment • Treatment of process air, liquids, and solids • Excludes material, energy, or releases associated with secondary ingot casting	Aluminum coils or plate shipped to component fabricators
Rolling cold mill	• Cold mill rolling of coils • Heat treating, leveling, anodizing, or painting • Finishing and packaging activities • Maintenance and repair of plant and equipment • Treatment of process air, liquids, and solids • Excludes material, energy, or releases associated with secondary ingot casting	Aluminum coils or plate shipped to component fabricators
Shape casting	• Pretreatment of hot metal (cleaning and auxiliary heating) • Recovery and handling of internal process scrap • Preparation and forming of cores and molds • Batching, metal treatment, and casting operations • Finishing and packaging activities • Maintenance and repair of plant and equipment • Treatment of process air, liquids, and solids	Semifabricated aluminum components

(continued)

Table 3.2 *(continued)*

Unit process	Operations of the unit process	Output
Manufacturing scrap transport	• Bailed, bundled, and segregated aluminum scrap on loading dock of auto part dismantler • Transportation by various modes	Scrap transported to aluminum recycling facility
Consumer scrap transport	• Containerized, bundled, and segregated aluminum scrap on loading dock of fabricator • Transportation by various modes	Scrap transported to aluminum recycling facility
Shredded aluminum transport	• Shredded and segregated aluminum scrap on loading dock of auto hulk shredder • Transportation by various modes	Scrap transported to aluminum recycling facility
Shredding and decoating	• Recovery, preparation, and handling of process materials • Shredding and decoating activities • Maintenance and repair of plant and equipment • Treatment of process air, liquids, and solids	Shredded aluminum to secondary ingot casting
Secondary ingot casting	• Recovery and handling of internal process scrap • Batching, metal treatment, and casting operations • Homogenizing, sawing, and packaging activities • Maintenance and repair of plant and equipment • Treatment of process air, liquids, and solids	Packaged ingots or hot metals transported to fabrication facilities

The only significant processing difference in bauxite mining is the need for beneficiation. Beneficiation occurs with ores from forested areas, while the grassland type typically does not require washing. The wastewater from washing is normally retained in a settling pond and recycled for continual reuse.

Electricity Production. The aluminum smelting process is energy-intensive, and as such, primary aluminum smelters are located close to large sources of reliable energy. The Canadian aluminum smelting industry is located in Quebec and British Columbia due to the abundant source of hydroelectric energy. Indeed, 100% of primary smelting capacity is supplied with hydroelectricity. Prior to entering the smelting process, the alternating current is converted into direct current through a series of rectifiers.

Alumina Refining. In alumina refining, bauxite is converted to aluminum oxide using the Bayer process. Most refineries blend bauxite to provide feedstock with consistent properties. Bauxite is ground and blended into a recycled liquor. The liquor contains sodium carbonate and sodium hydroxide plus supernatant liquor recycled from red mud holding ponds. The slurry is heated and pumped to digesters, which are heated pressure tanks. In digestion, iron and silicon impurities form insoluble oxides called red mud. The red mud settles out, and a rich concentration of sodium aluminate is filtered and seeded to form hydrate alumina crystals in precipitators. These crystals are then heated in a calcining process. The heat in the calciners drives off combined water, leaving alumina.

The major differences in processing are at the calcinations stage. Two types of kilns are used: rotary and fluid bed. The fluid bed or stationary kiln is newer and significantly more energy-efficient.

Significant reductions of energy per unit of alumina have been achieved over the last 15 years with the introduction of higher-pressure digesters and fluid flash calciners. Energy requirements have almost been halved in that period.

Anode Production. There are two types of aluminum smelting technologies that are distinguished by the type of anode that is used in the reduction process: Soderberg and prebake.

Soderberg design has a single anode that covers most of the top surface of a reduction cell (pot). Anode paste (briquettes) is fed to the top of the anode, and as the anode is consumed in the process, the paste feeds downward by gravity. Heat from the pot bakes the paste into a monolithic mass before it reaches the electrolytic bath interface.

The prebake design has prefired blocks of solid carbon suspended from axial busbars. The busbars both hold the anodes in place and carry the current for electrolysis.

The process for making the aggregate for briquettes or prebake blocks is identical. Coke is calcined, ground, and blended with pitch to form a paste that is subsequently extruded into blocks or briquettes and allowed to cool. While the briquettes are sent directly to the pots for consumption, the blocks are sent to a separate baking furnace.

Baking furnace technology has evolved from simple pits that discharged volatiles to the atmosphere during the baking cycle to closed-loop-type designs that convert the caloric heat of the volatile into a process fuel that reduces energy consumption for the process.

Aluminum Smelting. Molten aluminum is produced from alumina by the Hall-Heroult electrolytic process that dissolves the alumina in a molten cryolite bath and passes current through this solution, thereby decomposing the alumina into aluminum and oxygen. Aluminum is tapped out of the reduction cell (pot) at daily intervals, and the oxygen combines with the carbon to form carbon dioxide and carbon monoxide.

The pot consists of a steel shell lined with refractory insulation and with a hearth of carbon. This is known as the cathode. The cathode is filled with a cryolite bath and alumina, and an anode is suspended in the bath to complete the circuit for the pot. When started, a pot will run continuously for the life of the cathode, which may last in excess of 10 years. The current in a pot varies from 60,000 to over 300,000 A at a voltage drop of 4.2 to 5.0 V. Pots produce approximately 7.3 +/– 0.3 kg/day (16.2 +/– 0.6 lb/day) of aluminum for each kiloampere at an operating efficiency of 91 +/– 4%.

All North American aluminum smelters use some type of air pollution control system to reduce emissions. The primary system is typically a scrubber. Some plants use dry scrubbers with alumina as the absorbent, which is subsequently fed to the pots and allows for the recovery of scrubbed materials. Other plants use wet scrubbers, which recirculate an alkaline solution to absorb emissions. Unlike dry scrubbers, wet scrubbers absorb carbon dioxide, nitrogen oxide, and sulfur dioxide that are entrained in the wastewater liquor.

Ingot Casting. Molten metal siphoned from the pots is sent to a resident casting complex found in each smelter. In some cases, due to proximity, molten metal is transported directly to a shape casting foundry. Molten metal is transferred to a holding furnace, and the composition is adjusted to the specific alloy requested by a customer. In some instances, depending on the application and on the bath composition in the pots, some initial hot metal treatment to remove impurities may be done.

When the alloying is complete, the melt is fluxed to remove impurities and reduce gas content. The fluxing consists of slowly bubbling a combination of nitrogen and chlorine or carbon monoxide, argon, and chlorine through the metal. Fluxing may also be accomplished with an in-line degassing technology, which performs the same function in a specialized degassing unit.

Fluxing removes entrained gases and inorganic particulates by flotation to the metal surface. These impurities (typically called dross) are skimmed off. The skimming process also takes some aluminum, and as such, drosses are normally further processed to recover the aluminum content and to make products used in the abrasives and insulation industries.

Depending on application, metal is then processed through an in-line filter to remove any oxides that may have formed. Metal is then cast into ingots in a variety of methods: open molds (typically for remelt ingot), direct chill molds for various fabrication shapes, electromagnetic molds for some sheet ingots, and continuous casters for aluminum coils.

Aluminum Extruding. The extrusion process takes cast extrusion billet (round bar stock produced from direct chill molds) and produces extruded shapes. The process begins with an in-line preheat that takes the temperature of the billet to a predetermined level, depending on the alloy. The billet is then sheared, if not already cut to length, and deposited into a hydraulic press. The press squeezes the semiplastic billet through a heated steel die that forms the shape. The shape is extruded into lengths defined by the take-off tables and is either water quenched or air cooled. The shape is then clamped and stretched to form a solid, straightened length.

The straightened lengths are cut to final length multiples and placed in an aging furnace to achieve a desired temper. Lengths are then finished (drilled and shaped) and placed into a coating process. The types of coatings include anodized, painted, and lacquered finishes.

There are over 250 extrusion plants in North America, some with multiple presses. The technology is relatively mature, and variation in process efficiency is minor.

Depending on the shape and desired performance characteristics of the extrusion, some profiles are put through an impact extruding process using considerably higher pressures to form the final parts.

Aluminum hot rolling processes may begin with direct chill cast ingot or with continuous cast coils. Direct chill ingots are typically 45 to 66 cm (18 to 26 in.) thick and weigh from 13 to 27 metric tons (15 to 30 tons) each. The ingots are preheated to approximately 540 °C

(1000° F) and may be scalped on their rolling surfaces or directly fed to a reversing hot mill. In the reversing mill, the ingot passes back and forth between the rollers, and the thickness is reduced to between 25 to 50 mm (1 to 2 in.), with a corresponding increase in length. Following the reversing mill, the coils are fed to a continuous hot mill where the thickness is further reduced to less than 6.3 mm (0.25 in.). Continuous cast coils would also be included at this stage.

Aluminum Cold Rolling. Prior to the cold mill, the coils may be annealed to give the metal workability; some plants have moved toward self-annealing, which requires no additional energy input. Coils are then passed through a coil mill to further reduce the thickness to the customer's requirements and may also go through leveling, heat treating, slitting, cutting to length, and coating processes to meet customer requirements.

Shape Casting. Aluminum castings are produced in aluminum foundries that are designed to process a range of old and new scrap qualities that contain alloying elements.

Aluminum castings represent over 20% of total semifabricated production and are growing at an annualized rate of 4.8%. North American production is over 1 million tonnes. More than 80% of aluminum castings are used in the automotive industry.

The majority of castings are produced by secondary smelters, of which there are more than 70 production plants in North America. The technology employed is dependent on the shapes produced and uses both permanent and nonpermanent molds. Newly constructed plants include purpose-built facilities that supply the automotive market.

Manufacturing Scrap Transport. Manufacturing scrap is generated when semifabricated aluminum products are manufactured into final components or parts. The amount varies with application and will be defined by the component and final product manufacturers. This scrap is normally returned to the semifabricated aluminum producers.

Consumer Scrap. This material is typically generated by automotive dismantlers prior to the automotive hulks being sent to a hulk shredder. Given the high residual value of aluminum scrap and the premiums associated with segregated clean scraps, dismantlers invest the additional time and effort to remove easily distinguishable aluminum parts, such as body sheet, bumpers, wheels, brakes, and heat exchangers. Depending on market conditions, specific efforts to remove all aluminum engines, transmissions, and additional aluminum engine and suspension parts would also be included. In this study, the information on travel mode, load sizes and distances was defined by the secondary smelters.

Shredded Aluminum Scrap. All remaining aluminum automotive scraps are recovered from shredder residues using various media-separation technologies. These include flotation and eddy current methods. As noted, information on travel mode, load sizes, and distances was defined by the secondary smelters.

Shredding and Decoating. In some instances, a secondary aluminum smelter will process manufacturing and postconsumer scrap through a shredding and decoating process. Such processes generate smaller pieces, assist with the removal of impurities such as ferrous metals and other unwanted materials, and, in cases where the metal has been coated or finished, the organic cover may be removed through different types of decoating equipment (i.e., belt decoaters and fluidized bed decoaters). The main purpose for such processes is to improve the recovery and efficiency of recycling and to reduce environmental impacts of remelting coated materials.

Secondary Ingot Casting. The major distinction between secondary casters and primary casters is in the melting technology employed. Secondary smelters have a variety of purpose-built melting furnaces: top-loaded closed melters, rotary melters, and sidewell-feeding melting furnaces. These melting furnaces have different efficiencies and recoveries. Some of the melters also use a saltcake to cover the melt and reduce oxidation.

Inventory Analysis

Each unit process in the system is characterized and documented by a list of input materials, energy, and emissions, as shown in Fig. 3.4.

As noted, the primary source of data was obtained from questionnaires prepared and issued to the 13 participating companies. The evaluation of the data focused on the information received from the completed responses to these questionnaries.

The data categories (material inputs, energy inputs, environmental releases, material outputs) were predefined for each unit process.

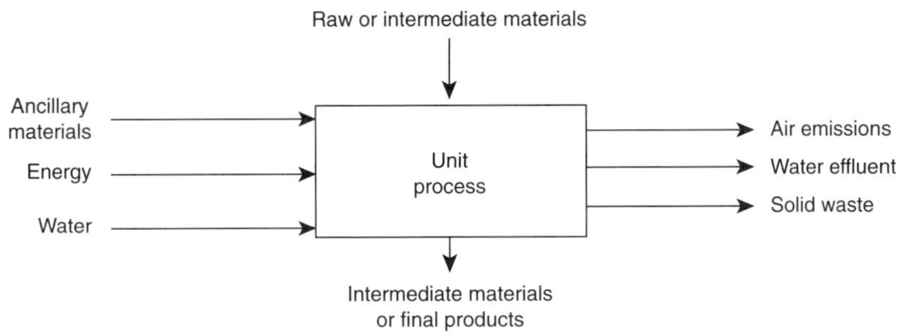

Fig. 3.4 Life-cycle inventory unit process template

These lists were identified by Weston, Inc. and the Aluminum Association and were completed based on the results of the survey.

The data categories used to present the results of this study are based on those specified by the USAMP methodology (Ref 3.1) document (Table 3.3). Additional data categories, at unit process level, were collected for Aluminum Association benchmarking purposes.

Calculation Procedures

In addition to the numerous assumptions that were made to simplify the data collection process, there were several special calculation procedures that were used to refine and integrate the information in the inventory. For example, all material and water consumption as well as environmental emissions were converted to metric units: mass in kilograms (kg); volume in liters (L); gaseous volume in cubic meters (m^3); surface area in square meters (m^2); and energy in mega joules (MJ). Other calculation methods address allocation procedures, deliberate omissions, and anomalies.

Allocation procedures are used to partition inputs and/or outputs within a specific product system. An allocation procedure is required when a unit process within a system shares a common pollution treatment infrastructure or where multiple products or co-products are produced in a common unit process.

For allocation of utilities and services common to several processes, allocations were made to reflect relative use of the service. For example, in instances where different unit processes (smelter and alumina refinery) were housed in the same effluent treatment facilities, allocation was determined by the percent of treatment load generated by each process.

Similar techniques based on mass of the unit process outputs were used to allocate some common solid waste streams, such as general refuse. When different unit processes shared a common energy source, for example, an electric grid, the metered energy to a specific unit process was used to allocate the common use items, such as plant and office lighting or space heating.

Deliberate Omissions. Capital assets (plant and equipment), transportation infrastructure (roads, street lights, service stations), and human activity (food and transport) associated with the production of materials are omitted from this study because the relative contribution to the product system for the functional unit being considered would be minor.

Table 3.3 Data categories specified by the United States Automotive Materials Partnership methodology document

1. Energy(a): Fossil Nonfossil Process (electrical and nonelectrical) Transportation Feedstock Total	4. Water emissions: Dissolved solids Suspended solids Heavy metals(c) Oils and greases Other organics Phosphates, ammonia
2. Water consumption: Ground Surface	5. Solid wastes: Total solid waste Sanitary/municipal
3. Air emissions: Dust and particulates (includes metals) Carbon dioxide Carbon monoxide Sulfur oxides Nitrogen oxides NMHC (includes halogenated organic compounds)(b) Methane Acid gases: HCl and HF Lead	6. Raw materials consumed: All significant inputs as defined by the decision rules

(a) All energies will include precombustion contributions.
(b) All organics other than methane are reported in the aggregate here.
(c) This includes arsenic, cadmium, chromium, cobalt, copper, lead, mercury, nickel, and zinc.

The manufacturing and postconsumer scrap is assumed to be ready for transport at the shipping location. Estimates for selective handling and processing (e.g., component dismantling) have not been included. These aspects are considered to be minor.

Treatment of Anomalies/Missing Data. Anomalies are extreme data values within a data set. Anomalies/missing data values are a result of misinterpreted requests for data input, misreported values, improper analysis, or simply not available from a reporting location. After an initial review of the data, additional information was requested by the facilities to verify certain misreported/missing values. The missing data feedback received from the individual plants was incorporated into the original data set. Where an anomaly was traced to process irregularities or accidental release, it was included in the data set. If an explanation could not be found, the anomaly was removed from the data set. When all attempts to secure actual inputs were exhausted, a calculated value was used based on the average of reported values from a unit process with similar technology.

Data Integration and Presentation

Life-cycle inventory modeling software calculates the LCI profile for each product system. Each of the unit processes and subsystems of the product system is aggregated to provide a summary cradle-to-gate LCI profile. Figure 3.5 shows the integration of unit process data with precombustion, transportation, fuel use, electricity, and ancillary LCI data to provide the product system LCI profile. Primary data for separate facilities concerning the same product or process were averaged and presented in a fashion that ensures confidentiality of individual company data. These aggregated results (weighted-average numbers normalized for each unit process) are given in the sections "Primary Aluminum Unit Processes" and "Secondary Aluminum Processes" in this article, along with unit process description, assumptions, and data quality assessment results.

Emissions from Fuel. The assumed energy values used are characteristically the same as fuels extracted in North America. The environmental releases (Table 3.4) from the combustion of these fuels include the precombustion energy for these fuels. Their emissions are added to the process emissions in the product system rollup.

Precombustion energy is defined as the energy associated with the acquisition, processing, and delivery of fossil fuels. Precombustion energy values were included for all fuels addressed within the scope of this LCI study.

Electrical Energy. The generation and delivery of electricity from electric utilities on a

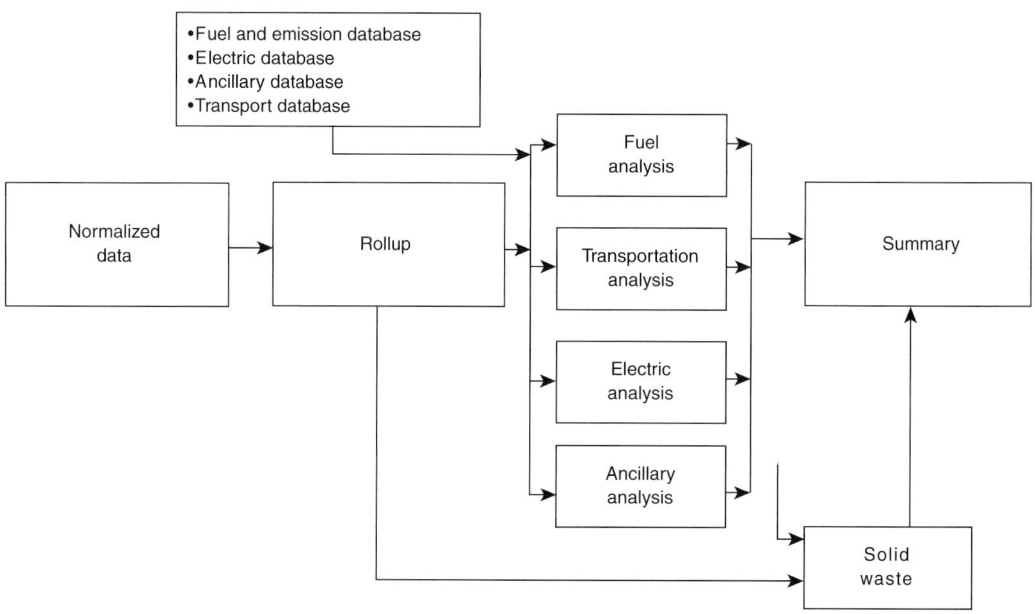

Fig. 3.5 Life-cycle inventory engine structure

Table 3.4 Typical industrial facility factors with fuel processing burden (per unit input)

		Heating fuel oil, kg	Diesel, L	Gasoline, L	Natural gas, m^3	Propane, L	Coal, kg
Mode		Industrial boiler	Industrial engine	Industrial engine	Industrial boiler	Industrial boiler	Industrial boiler
Water use	L	1.20×10^{-1}	1.02×10^{-1}	1.04×10^{-1}	2.32×10^{-7}	...	3.54×10^{-4}
Air emissions							
Particulate matter	kg/MJ	3.30×10^{-3}	1.58×10^{-3}	2.38×10^{-3}	4.92×10^{-5}	2.41×10^{-4}	1.30×10^{-3}
CO	kg/MJ	9.00×10^{-4}	1.27×10^{-2}	9.41×10^{-1}	6.64×10^{-4}	1.33×10^{-3}	6.87×10^{-3}
CO^2	kg/MJ	3.86	3.05	2.85	1.97	1.62	2.65
SO_X	kg/MJ	4.78×10^{-2}	3.71×10^{-3}	5.58×10^{-3}	1.86×10^{-2}	2.25×10^{-3}	1.53×10^{-2}
NO_X	kg/MJ	8.45×10^{-3}	4.75×10^{-2}	2.57×10^{-2}	9.01×10^{-3}	9.16×10^{-3}	7.74×10^{-3}
Nonmethane hydrocarbons	kg/MJ	1.03×10^{-3}	2.24×10^{-3}	4.72×10^{-2}	2.38×10^{-5}	2.77×10^{-3}	8.08×10^{-4}
Methane	kg/MJ	0	0	0	0	...	0
Lead	kg/MJ	0	0	0	0	...	0
HCl	kg/MJ	0	0	0	0	...	0
HF	kg/MJ	0	0	0	0	...	0
Water effluents							
BOD(a)	kg/MJ	4.07×10^{-4}	3.44×10^{-4}	7.85×10^{-4}	9.76×10^{-10}	...	1.17×10^{-6}
COD(b)	kg/MJ	0	0	0	0	...	0
Heavy metals	kg/MJ	0	0	0	0	...	0
Other organics	kg/MJ	...	...	...	...	...	...
Suspended solids	kg/MJ	1.85×10^{-3}	1.56×10^{-3}	3.56×10^{-3}	4.43×10^{-9}	...	5.30×10^{-6}
Dissolved solids	kg/MJ	0	0	0	0	...	0
Oil and grease	kg/MJ	2.34×10^{-4}	1.98×10^{-4}	3.60×10^{-4}	5.13×10^{-10}	...	6.57×10^{-7}
Phosphates	kg/MJ	...	...	...	...	...	...
NH^{4+}	kg/MJ	0	0	0	0	...	0
Solid waste	kg/MJ	1.71	1.45	1.28	3.48×10^{-6}	...	2.35×10^{-1}

(a) BOD, bacteriological oxygen demand. (b) COD, chemical oxygen demand

References

1. *Coal Industry Annual 1995*, DOE/EIA-0584(95), Energy Information Administration, 1996
2. *Petroleum Supply Annual 1995*, Vol 1, DOE/EIA-0340(95)/1, Energy Information Administration, May 1996
3. *Natural Gas Annual 1995*, DOE/EIA-0131(95), Energy Information Administration, Nov 1996
4. Stationary Point and Area Sources, *U.S. EPA Compilation of Air Pollutant Emission Factors*, AP-42, 5th ed., Vol 1

regional basis involves a mix of energy sources, electricity generation efficiency, transmission losses, and environmental releases. Data were determined for the ten census divisions in the U.S. and Canadian national grid based on information published by the Energy Information Administration (Ref 3.3) and the North American Electric Reliability Council (Ref 3.4). A hydropower profile was used to calculate LCI releases from facilities using hydropower. (A hydroelectric LCI profile was developed by Roy F. Weston, Inc. consistent with decision rules and data processing requirements with those of this study. The profile is based on confidential 1995 primary data from five hydropower plants in North America.) For smelter-based activities, a North American smelter electrical power mix, reported by the International Primary Aluminum Institute (IPAI) (Ref 3.5) was used. Where electrical grid information from offshore locations was not available, the U.S. average profile was used. Precombustion energy factors (for utilities) were included in this study.

Transportation Energy. The four modes of transportation identified in this study are truck, rail, barge, and ocean vessel. The environmental releases and the energy requirements (including precombustion factors) to deliver one metric tonne-kilometer (tonne-km) for each of the transport modes are presented (Table 3.5). Estimates of distance travel and mode of travel are provided by the reporting locations. Transportation has not been modeled independently in the life-cycle stages (primary or secondary stages) but rather has been incorporated into the product system rollup.

Feedstock Energy. The energy content of material resources, is calculated as the gross calorific value (high heat) of the energy resources removed from the Earth's energy reserves. It is the calorific value of the inputs to the system as opposed to the calorific value of the output.

Ancillary Material Analysis

The following is the decision rule process for completing this analysis:

1. Identify all potential ancillary material flows for a unit process that are greater than 1% by mass of the output of the unit process. When the ancillaries have been identified, a mass

Table 3.5 Transportation emission factors (includes precombustion burdens) per tonne-km

Mode	Unit	Heavy truck (18 tonne)	Rail (70 tonne)	Barge (12,000 tonne)	Ocean Vessel (35,000 tonne)
Fuel	...	Diesel	Diesel	Diesel	Fuel oil
Fuel consumption	MJ/tonne-km	1.04	3.35×10^{-1}	3.42×10^{-1}	9.72×10^{-2}
Water use	L/MJ	3.40×10^{-3}	3.40×10^{-3}	3.53×10^{-3}	3.53×10^{-3}
Air emissions					
Particulate matter	kg/MJ	8.25×10^{-5}	6.25×10^{-5}	6.50×10^{-5}	9.62×10^{-5}
CO	kg/MJ	6.04×10^{-4}	3.42×10^{-4}	3.55×10^{-4}	2.06×10^{-5}
CO_2	kg/MJ	1.01×10^{-1}	1.01×10^{-1}	1.05×10^{-1}	1.04×10^{-1}
SO_X	kg/MJ	1.28×10^{-4}	1.28×10^{-4}	1.33×10^{-4}	1.74×10^{-3}
NO_X	kg/MJ	6.24×10^{-4}	1.71×10^{-3}	1.78×10^{-3}	1.47×10^{-4}
Non-methane hydrocarbons	kg/MJ	1.58×10^{-4}	1.08×10^{-4}	1.12×10^{-4}	4.40×10^{-5}
Methane	kg/MJ	5.53×10^{-5}	5.27×10^{-5}	5.03×10^{-5}	5.03×10^{-5}
Lead	kg/MJ	1.60×10^{-9}	1.60×10^{-9}	1.67×10^{-9}	1.67×10^{-9}
HCl	kg/MJ	5.97×10^{-7}	5.97×10^{-7}	6.21×10^{-7}	6.21×10^{-7}
HF	kg/MJ	7.47×10^{-8}	7.47×10^{-8}	7.76×10^{-8}	7.76×10^{-8}
Water effluents					
BOD(a)	kg/MJ	1.15×10^{-5}	1.15×10^{-5}	1.20×10^{-5}	1.20×10^{-5}
COD(b)	kg/MJ	9.73×10^{-5}	9.73×10^{-5}	1.01×10^{-4}	1.01×10^{-4}
Heavy metals	kg/MJ	7.38×10^{-7}	7.38×10^{-7}	7.67×10^{-7}	7.67×10^{-7}
Other organics	kg/MJ	...	...	...	...
Suspended solids	kg/MJ	5.22×10^{-5}	5.22×10^{-5}	5.44×10^{-5}	5.44×10^{-5}
Dissolved solids	kg/MJ	1.38×10^{-3}	1.38×10^{-3}	1.44×10^{-3}	1.44×10^{-3}
Oil and grease	kg/MJ	6.63×10^{-6}	6.63×10^{-6}	6.89×10^{-6}	6.89×10^{-6}
Phosphates	kg/MJ	...	...	...	...
NH^{4+}	kg/MJ	1.68×10^{-6}	1.68×10^{-6}	1.75×10^{-6}	1.75×10^{-6}
Solid waste	kg/MJ	4.85×10^{-2}	4.85×10^{-2}	5.04×10^{-2}	5.04×10^{-2}

(a) BOD, bacteriological oxygen demand. (b) COD, Chemical oxygen demand

References

1. Emissions data derived from *EPA Compilation of Air Pollutant Emission Factors—Volume II: Mobile Sources*. Specific data used obtained from Table II-3-2 for steamship burning residual oil and distillate oil. Mode of operation assumed was cruising. Sulfur content assumed was 1%.
2. Emissions data derived from *EPA Compilation of Air Pollutant Emission Factors—Volume II: Mobile Sources*. Specific data used obtained from Table II-3-3 for diesel vessel with an internal combustion engine. Size of engine was assumed to be 3600 hp.
3. Data from *Transportation Energy Data Book*, 11th ed. Table 2.15 for Class I railroads and waterborne commerce
4. Emissions data derived from *EPA Compilation of Air Pollutant Emission Factors—Volume II: Mobile Sources*. Specific data used obtained from Table II-2-1
5. Average miles per gallon consumed by a heavy-duty truck obtained from "National Transportation Statistics, Annual Report," Sept 1993. Data for combination trucks in 1991
6. Emission factors developed from emission estimates in "EPA National Air Pollutant Emission Estimates, 1900–1991" and from energy consumption data for heavy-duty trucks in the *Transportation Energy Data Book*, 11th ed.

balance for the subsystems being analyzed is performed and normalized to the output from the subsystem.

2. Ancillary materials are then classified for inclusion in the scope of the study based on an analysis of their contribution to the mass of the subsystem, energy of the subsystem, and their environmental relevance.
3. All ancillary materials of a ranked ancillary list that have a cumulative mass contribution of up to 95% of the subsystem are included in the scope of the study.
4. A further decision rule is used to classify energy contribution. All ancillary materials that have a cumulative contribution of 99% of the total subsystem energy are included in the scope of the study, regardless of their mass ranking.
5. The remaining ancillary materials that are not selected as a result of the mass and energy criteria are analyzed for their environmental relevance. Specifically, an analysis is performed to determine if the ancillary material flow contributes more than 15% to an environmental release data category.

Data Quality

A comprehensive LCI involves the collection and integration of thousands of pieces of data regarding the product, process, or activity under study. To assess the data quality in this study, a number of data quality indicators (DQIs) were used.

Quantitative DQIs include mean and standard deviation values for measures of the variability of the data set values within each unit process. These precision measures provide the industry with an understanding of the variability of unit process performance and enable the industry to define opportunities for improvement from a benchmarking perspective.

Another DQI quantitative measure is data completeness, defined as a measure of the primary data values used in the analysis divided by the number of possible data points associated with a data category (within the domain). The

Table 3.6 Data completeness by unit process

Source	Unit process	Total reported	Total by unit process	Percent complete(a)
Primary	Bauxite mining	8	10	
	Electricity generation	7	8	
	Alumina refining	11	13	
	Anode production	25	27	
	Aluminum smelting	29	29	
	Primary ingot casting	22	23	
	Subtotal	**102**	**110**	**93%**
Manufacturing	Hot rolling	10	14	
	Aluminum extruding	16	21	
	Cold rolling	12	17	
	Aluminum casting	3	6	
	Subtotal	**41**	**58**	**71%**
Secondary	Manufactured scrap transport	9	9	
	Consumer scrap transport	12	12	
	Shredded aluminum transport	12	12	
	Shredding and decoating	9	13	
	Secondary ingot casting and melting	28	39	
	Subtotal	**70**	**85**	**82%**
	Total	**213**	**253**	**84%**

(a) The percentage complete is calculated by dividing the number of plants submitting data with the total number of plants for that unit operation within the domain.

completeness of reporting from the locations defined in the domain is highlighted in Table 3.6.

Qualitative DQIs include consistency and representativeness of the data. Consistency is a qualitative understanding of how uniformly the study methodology is applied to the various components of the study. This quality measure is one of the most important to manage in the inventory process. There are a number of steps that must be taken to ensure consistency. The most significant of these is communication. In a study that involves a number of different companies, which in turn collected data from over 200 different sites in different countries and continents, there must be a clear understanding of what data are being requested, how they are measured, how they are reported, and how they are used. For this study, a training session was conducted with company representatives to familiarize them with the key aspect of LCI analysis. The session focused on key requirements of the methodology and the data collection efforts. In addition, detailed instruction sheets were appended to each data input sheet, describing what was to be included under each data category. The reporting locations were encouraged to describe fully what they included in their data values and how it was measured, where possible, highlighting the sampling and analysis protocols that were employed. These descriptions were used to verify the consistency in data collection between the companies.

A further mechanism to enhance the consistency was the establishment of a hotline for reporting locations to obtain answers for specific questions or to address any difficulties they may be facing. Verbal and written communications flowed between the consultant team and the reporting locations, sometimes on several occasions to a single location, to verify the data inputs. This verification process also involved a critical review of each data category for each unit process.

Representativeness is a measure of the degree to which the data values used in the analysis present a true and accurate measurement of the average processes that the study is examining. The degree of representativeness is normally judged by the comparison of values determined in the study with existing reported values in other analyses or published data sources dealing with the subject matter. Any major variances identified should be examined and explained.

The results of the study and each unit process within the subsystems were compared with that of previous studies produced by the aluminum industry within both North America and Europe. Indeed, one of the purposes of the aluminum industry inventory was to quantify the improvements that have occurred since those analyses were completed. The comparison did not yield any major variances other than those related to the regional energy and energy-related emissions data. The other factor that leads to confidence in the representativeness of the data is the high level of completeness, as mentioned previously.

Primary Aluminum Unit Processes

The life-cycle stage of primary aluminum material processing includes the unit processes of bauxite mining, alumina refining, anode production, smelting, and primary ingot casting Fig. 3.1. As previously noted, transportation has not been modeled independently but rather has been incorporated into the product system rollup. For each life-cycle stage, the unit process is described, with assumptions made in the data aggregation. The aggregated data are presented as "mean," which is the weighted average of the normalized data for 1000 kg of the output from the unit process. Data quality measures, such as precision, completeness, consistency, and the qualitative DQIs (average DQI and DQI range), are also shown for the unit processes of primary ingot casting.

The energy consumption by unit process in the primary production chain is shown in Fig. 3.6 per 1000 kg of process output. The figure shows the minimum weighted average and maximum value of energy consumption for each process and clearly illustrates that smelting is the dominant energy-consuming operation. Alumina refining uses almost an order of magnitude less energy than smelting per unit of product output.

Figure 3.7 illustrates the CO_2 emissions per 1000 kg of output for the primary unit processes. With a close relationship between energy use and CO_2 releases, it is not surprising that smelting is the unit operation with the highest CO_2 emission. The baking of anodes is also a process with significant CO_2 emission. In this figure, the large variability in the data is due to major differences in production technology.

Bauxite Mining

Bauxite is composed largely of hydrated aluminum oxides, normally including significant impurities. The major impurities in commercially mined bauxite include silicon, titanium, iron compounds, and other materials. The most valuable bauxite sources tend to be found near the equator. Some of the more important sources are found in Australia, Guinea, Brazil, and the Caribbean.

Bauxite is typically found at a depth of 0 to 180 m (600 ft) beneath the surface, with an average of 4 m (80 ft). The overburden is removed and stockpiled for use in restoring the site after the bauxite has been removed. The bauxite deposit may be loosened with explosives, depending on its hardness and other local conditions. In some

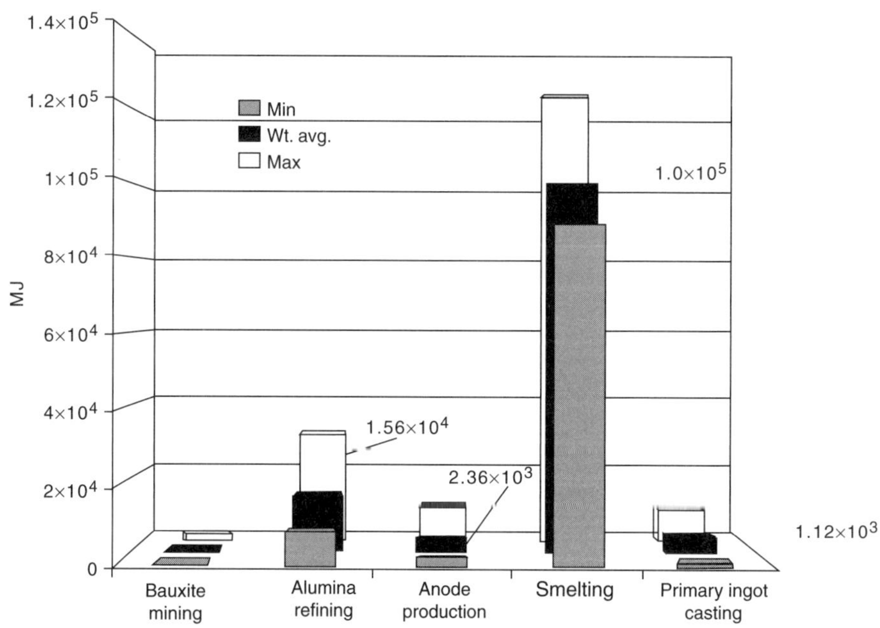

Fig. 3.6 Primary unit processes energy consumption per 1000 kg of output. Variation due to difference in technology is not accounted for.

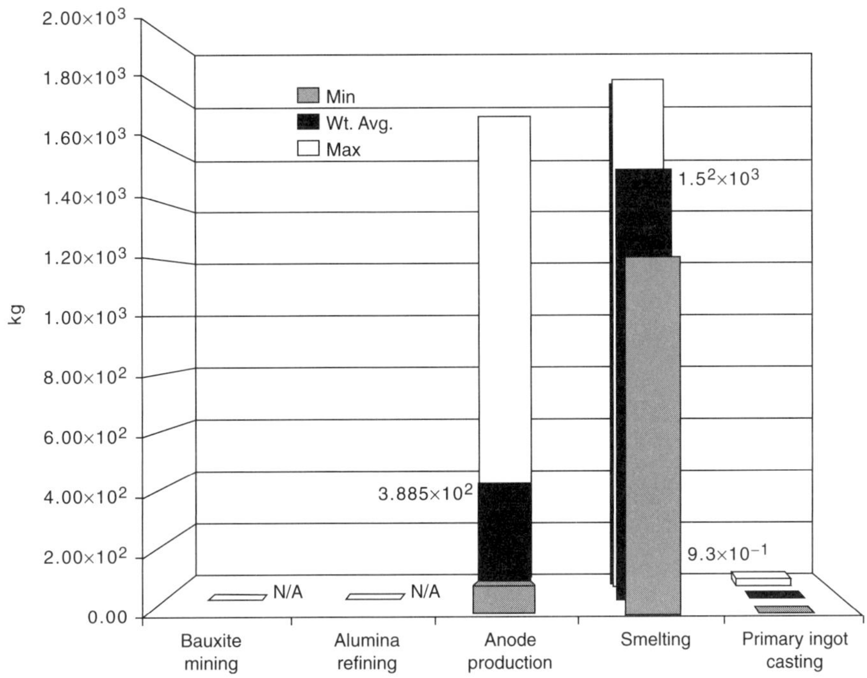

Fig. 3.7 Primary unit processes CO_2 emission per 1000 kg of output. Variation due to difference in technology is not accounted for. Note: CO_2 values presented here represent process-related emissions only.

cases, the bauxite is crushed with dust-control equipment and/or treated with water to remove impurities before it is shipped. The washing process is called beneficiation. Beneficiated bauxite typically is dried prior to shipment to the refinery. Ten bauxite mines supplied the major sources of ore for the thirteen alumina refineries providing data for this study. The operations associated with this unit process include:

- Extraction of bauxite-rich minerals from the site
- Beneficiation activities, such as washing, screening, or drying
- Treatment of mining site residues and waste
- Site restoration activities, such as grading, dressing, and planting

The output of this unit process is the bauxite that is transported to an alumina refinery. The resource consumption and environmental releases for bauxite mining are shown in Fig. 3.8.

Alumina Refining

In alumina refining, bauxite is converted to aluminum oxide (alumina) using the Bayer process. Most refineries use a mixture of blended bauxite to provide feedstock with consistent properties. The mixture is ground and blended with recycled plant liquor. This liquor contains dissolved sodium carbonate and sodium hydroxide recovered from previous extraction cycles plus supernatant liquor recycled from the red mud holding ponds. The aluminum trihydrate is recovered by precipitation and is filtered, dried, and calcined to form alumina.

Following refining, the alumina is transported to the smelter. There were a total of 13 refineries that participated in this study. Four of these refineries are in North America and nine were offshore. This introduces a large variation in both the bauxite and alumina transportation distance reported by the operation locations. The boundaries for this study begin with the unloading of process materials to their storage areas on site. The operations associated with this unit process include:

- Bauxite grinding, digestion, and processing of liquors
- Alumina precipitation and calcination
- Maintenance and repair of plant and equipment
- Treatment of process air, liquids, and solids

The output of this unit process is smelter-grade alumina transported to primary aluminum

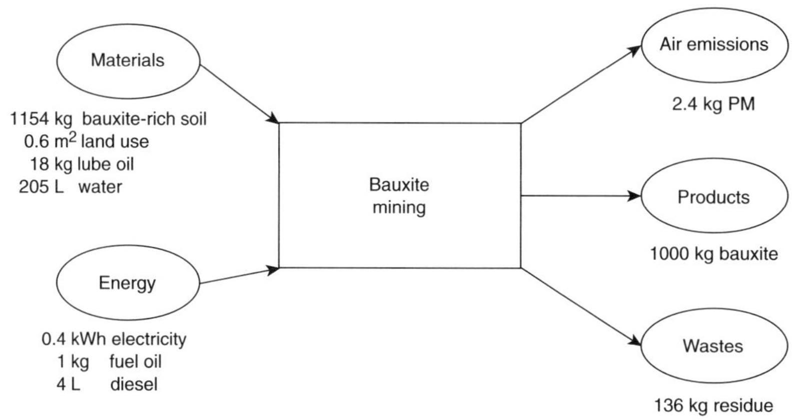

Fig. 3.8 Bauxite mining—resource consumption and environmental releases (per 1000 kg of unit process output). PM, particulate matter

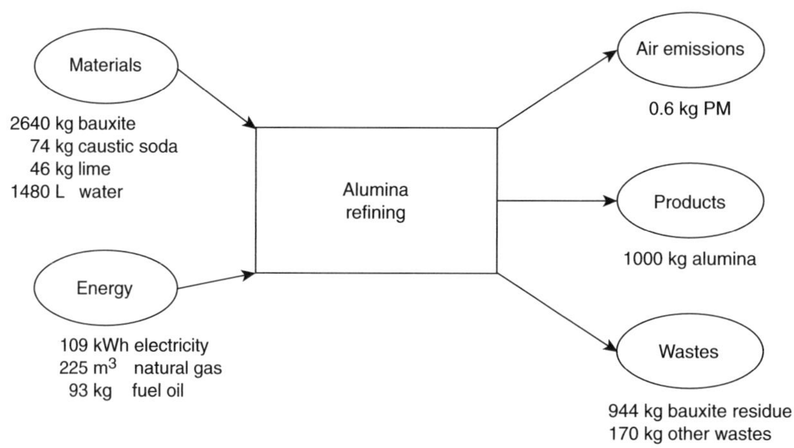

Fig. 3.9 Alumina refining—resource consumption and environmental releases (per 1000 kg of unit process output). Note: Difference in total mass of inputs and outputs is attributable to free and bound water in bauxite and residues. PM, particulate matter

smelters in North America. The resource consumption and environmental releases for alumina refining are shown in Fig. 3.9.

Anode Production

There are two types of aluminum smelting technologies that are distinguished by the type of anode that is used in the reduction process: Soderberg and prebake. Soderberg design has a single anode that covers most of the top surface of a reduction cell (pot). Anode paste (briquettes) is fed to the top of the anode, and as the anode is consumed in the process, the paste feeds downward by gravity. Heat from the pot bakes the paste into a monolithic mass before it gets to the electrolytic bath interface. The prebake design has prefired blocks of solid carbon suspended from axial busbars. The busbars both hold the anodes in place and carry the current for electrolysis.

The process for making the aggregate for briquettes or prebake blocks is identical. Coke is calcined, ground, and blended with pitch to form a paste that is subsequently extruded into blocks or briquettes and allowed to cool. While the briquettes are sent directly to the pots for consumption, the blocks are sent to a separate baking furnace. Baking furnace technology has evolved from simple pits that discharged volatiles to the atmosphere during the baking cycle to closed-loop-type designs that convert the caloric heat of the volatile into a process fuel that reduces energy consumption for the process.

Twenty-seven anode facilities have participated in this study. This unit process (as applied to this study) begins with the unloading of process materials to their storage areas on site. The operations associated with this unit process include:

- Recovery of spent anode materials
- Anode mix preparation; block or briquette forming and baking
- Rodding of baked anodes
- Maintenance and repair of plant and equipment
- Treatment of process air, liquids, and solids

The output of this unit process is rodded anodes or briquettes transported to a primary aluminum smelter. The resource consumption and environmental releases for anode production are shown in Fig. 3.10.

Aluminum Smelting

Molten aluminum is produced from alumina by the Hall-Heroult electrolytic process (Ref 3.6). This involves two steps: dissolving the alumina in a molten cryolitic bath, and passing electric current through this solution, thereby decomposing the alumina into aluminum and carbon dioxide.

There are two generic types of reduction cells: prebake and Soderberg. The prebake technology has two variants referring to how alumina is added, namely, *center-worked* prebake and *side-worked* prebake. Also, the Soderberg technology has two variants based on how the electricity is introduced to the cell, namely, *vertical stud* Soderberg and *horizontal stud* Soderberg. All plants built since approximately 1970 are of the prebake type. The two technologies are differentiated by the type of anodes they consume.

The prebake design has prefired blocks of solid carbon suspended from axial busbars. The busbars both hold the anodes in place and carry the current required for electrolysis. The anode blocks are consumed until they can no longer be safely held in place without attack on the stubs and are then replaced. The butts of the anodes are returned to the anode plant for recycling into new anodes by way of adding calcined petroleum coke and pitch.

The Soderberg design has a single anode that covers most of the top surface of the pot. The anode is baked in place in a large rectangular steel shell.

All aluminum smelters in North America use some type of air pollution control system to reduce emissions. The primary system is typically a scrubber. Most plants use dry scrubbers, with alumina as the absorbent. The alumina is then returned to the pots as feedstock. This allows recovery of the scrubbed materials. Other plants use wet scrubbers, which recirculate an alkaline solution to absorb emissions. Unlike dry scrubbers, wet scrubbers absorb the emitted carbon dioxide, nitrogen oxide, and sulfur dioxide. Plants with wet scrubbers have a wastewater stream from the wet scrubber blowdown.

All 29 North American smelters (within the domain of this study) have participated in this

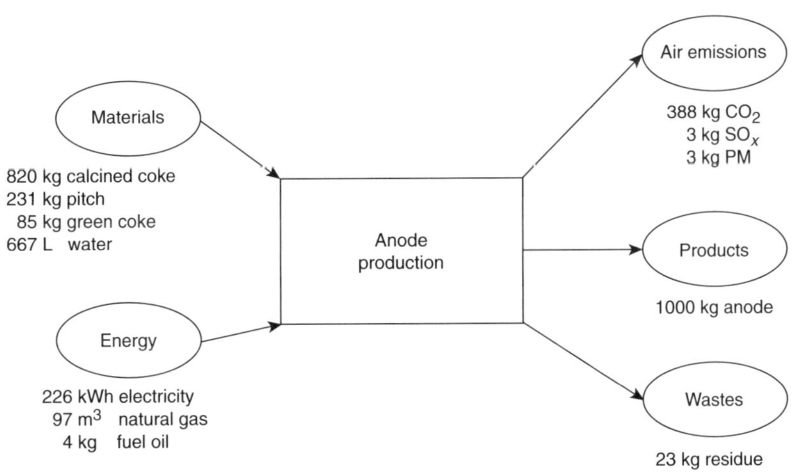

Fig. 3.10 Anode production—resource consumption and data quality (per 1000 kg of unit process output). PM, particulate matter

study. The unit process (as defined for this study) begins with the unloading of process materials to their storage areas on site. The operations associated with this unit process include:

- Recovery, preparation, and handling of process materials
- Manufacture of major process equipment (e.g., cathode shells)
- Process of control activities (metal, bath, heat)
- Maintenance and repair of plant and equipment
- Treatment of process air, liquids, and solids

The output of this unit process is hot metal transported to an ingot casting facility. This unit process excludes material, energy, and environmental releases associated with anode production and casting operations. The resource consumption and environmental releases for aluminum smelting are shown in Fig. 3.11. In addition, environmental releases of perfluoro carbon and carbonyl sulfide generated in aluminum smelting are summarized in the following sections.

Perfluoro Carbon (PFC) Generation in Aluminum Smelting. The PFC values are calculated from the 1990 and 1996 values reported in the IPAI-PFC report (Ref 3.7). Table 3.7 summarizes the average PFC emissions from aluminum smelting based on IPAI estimates and actual 1995 production rates for primary aluminum taken from the domain analysis. The PFC calculation is based on the assumption that 92% CF_4 (perfluoromethane) and 8% C_2F_6 (perfluoroethane) are emitted. CO_2 equivalents are calculated based on 6500 (kg) global-warming potential for CF_4 and 9200 (kg) global-warming potential for C_2F_6. Global warming potential is defined as the total impact over time of adding a unit of a greenhouse gas to the atmosphere. For the production of every 1000 kg of primary aluminum, 0.38 kg of PFC is emitted to the atmosphere.

Carbonyl Sulfide (COS) Generation in Aluminum Smelting. The COS values are calculated from the reported 1995 values (Ref 3.8). For the production of every 1000 kg of primary aluminum, 1.12 kg of COS is emitted to the atmosphere.

Primary Ingot Casting

This unit process begins with molten primary aluminum and ends with sheet ingot suitable for rolling, extruding, or shape casting. There are other types of ingot, called T-ingot, as well as pigs or sows. Aluminum is commonly shipped in these solid forms if it is necessary to transport it a long way from a smelter or reclaimer to an ingot casting facility. This process description does not include information on any type of ingot besides sheet ingot from primary metal and additives. Ingots that contain manufacturing or recycling aluminum are evaluated in the recycling life-cycle stage.

Twenty-two North American primary ingot casting facilities participated in the study. The unit process (as defined for this study) begins

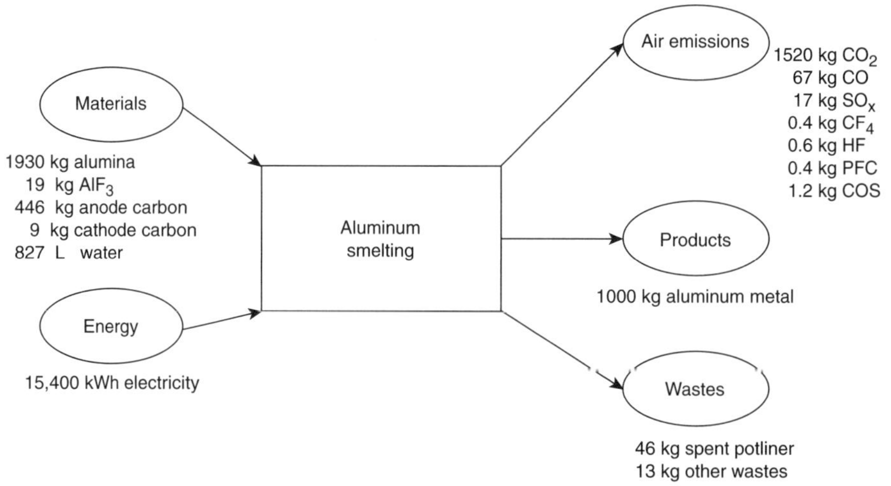

Fig. 3.11 Aluminum smelting—resource consumption and environmental releases (per 1000 kg of unit process output). PFC, perfluoro carbon; COS, carbonyl sulfide

Table 3.7 Perfluoro carbon (PFC) summary for 1996

PFC generation rates (kg of PFC per tonne of aluminum)	
CWPB	0.15
SWPB	1.00
VSS	1.50
HSS	0.45
Average North American	**0.38**
PFC generation (kg of PFC)	
CWPB	691,350
SWPB	146,000
VSS	1,254,000
HSS	310,014
Total	**2,401,364**
PFC generation (tonnes of CO_2 equivalents)	
	16,127,561
Aluminum production	
Smelter type	
CWPB	4609
SWPB	146
VSS	836
HSS	689
North America	**6280**

CWPB, center-worked prebake; SWPB, side-worked prebake; VSS, vertical stud Soderberg; HSS, horizontal stud Soderberg

with the unloading of process materials to their storage areas on site. The operations associated with this unit process include:

- Pretreatment of hot metal (cleaning and auxiliary heating)
- Recovery and handling of internal process scrap
- Batching, metal treatment, and casting operations
- Homogenizing, sawing, and packaging and casting operations
- Maintenance and repair of plant and equipment
- Treatment of process air, liquids, and solids

The output of this unit process is packaged aluminum ingots transported to an aluminum fabricating facility. The resource consumption and environmental releases for primary ingot casting are shown in Fig. 3.12. Data precision and completeness for resource consumption and environmental releases data for primary ingot casting are shown in Tables 3.8 and 3.9.

Secondary Aluminum Processing

The life-cycle stage of secondary aluminum processing (Fig. 3.2) includes the unit processes of manufacturing scrap transport, consumer scrap transport, shredding and decoating, and secondary ingot casting. Like the LCI assessment of primary processing, transportation has not been modeled independently but rather has been incorporated into the product system rollup. The aggregated data are presented as "mean," which is the weighted average of the normalized data for 1000 kg of the output from the unit process. Data quality measures, such as precision, completeness, consistency, and the qualitative DQIs (average DQI and DQI range), are also shown for secondary ingot casting.

Secondary Aluminum Transport

Secondary aluminum transport includes manufactured scrap, consumer scrap, and shredded

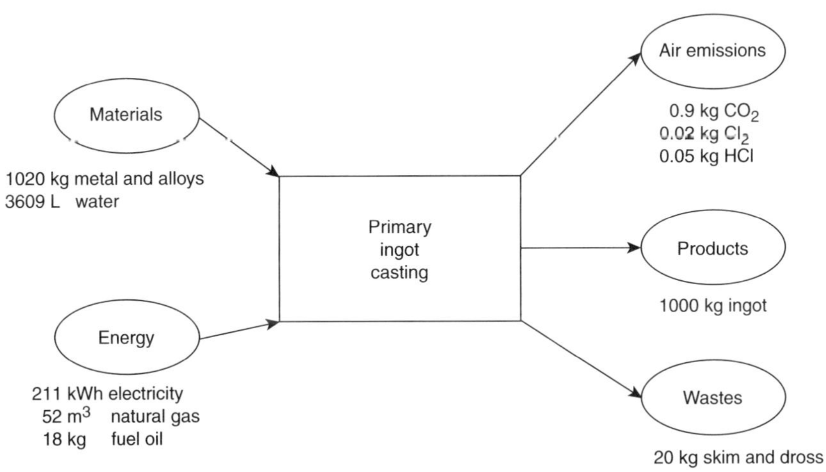

Fig. 3.12 Primary ingot casting—resource consumption and environmental releases (per 1000 kg of unit process output)

Table 3.8 Primary ingot casting—resource consumption and data quality (per 1000 kg of unit process output)

Inputs	Units	Mean	Standard deviation	Completeness,%	DQI average(a)	DQI range(a)
Material						
Lube oil	kg	5.35×10^{-3}	2.45×10^{-2}	13	3.8	1–5
Filter media	kg	1.37	3.46	33	3.6	1–5
Primary metal	kg	1.00×10^{3}	1.19×10	100	4.8	3–5
Alloying additives	kg	1.74×10	1.25×10	100	4.0	1–5
Grain refiners	kg	2.27	4.18	63	3.7	1–5
Surface water	L	2.86×10^{3}	4.59×10^{3}	67	2.4	1–5
Ground water	L	7.49×10^{2}	2.40×10^{3}	17	3.9	1–5
Energy						
Light fuel oil	kg	1.74×10	3.04×10	100	4.3	3–5
Diesel	L	1.84×10^{-1}	2.96×10^{-1}	100	3.0	1–5
Gasoline	L	7.46×10^{-2}	1.19×10^{-1}	100	3.2	1–4
Natural gas	m^3	5.18×10	4.82×10	100	4.3	1–5
Propane	L	7.98×10^{-1}	1.24	100	4.0	2–5
Purchased/self-generated electricity	kWh	2.11×10^{2}	1.87×10^{2}	100	3.1	1–5

(a) Ranking and definitions of data quality indicators (DQIs).

aluminum. Because data for scrap transport operations were provided by less than three companies, data for these unit operations are not presented here, but normalized data from these operations were given to the individual (participating) companies.

Manufactured Scrap Transport. The unit process (as defined for this study) begins with the containerized, bailed or bundled, and segregated aluminum manufacturing scrap on the loading dock of a component or final product manufacturer. The unit process includes the transport of manufacturing scrap by various transport modes. The output of the unit process is manufacturing scrap transported to an aluminum recycling facility.

Consumer Scrap Transport. This unit process begins with disassembled aluminum components containerized, bailed or bundled, and segregated on the loading dock of an automobile part dismantler. The unit process includes the transport of consumer scrap by various transport modes. The output of the unit process is consumer scrap transported to an aluminum recycling facility.

Shredded Aluminum Transport. This unit process begins with the containerized, bailed or bundled, and segregated aluminum shredded scrap on the loading dock of a component or final product manufacturer. The unit process includes the transport of shredded scrap by various transport modes. The output of the unit process is shredded scrap transported to an aluminum recycling facility.

Shredding and Decoating

This unit process begins with the unloading of segregated aluminum scrap to designated storage areas on site. The operations associated with this unit process include:

- Recovery, preparation, and handling of process materials
- Shredding and decoating activities
- Maintenance and repair of plant and equipment
- Treatment of process air, liquids, and solids

The output of this unit process is shredded aluminum transported to a secondary ingot casting facility. As for scrap transportation, data for the shredding and decoating operation were provided by less than three companies and are not presented here. Normalized data from these operations were provided to participating companies.

Secondary Ingot Casting

The major distinction between secondary casters and primary casters is in the melting technology employed. Secondary smelters have a variety of melting furnaces: top-loaded closed melters, rotary melters, and sidewell-feeding melting furnaces. These melting furnaces have different efficiencies and recoveries. Some of the melters also use a saltcake to cover the melt and reduce oxidation.

Thirty-nine secondary ingot casting facilities participated in this study. This unit process begins with the unloading of process materials to their storage areas on site. The operations associated with this unit process include:

- Recovery and handling of internal process scrap
- Batching, metal treatment, and casting operations

Table 3.9 Primary ingot casting—environmental releases and data quality (per 1000 kg of unit process output)

Releases	Units	Mean	Standard deviation	Completeness, %	DQI average(a)	DQI range(a)
Air emissions						
Particulate matter	kg	5.77×10^{-2}	5.41×10^{-2}	71	3.6	1–5
CO	kg	6.01×10^{-3}	1.31×10^{-2}	21	3.4	1–5
CO_2	kg	9.34×10^{-1}	4.52	13	3.0	3–3
SO_X	kg	1.10×10^{-2}	2.85×10^{-2}	17	3.5	1–5
NO_X	kg	2.62×10^{-2}	5.53×10^{-2}	25	3.2	1–5
Cl_2	kg	1.76×10^{-2}	4.96×10^{-2}	58	3.4	1–5
HCl	kg	5.27×10^{-2}	6.37×10^{-2}	58	3.6	1–5
F	kg	1.92×10^{-2}	8.56×10^{-2}	8	5.0	5–5
HF	kg	2.30×10^{-3}	9.57×10^{-3}	13	4.7	4–5
Nonmethane hydrocarbons	kg	3.82×10^{-3}	3.00×10^{-2}	17	3.0	1–5
Organics	kg	1.05×10^{-2}	NA	4	3.0	3–3
Lead	kg	9.37×10^{-6}	NA	4	5.0	5–5
Metals	kg	1.63×10^{-3}	4.26×10^{-3}	17	3.3	1–5
Water effluents						
COD(b)	kg	2.11×10^{-1}	2.93×10^{-1}	58	2.9	1–5
BOD(c)	kg	4.14×10^{-2}	5.41×10^{-2}	29	3.0	1–5
Lead	kg	3.18×10^{-6}	2.88×10^{-5}	17	2.7	1–5
Iron	kg	8.59×10^{-4}	1.19×10^{-2}	17	2.5	2–3
NO_3	kg	4.71×10^{-4}	4.46×10^{-3}	17	3.0	2–5
Mercury	kg	1.36×10^{-8}	NA	4	3.0	5–5
Metals	kg	2.55×10^{-3}	9.57×10^{-3}	29	3.4	1–5
NH^{4+}	kg	3.63×10^{-4}	3.65×10^{-3}	17	3.0	2–5
Cl^-	kg	7.80×10^{-3}	NA	4	3.0	3–3
CN^-	kg	1.73×10^{-6}	1.49×10^{-5}	13	3.3	3–4
F^-	kg	2.66×10^{-3}	6.09×10^{-3}	33	3.8	3–5
S	kg	5.70×10^{-4}	NA	4	3.0	3–3
Dissolved organics	kg	1.26×10^{-2}	5.76×10^{-2}	17	3.0	1–5
Suspended solids	kg	6.66×10^{-2}	8.36×10^{-2}	71	3.1	1–5
Detergents	kg	1.47×10^{-5}	1.91×10^{-4}	8	2.5	2–3
Dissolved chlorine	kg	8.78×10^{-5}	1.14×10^{-3}	13	2.7	2–4
Phenol	kg	1.20×10^{-6}	1.47×10^{-5}	13	2.5	1–4
Dissolved solids	kg	1.80×10^{-1}	4.50×10^{-1}	42	2.0	1–4
Oil and grease	kg	2.33×10^{-2}	7.42×10^{-2}	75	3.1	1–5
Solid waste						
Process Related	kg	1.89×10	1.22×10	100	3.9	1–5
Environmental abatement	kg	7.16×10^{-1}	2.62	25	3.5	1–5
Municipal	kg	1.61	2.01	88	2.9	1–5

(a) See preceding table for ranking and definitions of data quality indicators (DQIs). (b) COD, chemical oxygen demand. (c) BOD, bacteriological oxygen demand

- Homogenizing, sawing, and packaging and casting operations
- Maintenance and repair of plant and equipment
- Treatment of process air, liquids, and solids

The output of this unit process is packaged aluminum ingots transported to an aluminum fabricating facility. The resource consumption and environmental releases for secondary ingot casting are shown in Fig. 3.13. Data precision and completeness for resource consumption and environmental releases data for secondary ingot casting are shown in Tables 3.10 and 3.11.

Manufacturing Unit Processes

The life-cycle stage of manufacturing includes the unit processes of shape casting, aluminum extruding, hot rolling, and cold rolling. Ingots are provided to this life-cycle stage from both primary and secondary aluminum processing. As previously noted for primary and secondary processing, transportation has not been modeled independently but rather has been incorporated into the product system rollup. The aggregated data are presented as "mean," which is the weighted average of the normalized data for 1000 kg of the output from the unit process.

Shape Casting. Aluminum castings are produced in aluminum foundries that are designed to process a range of old and new scrap qualities that contain alloying elements. They represent over 20% of total semifabricated production and are growing at an annualized rate of 4.8%. North American cast aluminum production is over a million metric tons. More than 80% of aluminum castings are used in the automotive

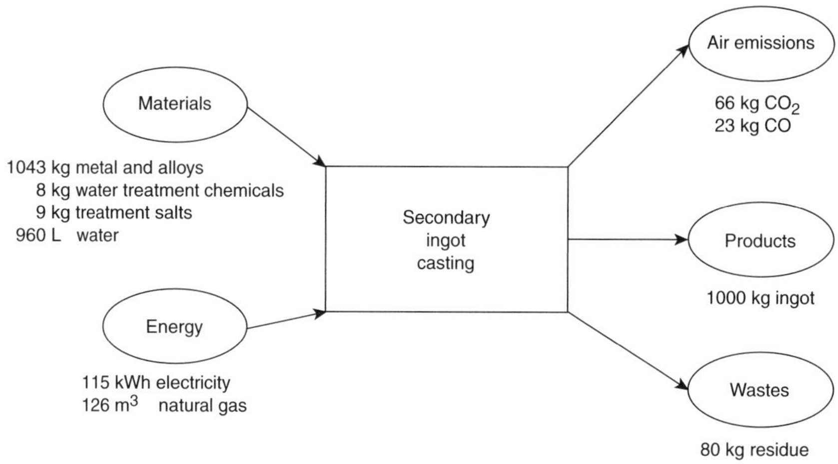

Fig. 3.13 Secondary ingot casting—resource consumption and environmental releases (per 1000 kg of unit process output)

Table 3.10 Secondary ingot casting—resource consumption and data quality indicator (DQI) (per 1000 kg of unit process output)

Inputs	Units	Mean	Standard deviation	Completeness, %	DQI average	DQI range
Material						
Lube oil	kg	5.35×10^{-3}	2.45×10^{-2}	13	3.8	1–5
Filter media	kg	1.37	3.46	33	3.6	1–5
Primary metal	kg	1.00×10^{3}	1.19×10	100	4.8	3–5
Alloying additives	kg	1.74×10	1.25×10	100	4.0	1–5
Grain refiners	kg	2.27	4.18	63	3.7	1–5
Surface water	L	2.86×10^{3}	4.59×10^{3}	67	2.4	1–5
Ground water	L	7.49×10^{2}	2.40×10^{3}	17	3.9	1–5
Energy						
Light fuel oil	kg	1.74×10	3.04×10	100	4.3	3–5
Diesel	L	1.84×10^{-1}	2.96×10^{-1}	100	3.0	1–5
Gasoline	L	7.46×10^{-2}	1.19×10^{-1}	100	3.2	1–4
Natural gas	m^3	5.18×10	4.82×10	100	4.3	1–5
Propane	L	7.98×10^{-1}	1.24	100	4.0	2–5
Purchased/self-generated electricity	kWh	2.11×10^{2}	1.87×10^{2}	100	3.1	1–5

industry. The majority of castings are produced by secondary smelters, of which there are more than 70 production plants in North America. The technology employed is dependent on the shapes produced and employs both permanent and nonpermanent molds.

This unit process (as defined for this study) begins with the unloading of process materials to their storage areas on site. The operations associated with this unit process include:

- Pretreatment of hot metal (cleaning and auxiliary heating)
- Recovery and handling of internal process scrap
- Preparation and forming of cores and molds
- Batching, metal treatment, and casting operations
- Finishing and packaging activities
- Maintenance and repair of plant and equipment
- Treatment of process air, liquids, and solids

The output of this unit process is semifabricated aluminum components that are transported to a component or final product manufacturer. The resource consumption and environmental releases for aluminum shape casting are shown in Fig. 3.14.

Extruding. The extrusion process takes cast extrusion billet (round bar stock produced from direct chill molds) and produces extruded shapes. The process begins with an in-line preheat that takes the temperature of the billet to a predetermined level, depending on the alloy. The billet is then sheared, if not already cut to

Table 3.11 Secondary ingot casting—environmental release and data quality indicator (DQI) (per 1000 kg of unit process output)

Releases	Units	Mean	Standard deviation	Completeness, %	DQI average	DQI range
Air emissions						
Particulate matter	kg	2.38×10^{-1}	3.43×10^{-1}	81	2.9	1–5
CO	kg	2.35×10	4.91×10	70	3.0	1–5
CO_2	kg	6.59×10	6.08×10^{2}	15	2.0	1–4
SO_x	kg	1.02×10^{-2}	3.65×10^{-2}	56	3.0	1–5
NO_x	kg	2.02×10^{-1}	7.58×10^{-1}	67	2.9	1–5
Cl_2	kg	6.34×10^{-2}	8.55×10^{-2}	41	3.5	1–5
HCl	kg	1.71×10^{-1}	1.51×10^{-1}	44	3.3	1–5
HF	kg	1.07×10^{-2}	1.44×10^{-2}	15	4.3	4–5
Nonmethane hydrocarbons	kg	8.91×10^{-2}	1.80×10^{-1}	41	2.8	1–5
Organics	kg	1.71×10^{-2}	5.21×10^{-2}	19	3.8	2–5
Lead	kg	1.99×10^{-4}	2.43×10^{-4}	33	3.4	3–5
Metals	kg	1.74×10^{-4}	4.14×10^{-4}	11	3.5	1–5
Water effluents						
COD(a)	kg	2.92×10^{-2}	5.80×10^{-2}	100	3.7	2–5
BOD(b)	kg	3.76×10^{-2}	6.74×10^{-2}	100	3.6	2–5
Lead	kg	8.06×10^{-7}	1.07×10^{-5}	22	3.3	1–5
Iron	kg	9.31×10^{-6}	2.58×10^{-4}	15	3.5	2–5
NO_3	kg	2.12×10^{-4}	3.19×10^{-3}	11	2.7	2–3
Metals	kg	2.12×10^{-4}	3.37×10^{-3}	33	2.8	1–5
NH^{4+}	kg	1.42×10^{-5}	4.95×10^{-4}	15	3.2	2–5
F^-	kg	1.06×10^{-4}	1.48×10^{-3}	22	2.8	1–5
Dissolved organics	kg	3.63×10^{-4}	5.06×10^{-3}	15	3.3	1–5
Suspended solids	kg	2.25×10^{-2}	1.14×10^{-1}	100	3.5	1–5
Detergents	kg	3.65×10^{-6}	7.23×10^{-5}	11	2.7	2–3
Hydrocarbons	kg	4.65×10^{-4}	1.08×10^{-3}	11	1.3	1–2
Dissolved solids	kg	1.29×10^{-3}	NA	4	4.5	4–4
Phosphates	kg	9.61×10^{-6}	NA	4	2.0	3–3
Other nitrogen	kg	1.86×10^{-4}	2.38×10^{-3}	19	3.6	2–5
Oil and grease	kg	2.56×10^{-2}	1.93×10^{-2}	52	3.6	1–5
Solid waste						
Process related	kg	7.87×10	7.54×10	85	2.9	1–5
Environmental abatement	kg	6.54×10^{-1}	1.82	19	3.7	1–5
Municipal	kg	4.81	2.23×10	67	2.6	1–5

(a) COD, chemical oxygen demand. (b) BOD, bacteriological oxygen demand

length, and deposited into a hydraulic press. The press squeezes the semiplastic billet through a heated steel die that forms the shape. The shape is extruded into lengths defined by the take-off tables and is either water quenched or air cooled. The shape is then clamped and stretched to form a solid, straightened length. The straightened lengths are cut to final length multiples and placed in an aging furnace to achieve a desired temper. Lengths are then finished (drilled and shaped) and placed into a coating process. The types of coatings include anodized, painted, and lacquered finishes.

There are over 250 extrusion plants in North America, some with multiple presses. The technology is relatively mature, and variation in process efficiency is minor. Depending on the shape and desired performance characteristics of the extrusion, some profiles are put through an impact extruding process that uses considerably higher pressures to form the final parts.

The operations associated with this unit process include:

- Preheating and cutting or shearing of billet lengths
- Extruding of shapes, cooling, stretching, and cutting
- Heat treating, aging, anodizing, or painting
- Finishing and packaging activities
- Maintenance and repair of plant and equipment and treatment of process air, liquids, and solids

The output of this unit process is semifabricated or finished extruded profiles transported to a component or final product manufacturer. The unit process excludes materials, energy, and environmental releases associated with secondary ingot casting operations, if any. The resource consumption and environmental releases for aluminum extruding are shown in Fig. 3.15.

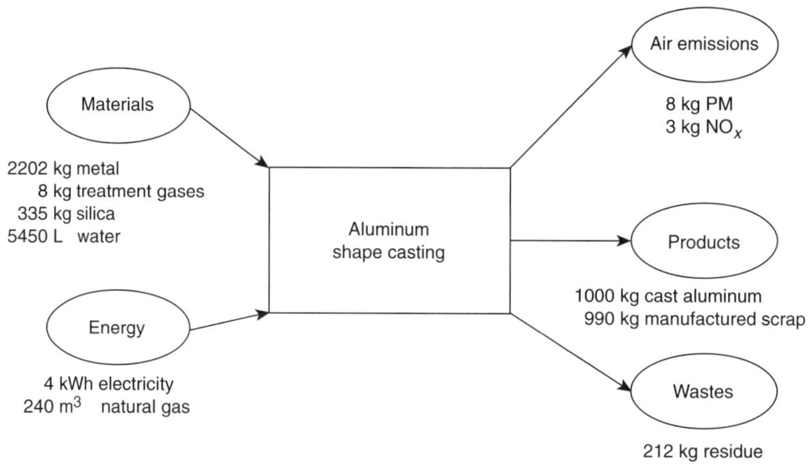

Fig. 3.14 Shape casting—resource consumption and environmental releases (per 1000 kg of unit process output). PM, particulate matter

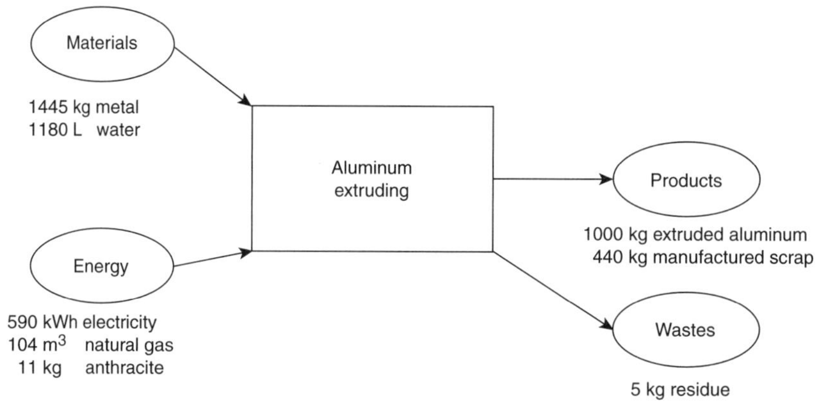

Fig. 3.15 Aluminum extruding—resource consumption and environmental releases (per 1000 kg of unit process output)

Hot and Cold Mill Rolling. The ingots are typically 45 to 66 cm (18 to 26 in.) thick and weigh 13 to 27 $\times$ 10^3 kg (15 to 30 tons). The ingots are preheated to approximately 540 °C (1000 °F) and fed through a hot reversing mill. In the reversing mill, the coil passes back and forth between rollers, and the thickness is reduced from the initial thickness to 25 to 50 mm (1 to 2 in.), with a corresponding increase in length. Following the reversing mills, the slabs are fed to a continuous hot mill, where the thickness is further reduced to less than 6 mm (1/4 in.) in thickness. The metal, called reroll or hot coil, is rolled into a coil and is ready to be transferred to the cold mill.

Prior to the cold mill, the coils may be annealed to give the metal the workability for downstream working. The coils are then passed through multiple sets of continuous rollers to reduce the gage. The coils are cut to the width and length as required by the customer. The coils are packaged to prevent damage to the metal in shipping.

The resource consumption and environmental releases for aluminum rolling are shown in Fig. 3.16 (hot rolling) and 3.17 (cold rolling).

Results by Product System

Figures 3.18 and 3.19 summarize the cradle-to-gate profiles of primary and secondary product systems. The product systems include the raw material acquisition, processing, and product

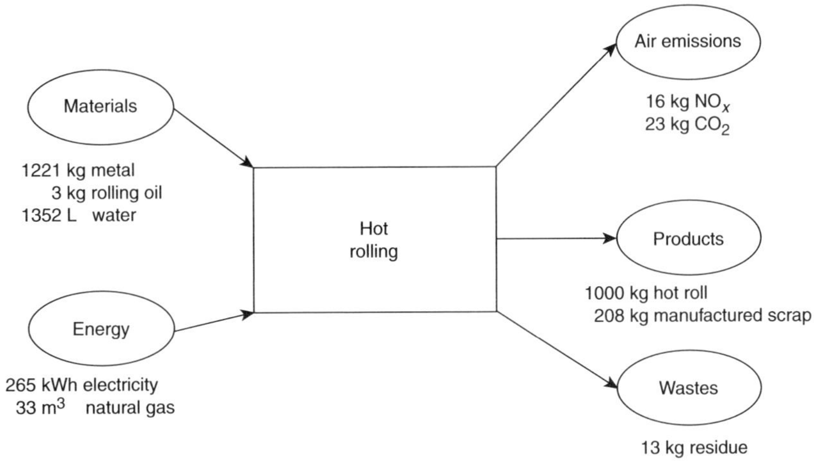

Fig. 3.16 Hot rolling—resource consumption and environmental releases (per 1000 kg of unit process output)

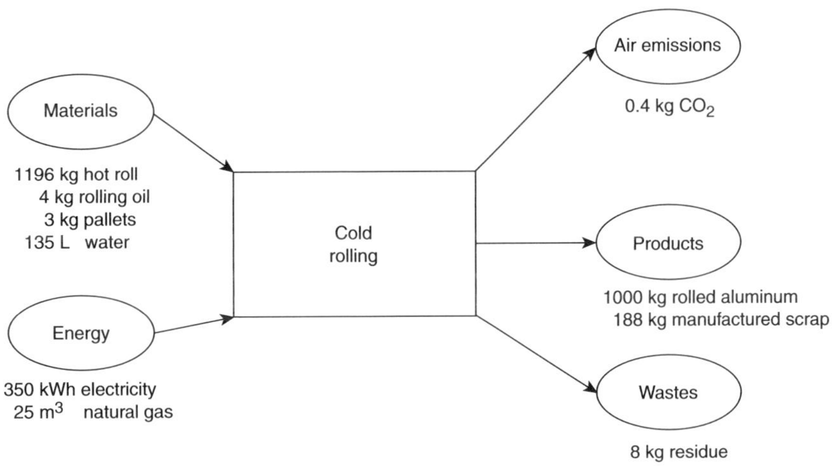

Fig. 3.17 Cold rolling—resource consumption and environmental releases (per 1000 kg of unit process output)

manufacture life-cycle stages. Material consumption includes the consumption of minerals, water, and petrochemical feedstocks. Energy consumption includes the use of energy at the plant (purchased/self-generated); energy used in transportation of raw materials, products and solid waste; and precombustion energy (energy expended to extract, refine, and deliver a fuel).

The outputs of the product systems are products, air emissions, water effluents, and solid waste. The air emissions quantified include dust and particulates (includes metals), CO_2, CO, SO_X, NO_X, nonmethane hydrocarbons (includes halogenated organic compounds), CH_4, acid gases (HCl and HF), and lead. The water effluents quantified include dissolved solids, suspended solids, heavy metals, oils and greases, other organics, and phosphates and ammonia. The solid waste produced by the product system was aggregated based on the waste management method used. Waste management options include landfilling, recycling, incineration, and beneficial use.

Automotive Aluminum Recycling. Figure 3.20 portrays the mass flow of extruded, rolled, and shape cast aluminum products through the various steps of vehicle assembly, use, dismantling, and recycling. Secondary aluminum recovered during recycling is returned for the

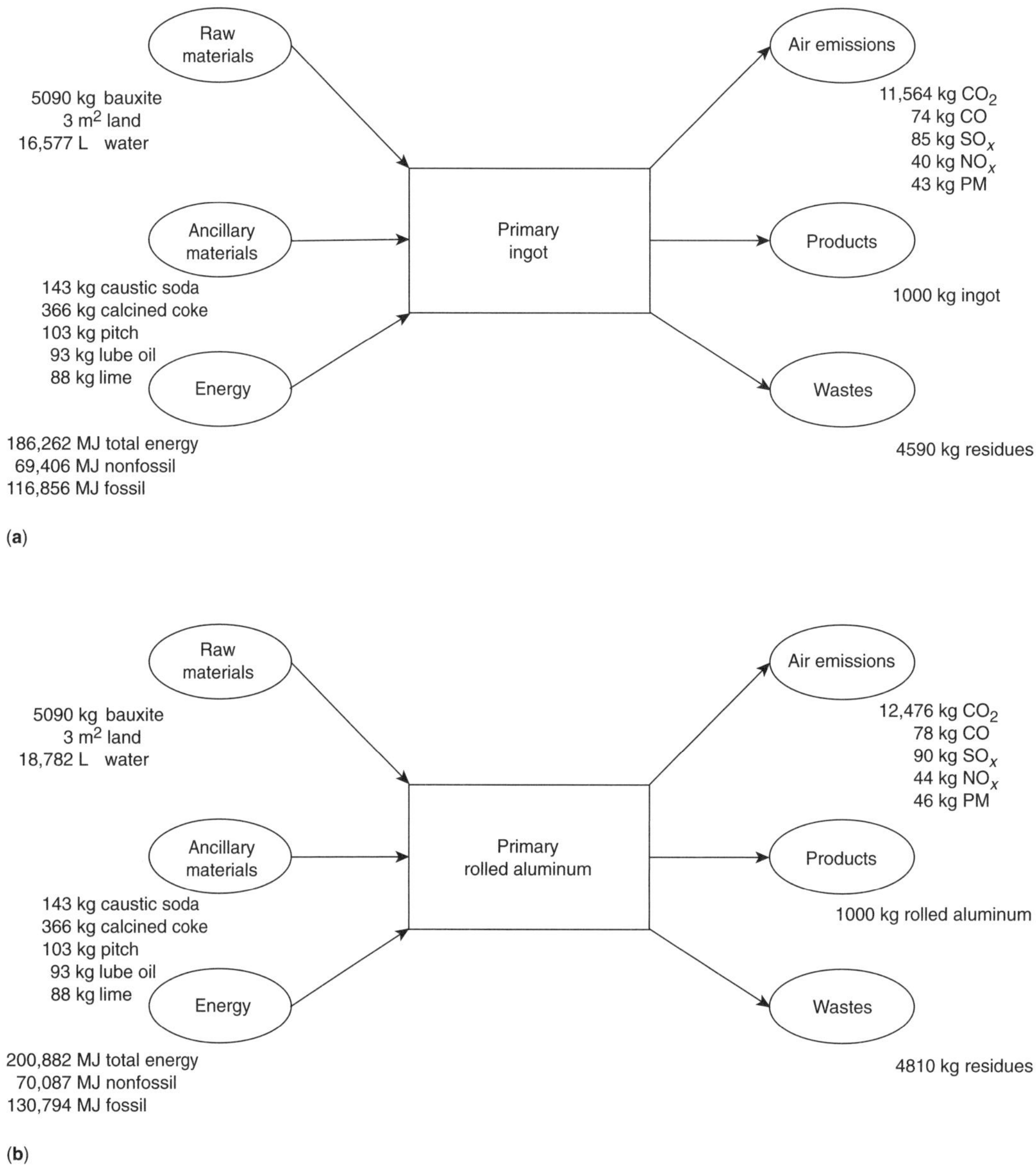

Fig. 3.18 Resources and environmental releases (per 1000 kg output) for primary aluminum products. (a) Primary ingot. (b) Primary rolled product. (c) Primary extrusions. (d) Primary shape castings. PM particulate matter

manufacture of new components. This chart was developed through the combined efforts of representatives of the Auto and Light Truck Committee, the Recycling Committee, and the Life-Cycle Task Group of the Aluminum Association. Figure 3.20(a) shows the situation for the 1995 calendar year, and Fig. 3.20(b) portrays the future steady-state situation, when a fleet of more aluminum-intensive vehicles is recycled. Currently, because the amount of aluminum, especially wrought aluminum, used in each vehicle is growing, there is more aluminum being consumed in each new vehicle than is now being recovered from used vehicles. However, as the more aluminum-intensive vehicles reach the end of their useful life of approximately 10 to 12 years, larger quantities of aluminum will be recycled, and a higher steady-state situation should be reached for wrought products. This assertion is based on the economical and technical performance of the recycling infrastructure. This steady-state

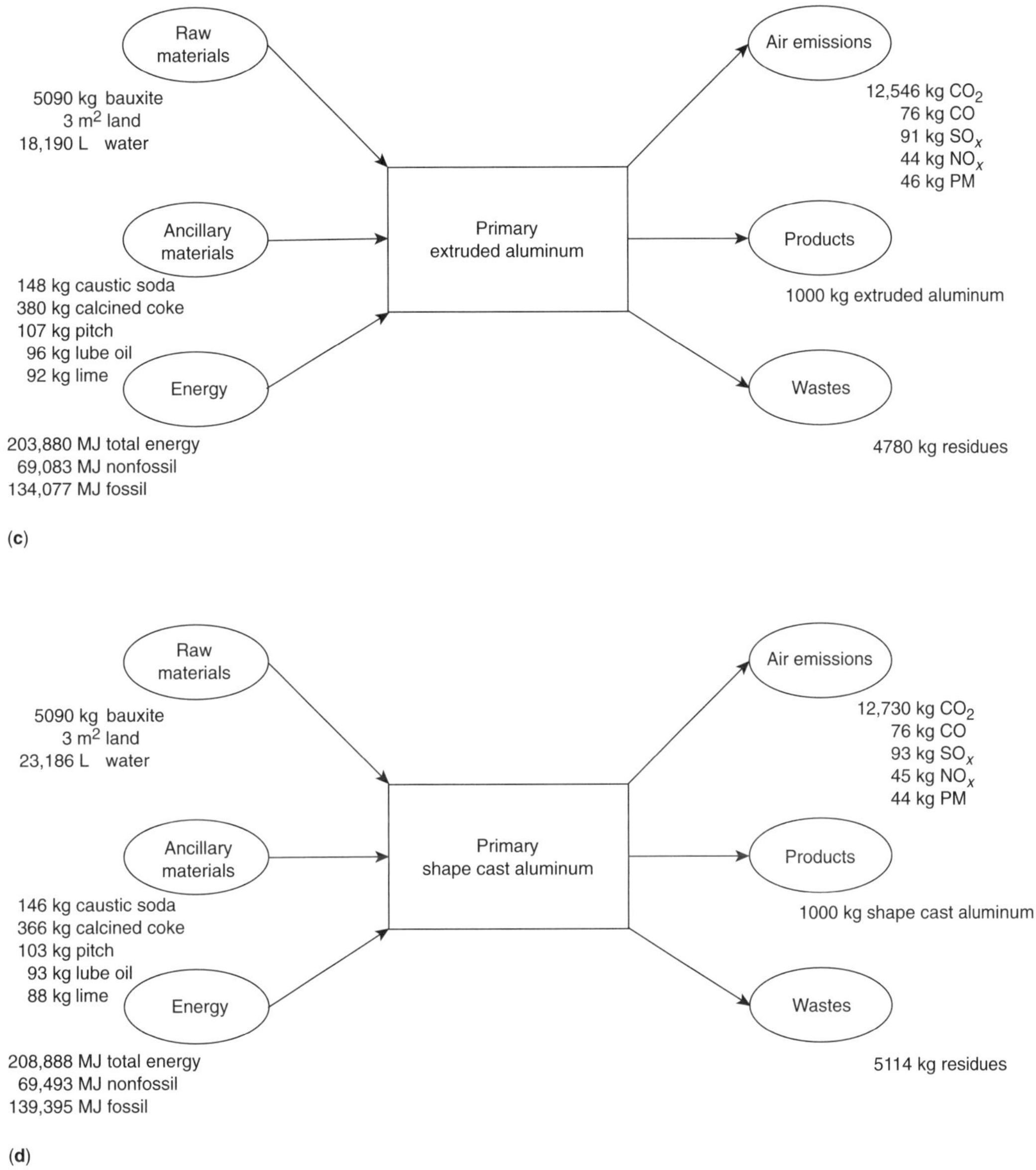

Fig. 3.18 (continued) (c) Primary extrusions. (d) Primary shape castings. PM, particulate matter

situation will probably be attained during the second decade of the next century. Some estimates for the use of aluminum in vehicles at that time range from 180 to 315 kg (400 to 700 lb) per vehicle.

The results released to USAMP will contribute in achieving the objectives of the generic automobile LCI effort. The energy and environmental profiles modeled and submitted to the automakers are based on the present industry practice, where only 11% recycled aluminum is used in wrought products and 85% is used in cast aluminum products. Currently, the distribution of aluminum use by product form in automotive applications is made up of 73.8% cast aluminum, 22.8% extruded aluminum, and 3.4% rolled aluminum. Accordingly, using these percentages, a composite energy and environmental profile for the generic USAMP vehicle can be derived for automotive aluminum. The results for the three product systems are presented in Fig. 3.21.

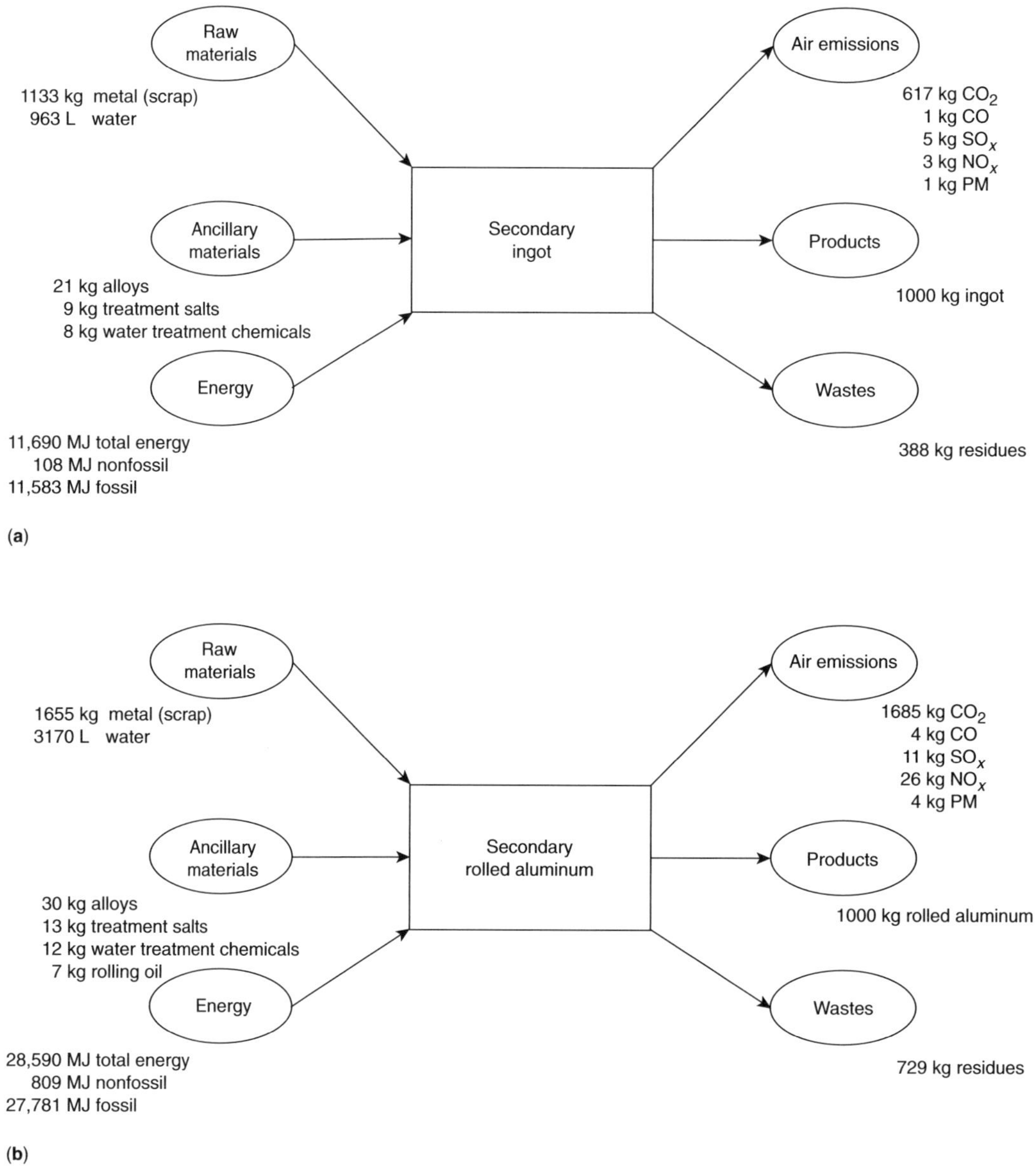

Fig. 3.19 Resource consumption and environmental releases (per 1000 kg output) for products from secondary aluminum. (a) Ingot. (b) Rolled aluminum. (c) Extruded aluminum. (d) Shape castings. PM, particulate matter

Interpretation of LCI Results

The aluminum LCI results described here are based on 1995* actual average data collected from 213 facilities (13 companies and 15 primary and secondary unit operations). Information on energy used for the primary ingot (Fig. 3.3) product system revealed that 84% of the total energy along the life cycle was from process energy (nonelectric and electric). Electric process energy accounted for 57% of the total energy. Production of secondary ingot requires 92% of its energy from process (nonelectric and electric), with electric process energy accounting for 21% of the process energy.

With regard to recycling, when the energy associated with transportation is discounted, the relative energy consumption between secondary

*There was one location where actual 1992 to 1993 data were used in this analysis.

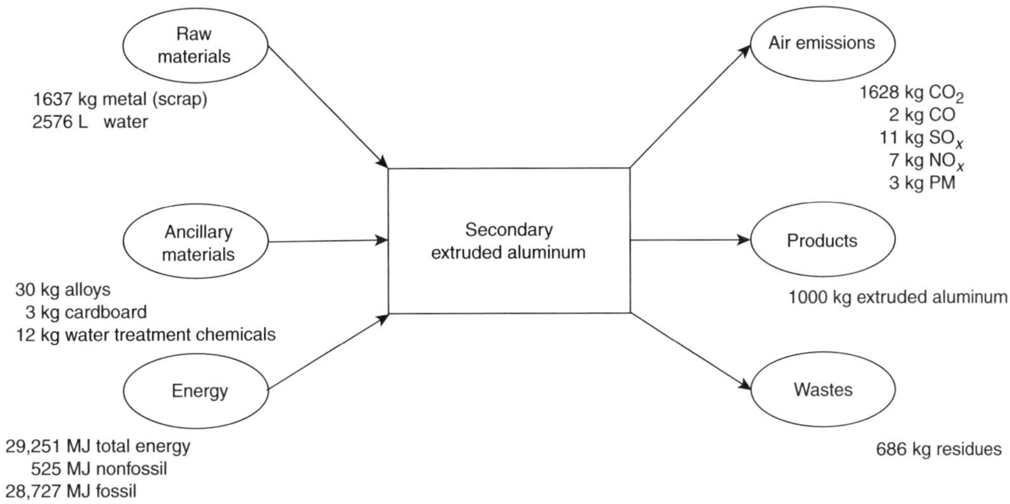

(c)

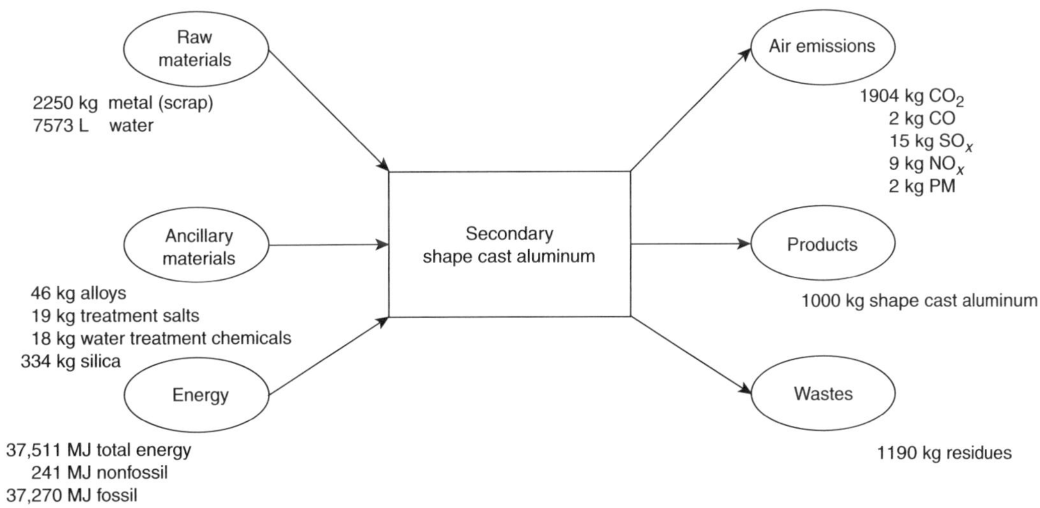

(d)

Fig. 3.19 (continued) (c) Extruded aluminum. (d) Shape castings. PM, particulate matter

(recycled) and primary ingot indicates that only 6.3% of the energy is required to produce secondary ingot as compared to primary ingot (as shown in Fig. 3.22). Accordingly, the energy efficiency and environmental performance of the industry as a whole will continue to improve as the extent of recycling increases.

Areas for Process Improvement. The USAMP partners in the LCI study consider the database suitable for benchmarking and process improvement, but it should not be used for the purposes of materials selection. For materials selection in actual application, other factors, such as cost, materials availability, and performance, must also be considered.

This LCI study has provided opportunities for benchmarking and subsequent process improvement. For example, each company contributing data to the study has received its own data, together with the industry average data, in the same format. Thus, each company can directly compare its performance in a given unit process

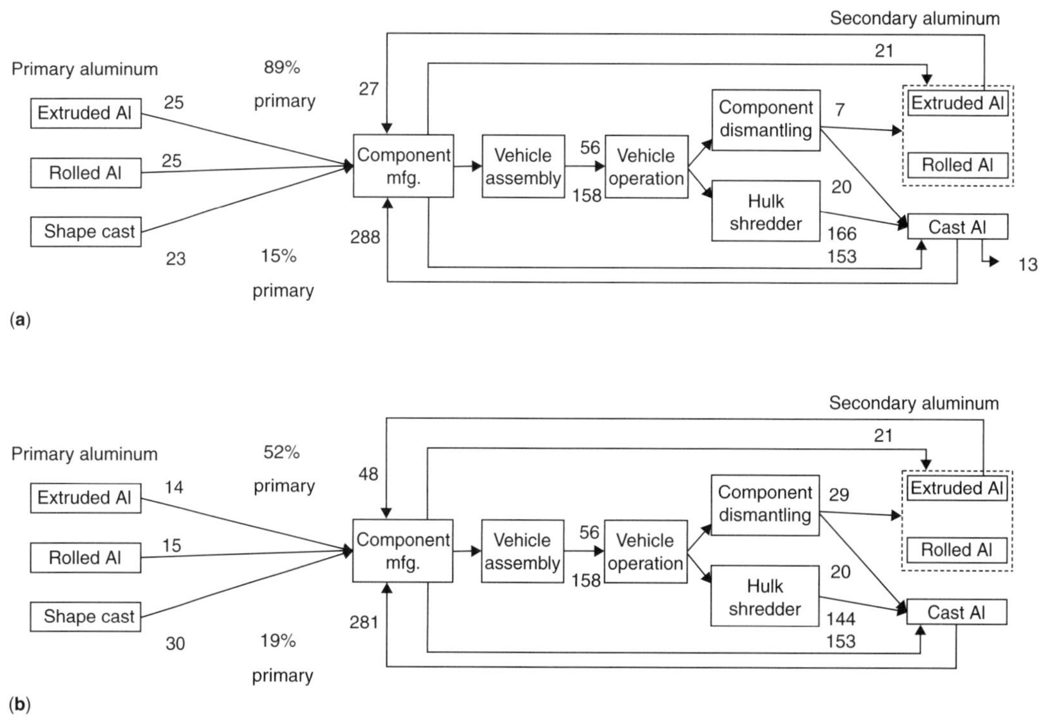

Fig. 3.20 Automotive aluminum recycling. (a) Current practice. (b) Steady-state future practice

to the benchmark performer and to the industry average. It is believed that this study will help identify several areas where reductions in raw materials and energy consumption, as well as environmental loadings, can be achieved in aluminum production processes.

To facilitate the benchmarking process, the use of an improvement potential index has been explored. The improvement potential has been defined as:

$$\text{Improvement potential} = \text{Standard deviation} \div \text{Mean} \times \text{Contribution}$$

For each data category, the standard deviation of the input or output was divided by the mean of the input or output and then multiplied by the contribution that the input or output value had on the data category. Conceptually, the improvement potential could be used as a means of prioritizing or selecting the process improvement efforts or investments.

It is critical to recognize, however, that this LCI database is enormous in its scope, and in each unit process, it arbitrarily compares data from a wide range of operations in different locations, using raw materials of varying quality, that is, bauxite type and alumina quality. Most importantly, the report often compares older technologies with newer variations of the same technology or even with radically different technologies that employ different operating principles, that is, fluid bed calcination versus rotary calcination of alumina trihydrate.

Given that the input data were provided under conditions of anonymity, no attempt is made to suggest specific areas of process improvement in view of the enormous number of raw material, process, and technology variables. It is considered that specific process improvements will emerge when knowledgeable scientists and engineers sift through the detailed database and, using it as a background, can apply their own unique knowledge of their companies' specific processes to define the best opportunities for process improvement.

However, with the aforementioned caveats in mind, it was considered instructive to evaluate the improvement potential for the category of energy use. Figure 3.23 shows the improvement potential, expressed as a percentage, for energy use as a function of the unit process. The relatively high value for alumina refining (39.1%) is probably due to a combination of factors. For

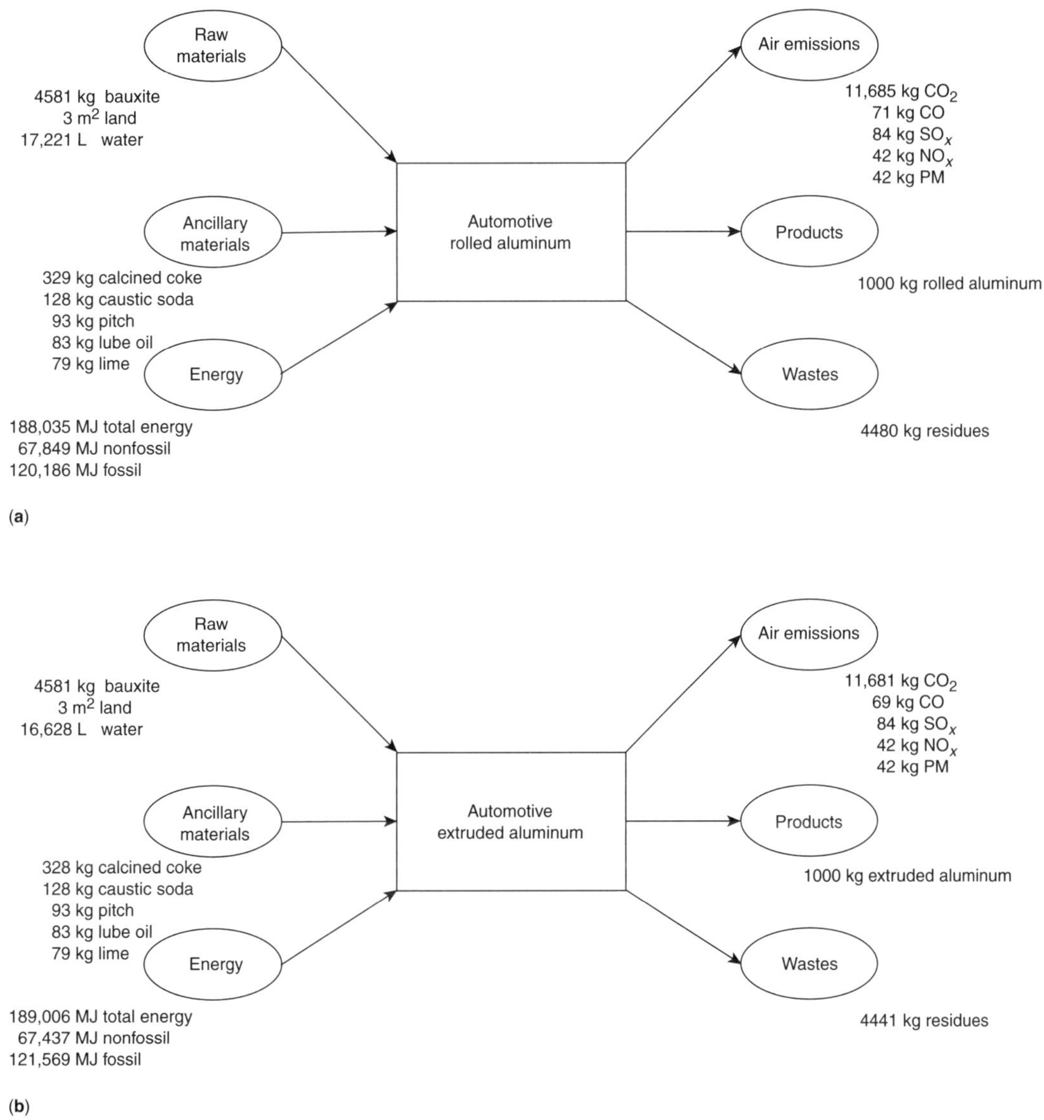

Fig. 3.21 Resource consumption and environmental releases (per 1000 kg output) for aluminum automotive products. (a) Rolled. (b) Extruded. (c) Shape cast. (d) Composite. PM, particulate matter

example, in alumina refining, the quality of the bauxite ore dictates whether low- or high-temperature digestion conditions must be used to dissolve the alumina. Also, the calcination of the resulting aluminum trihydrate intermediate product can be conducted using either the traditional rotary kiln or the more recent and energy-efficient fluid bed process.

The relatively low value for smelting (2.1%) is believed to reflect the fact that smelting has already been extensively optimized through research and development and plant practice, because this process step has long been recognized as energy-intensive (Fig. 3.6, 3.7). In smelting, energy improvements to some new level of efficiency will require some fundamental development in advanced electrode technology, for example, wettable cathode and/or inert anode technology and related cell design improvements. This is consistent with the fact that the highest long-term research and development priorities in the technology roadmap of the industry relate to the development of advanced electrode technology.

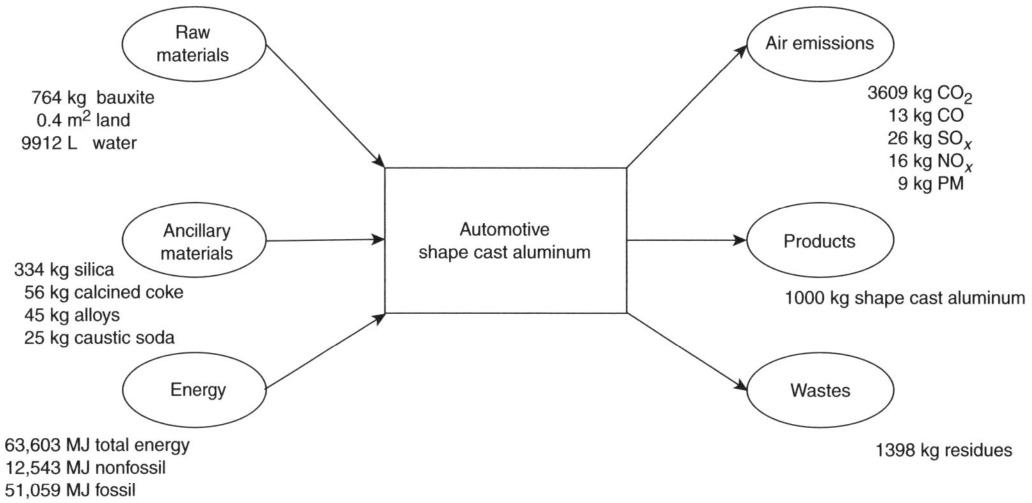

(c)

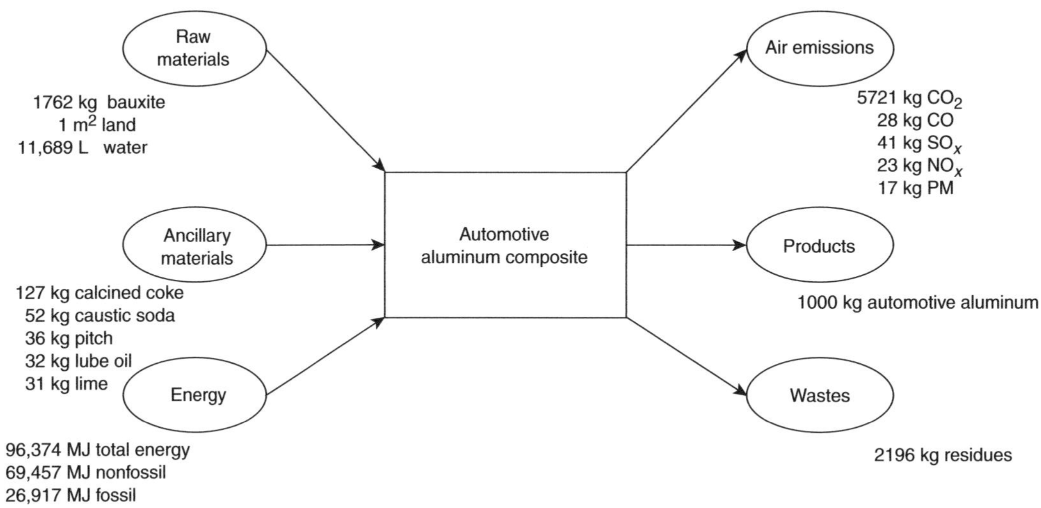

(d)

Fig. 3.21 **(continued)** Resource consumption and environmental releases (per 1000 kg output) for aluminum automotive products. (a) Rolled. (b) Extruded. (c) Shape cast. (d) Composite. PM, particulate matter

Figure 3.24 shows the improvement pontential data for CO_2 emissions by unit process. Neither bauxite mining nor alumina refining has any process-related CO_2 emissions; here, the CO_2 emissions are derived from the fuels used, for example, during calcination. However, during anode production and smelting, there appears to be significant potential for improvement. Again, raw material variations and technology differences may account for a large portion of the variation. For example, during anode baking, the newest furnace technology captures the hydrocarbon offgases from the baking process and recycles them as a fuel supplement. In the case of smelting, technology differences probably account for a significant portion of the 10.8% improvement potential value. Note that the smelter data incorporate data from both vertical stud and horizontal stud Soderberg processes as well as side-worked

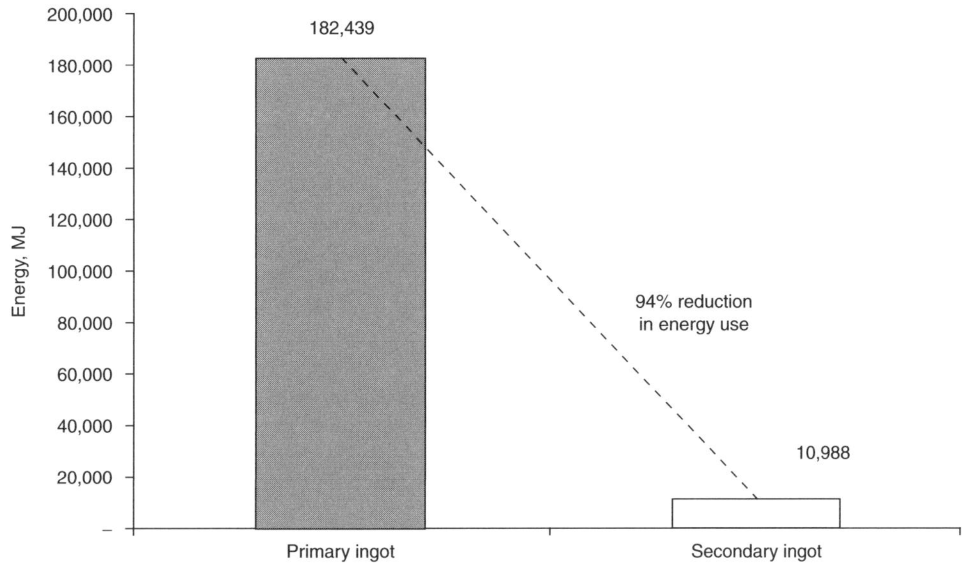

Fig. 3.22 Energy consumption comparison of recycled (secondary) ingot to primary ingot. Excludes transportation energy

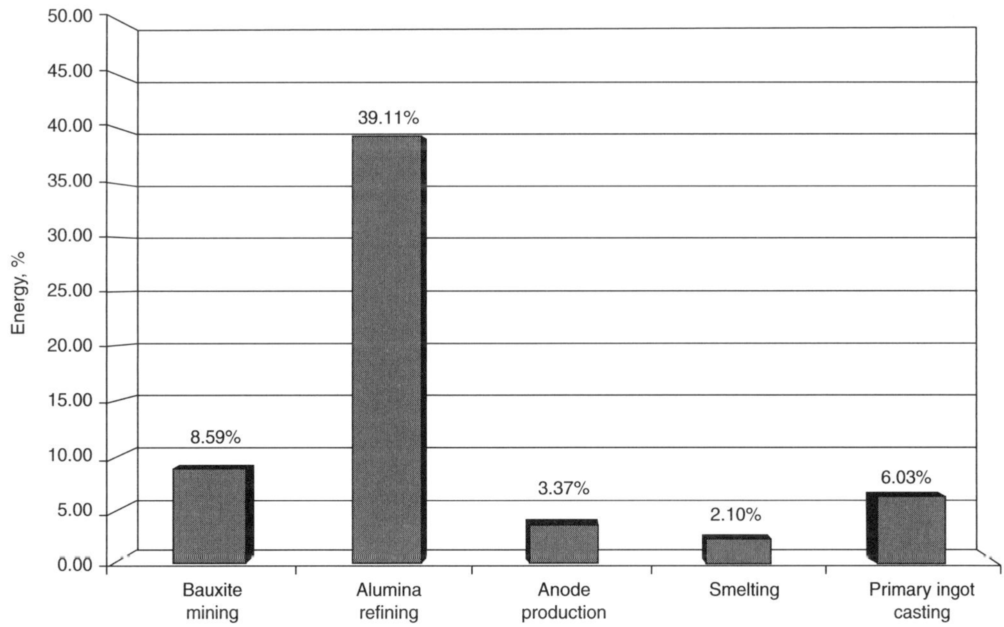

Fig. 3.23 Improvement potential for energy consumption in primary aluminum processing

and center-worked prebake cells. Emissions of CO_2 can also be influenced by materials variability, for example, coke quality, and by cell operating practices.

In summary, it is anticipated that energy and environmental improvements will continue to be made in aluminum production by dedicated plant and process engineers who exploit this LCI database as a tool, together with their own specific insights of different production processes, to achieve overall performance improvements.

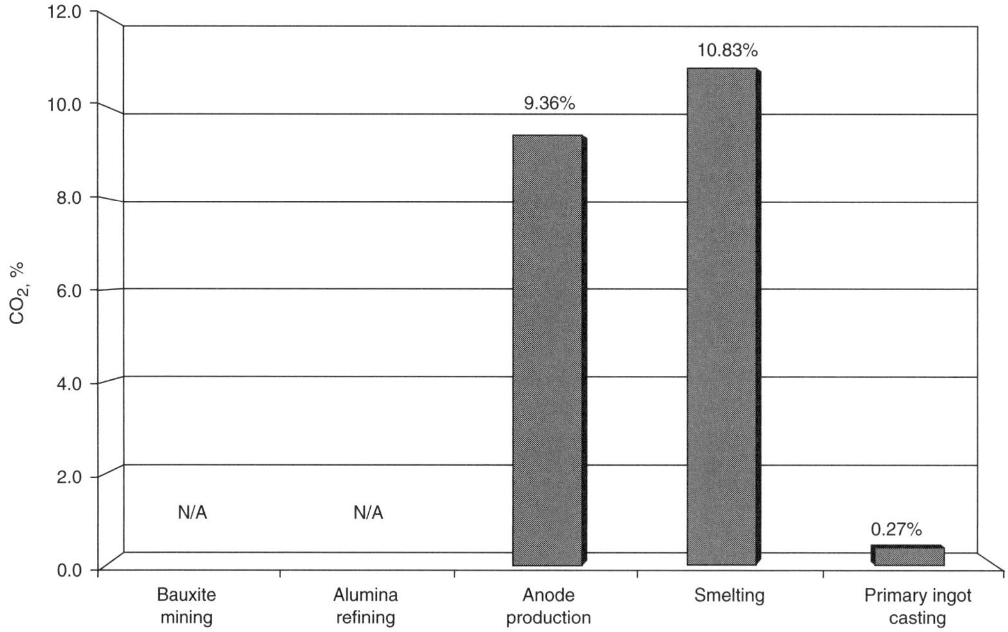

Fig. 3.24 Improvement potential for CO_2 emissions in primary aluminum processing

REFERENCES

3.1. "USCAR/USAMP Generic Automobile Life Cycle Inventory Methodology Document", Roy F. Weston, Inc., 1997
3.2. "Environmental Management—Life Cycle Assessment—Goal and Scope Definition and Inventory Analysis", Draft International Standard ISO/DIS 14041, International Standards Organization, 1997
3.3. *Electric Power Annual 1995*, Vol 1, DOE/EIA-0348(95)/1, Energy Information Administration, July 1996
3.4. "Electricity Supply and Demand 1995–2004, Summary of Electric Utility Supply and Demand Projections," North American Electric Reliability Council, June 1995
3.5. "Electric Power Used in Primary Aluminum Production," International Primary Aluminum Institute (IPAI) Form ES 002
3.6. K. Grjotheim, et al., *Aluminum Electrolysis, Fundamentals of the Hall-Heroult Process*, Chap. 4, 5, Aluminum-Verlag, Dusseldorf, 1982
3.7. "PFC Generation in 1996," International Primary Aluminum Institute
3.8. COS Generation in 1995, *Summary of Major Air Emissions of U.S. Aluminum Smelters* (1995 reporting year), Environmental Defense Fund; http:// www.scorecard.org (accessed June 2007)

CHAPTER 4

Life-Cycle Assessment of Aluminum: Inventory Data for the Worldwide Primary Aluminum Industry*

THE INTERNATIONAL ALUMINUM INSTITUTE (IAI) has been involved in the worldwide collection of aluminum data to be used in life-cycle assessments in order to "develop as complete an understanding as possible of the positive contributions that aluminum makes to the environmental and economic well-being of the world's population; of any negative economic or environmental impacts that its production may cause; and of the balance between these positives and negatives during the entire life cycle of the material" (Ref 4.1). This chapter contains results from a March 2003 report (Ref 4.2) with the purpose of collecting all significant life-cycle inventory data for primary aluminum (i.e., raw materials and energy use, air and water emissions, solid waste generated), with worldwide coverage (except Russia and China). Cumulative inputs and outputs of environmental significance (air emissions, waste generation, resource consumption) are defined for producing primary aluminum ingot from bauxite ore. It does not include the inputs and outputs associated with the further processes related to the production of final products from ingots and the recycling of the end-of-life product to obtain recycled aluminum ingots.

The intended purpose of this life-cycle inventory is to accurately characterize resource consumption and significant environmental aspects associated with the worldwide production of primary aluminum as a globally traded commodity. The primary aluminum production covered by this life-cycle inventory includes the following unit processes:

- Bauxite mining
- Alumina production
- Anode production: production of prebaked anodes, production of Soderberg paste
- Electrolysis
- Ingot casting

The collected data will serve as a credible basis for subsequent life-cycle assessments of aluminum products. In addition, the last section provides guidance on key features about how to treat aluminum in life-cycle assessments where recycled aluminum is used for aluminum products. No specific additional unit processes, in particular about energy production, transport, petroleum coke and pitch production, caustic soda production, and so on, have been added to the process in order to avoid nonelementary flows. Practitioners of life-cycle assessment who will use the data of this report may include such additional unit processes from their own databases (see also the section "Data Interpretation Items" in this chapter).

Data Coverage, Reporting, and Interpretation

Data from worldwide aluminum producers were requested for the calendar year 2000. Data for the life-cycle survey were obtained from:

- Eighty-two worldwide aluminum electrolysis plants producing 14.7 million metric tons

*Adapted with permission of the International Aluminum Institute from Ref 4.2

(tonnes, t) of primary aluminum, representing approximately 60% of the worldwide aluminum smelting operations. Base: primary aluminum of 24,464,400 tonnes (t) from the World Bureau of Metal Statistics (WBMS).

- Twenty-three worldwide alumina facilities producing 30.8 million metric tons of alumina, representing approximately 59% of the worldwide alumina operations. Base: world alumina production of 52,419,000 t, as 48,119,000 t from IAI plus 4,300,000 t estimated for China.
- Seventy-two worldwide aluminum cast houses producing 14.0 million tonnes (metric tons) of primary aluminum ingot, representing approximately 57% of the worldwide aluminum ingot casting operations. Total world aluminum production in the year 2000 from WBMS was 32,623,800 t, of which 24,464,400 t primary aluminum was taken as the cast house production base (plus 8,159,400 t secondary aluminum not relevant here).

Overall, primary aluminum energy returns represented approximately 69% of the world primary aluminum production; alumina energy returns represented approximately 70% of the world alumina production; and perfluorocarbon returns represented approximately 66% of the world primary aluminum production. The data were reasonably evenly distributed on a worldwide basis, with the nonavailability of data from China and Russia being mainly responsible for the comparatively poor coverage of Asia and Europe. For example, the survey's coverage of electrolysis plants in terms of reported primary aluminum production as a percentage of total primary aluminum production was approximately 82% in Africa, 75% in North America, 58% in Latin America, 34% in Asia (data for China not available), 52% in Europe (data for Russia and other Commonwealth of Independent States countries not available), and 100% in Oceania.

Collected data were combined together into inventory summaries of input and output (Tables 4.1 and 4.2) resources to produce 1000 kg of aluminum ingot (see also the section "Unit Processes and Results by Process" in this chapter). Selection of variables (Table 4.3) for this inventory was based on their environmental relevance, either specific to the primary aluminum production or as generally acknowledged environmental issues. It should also be noted that only direct energy-consumption figures were documented for this inventory and that CO_2 emission data were not included. Comprehensive energy data and CO_2 emission data, including those associated with the generation of electricity, the production of fuel (precombustion), and the combustion of fuel, are covered in Ref 4.3 and summarized in Table 4.4.

Data Quality

Data reporting includes quantitative and qualitative data quality indicators calculated for each data item. All values presented represent production-weighted mean values. Completeness is a quality indicator that covers the possibility for when not all respondents provided an answer. It is a ratio of the number of responses for the data item versus the total number of responses to the survey. Data statistics assume a normal distribution of results.

Data Consistency. During data review, a significant scatter in individual data items was noticed. Beyond obviously unrealistic outliers, high- and low-value outlier responses appeared likely to wrongly influence the final weighted average values. To address this issue, all individual answers beyond two standard deviations from the average value have been considered as outliers. Every individual outlier respondent has been queried accordingly with a request to check the response item for correction or confirmation. All outliers were adjusted according to answers received. If no answer had been received to the outlier query (approximately 60% of outliers), the individual outlier item was removed from the survey. The effect of this correction from outliers is summarized in Table 4.5. The table shows the percent difference in inventory results obtained by reintroducing the outliers (individual answers beyond two standard deviations from the average value, identified in the initial survey results and not commented on query). Reasons for the occurrence of outliers range from reporting mistakes (e.g., one nonrealistic low alumina consumption report, bringing in the same percent difference in all alumina-production-related data) to allocation issues between operations within plants (e.g., water input and output, steel input) and to measurement issues (e.g., fluoride, polycyclic aromatic hydrocarbons, mercury air and water emissions, benzo-a-pyrene air emissions, and other by-products for external recycling), the latter being generally responsible for the largest differences.

Table 4.1 Resource inputs from inventory data to produce 1000 kg of primary aluminum

	Process					
Inputs	**Bauxite mining**	**Alumina production**	**Anode production**	**Electrolysis**	**Ingot casting**	**Total**
Raw materials						
Bauxite	5168	...	...	...	...	5168 kg
Caustic sode	...	159	...	...	...	159 kg
Calcined lime	...	86	...	...	...	86 kg
Alumina						1925 kg
Petroleum coke	...	...	349	...	...	349 kg
Pitch	...	...	92	...	...	92 kg
Anode						441 kg
Aluminum fluoride	...	...	...	17.4	...	17.4 kg
Cathode carbon	...	...	...	6.1	...	6.1 kg
Aluminum (liquid metal)						1000 kg
Alloy additives	...	...	...	...	20	20 kg
Chlorine	...	...	...	...	0.068	0.068 kg
Cast ingot						1000 kg
Other raw material inputs						
Fresh water	...	6.4	0.5	2.95	3.15	13.0 m^3
Sea water	...	6.5	0.001	20.8	0.2	27.5 m^3
Refractory materials	...	...	5.5	6	...	11.5 kg
Steel (for anodes)	...	...	1.4	...	...	1.4 kg
Steel (for cathodes)	...	...	...	5.5	...	5.5 kg
Fuels and electricity						
Coal	...	185	0.9	...	...	186 kg
Diesel oil	10.3	1.2	1.4	...	0.1	13.0 kg
Heavy oil	...	221.4	6.2	...	10	238 kg
Natural gas	...	233	23	...	52	308 m^3
Electricity	...	203	62	15,365	81	15,711 kWh

Table 4.2 Resource outputs from inventory data when producing 1000 kg of primary aluminum

	Process					
Outputs	**Bauxite mining**	**Alumina production**	**Anode production**	**Electrolysis**	**Ingot casting**	**Total**
Air emissions						
Fluoride gaseous (as F)	...	...	0.02	0.55	...	0.57 kg
Fluoride particulate (as F)	...	...	0.004	0.5	...	0.50 kg
Particulates	12.2	1.2	0.1	3.3	0.08	16.9 kg
NO_x (as NO_2)	...	2.24	0.13	0.35	0.12	2.8 kg
SO_2	...	10.2	0.7	13.6	0.2	24.7 kg
Total polycyclic aromatic hydrocarbons (PAH)	...	...	0.02	0.13	...	0.15 kg
Benzo-a-pyrene (BaP)	...	...	0.1	5.0	...	5.1 g
CF_4	...	...	...	0.22	...	0.22 kg
C_2F_6	...	...	...	0.021	...	0.021 kg
Hydrogen chloride (HCl)	...	...	...	...	0.067	0.067 kg
Mercury	...	0.00020	...	...	...	0.00020 kg
Water emissions						
Fresh water	...	6.4	...	3.2	3.8	13.4 m^3
Sea water	...	6.6	...	20.9	...	27.5 m^3
Fluoride (as F)	...	...	...	0.2	...	0.20 kg
Oil/grease	...	0.13	...	0.008	0.009	0.15 kg
PAH (six Borneff components)	...	...	...	3.77	...	3.77 g
Suspended solids	...	1.43	...	0.21	0.02	1.66 kg
Mercury	...	0.0018	...	...	...	0.0018 kg
By-products for external recycling						
Bauxite residue	...	2.3	...	...	...	2.3 kg
Dross	...	...	...	...	13.0	13.0 kg
Filter dust	...	...	...	...	0.57	0.57 kg
Other by-products	...	3.5	2.8	5.1	...	11.4 kg
Refractory material	...	...	3.1	0.5	0.5	4.1 kg
Scrap sold	...	...	...	...	2.2	2.2 kg

(continued)

Table 4.2 *(continued)*

	Process					
Outputs	**Bauxite mining**	**Alumina production**	**Anode production**	**Electrolysis**	**Ingot casting**	**Total**
Spent potliner (SPL) carbon fuel—reuse	...	...	...	9.9	...	9.9 kg
SPL refractory bricks—reuse	...	...	...	5.5	...	5.5 kg
Steel	...	...	1.7	6.9	...	8.6 kg
Solid waste						
Bauxite residue (red mud)	...	1905	...	...	...	1905 kg
Carbon waste	...	...	2.4	4.6	...	7.0 kg
Dross—landfill	...	...	...	...	7.7	7.7 kg
Filter dust—landfill	...	...	...	...	0.4	0.40 kg
Other landfill wastes	703	47.5	2.7	7.3	1.3	762 kg
Refractory waste—landfill	...	...	2.5	1.2	0.7	4.4 kg
Scrubber sludges	...	...	0.8	13.7	...	14.5 kg
SPL—landfill	...	...	...	17.3	...	17.3 kg
Waste alumina	...	...	...	4.7	...	4.7 kg

Table 4.3 Units of input and output data for inventory of primary aluminum unit process

Inputs	Unit	Outputs	Unit
Raw materials		**Air emissions**	
Bauxite	kg	Fluoride gaseous (as F)	kg
Caustic soda (for alumina production)	kg	Fluoride particulate (as F)	kg
Calcined lime (for alumina production)	kg	Particulates	kg
		NO_x (as NO_2)	kg
Petroleum coke (for anode production)	kg	SO_2	kg
Pitch (for anode production)	kg	Total Polycyclic aromatic hydrocarbons (PAH)	kg
		Benzo-a-pyrene (BaP)	kg
Aluminum fluoride (for electrolysis)	kg	CF_4	kg
Cathode carbon (for electrolysis)	kg	C_2F_6	kg
		Hydrogen chloride (HCl)	kg
Alloy additives (for ingot casting)	kg	Mercury	kg
Chlorine (for ingot casting)	kg		
Other raw material inputs		**Water emissions**	
Fresh water	m^3	Fresh water	m^3
Sea water	m^3	Sea water	m^3
Refractory materials	kg	Fluoride (as F)	kg
Steel (for anodes)	kg	Oil/grease	kg
Steel (for cathodes)	kg	PAH (six Borneff components)	g
		Suspended solids	kg
		Mercury	kg
Fuels and electricity		**By-products for external recycling**	
Coal	kg	Bauxite residue	kg
Diesel oil	kg	Dross	kg
Heavy oil	kg	Filter dust	kg
Natural gas	m^3	Other by-products	kg
Electricity	kWh	Refractory material	kg
		Scrap sold	kg
		Spent potliner (SPL) carbon fuel—reuse	kg
		SPL refractory bricks—reuse	kg
		Steel	kg
		Solid waste	
		Bauxite residue (red mud)	kg
		Carbon waste	kg
		Dross—landfill	kg
		Filter dust—landfill	kg
		Other landfill wastes	kg
		Refractory waste—landfill	kg
		Scrubber sludges	kg
		SPL—landfill	kg
		Waste alumina	kg

Table 4.4 Average greenhouse gas (GHG) emissions for primary aluminum unit process included in this study

	Bauxite	Refining	Anode	Smelting	Casting
GHG emissions	[kg of CO_2 equivalents per 1000 kg of process output]				
Process	0	0	388	1626	0
Electricity	0	58	63	5801	77
Fossil fuel	16	789	135	133	155
Transport	32	61	8	4	136
Ancillary	0	84	255	0	0
Perfluorocarbons	0	0	0	2226	0
Total	**48**	**991**	**849**	**9789**	**368**

Source: Ref 4.3

Table 4.5 Difference on the inventory results from outlier reintroduction

Inputs	Difference(a),%	Outputs	Difference(a),%
Raw material inputs		**Air emission outputs**	
Bauxite	– 0.7	Fluoride gaseous (as F)	16
Caustic soda	– 0.7	Fluoride particulate (as F)	17
Calcined lime	– 0.7	Particulates	20
Alumina	– 0.7	NO_x (as NO_2)	14
		SO_2	2
Petroleum coke (for anode production)	0	Total polycyclic aromatic hydrocarbons (PAH)	23
Pitch (for anode production)	0	Benzo-a-pyrene (BaP)	6
Anode	0	CF_4	0
		C_2F_6	0
Aluminum fluoride	3	Hydrogen chloride (HCl)	8
Cathode carbon	6	Mercury	139
Aluminum (liquid metal)			
Alloy additives (for ingot casting)	0	**Water emission outputs**	
Chlorine (for ingot casting)	6	Fresh water	28
Cast ingot		Sea water	–0.2
		Fluoride (as F)	39
		Oil–grease	62
Other raw material inputs		PAH (six Borneff components)	102
Fresh water	46	Suspended solids	73
Sea water	–0.2	Mercury	879
Refractory materials	38		
Steel (for anodes)	12	**By-products for external recycling**	
Steel (for cathodes)	0.7	Bauxite residue	–0.7
		Dross	32
Fuel and electricity inputs		Filter dust	–2
Coal	–0.7	Other by-products	123
Diesel oil	–0.06	Refractory material	33
Heavy oil	–0.6	Scrap sold	20
Natural gas	–0.5	Spent potliner (SPL) carbon fuel—reuse	0
Electricity	–0.01	SPL refractory bricks—reuse	0
		Steel	10
		Solid waste outputs	
		Bauxite residue (red mud)	7
		Carbon waste	17
		Dross—landfill	–2
		Filter dust—landfill	7
		Other landfill wastes	14
		Refractory waste—landfill	26
		Scrubber sludges	45
		SPL—landfill	3
		Waste alumina	53

(a) This table shows the percent difference in inventory results obtained by reintroducing the outliers (individual answers beyond two standard deviations from the average value, identified in the initial survey results and not commented on query). Reasons for the occurrence of outliers range from reporting mistakes (e.g., one nonrealistic low alumina consumption report, bringing in the same percent difference in all alumina-production-related data) to allocation issues between operations within plants (e.g., water input and output, steel input) and to measurement issues (e.g., fluoride, polycyclic aromatic hydrocarbons, mercury air and water emissions, other by-products for external recycling), the latter being generally responsible for the largest differences.

Missing Process Data Supplemented. During the course of the survey, it was realized that the life-cycle survey forms distributed to companies had omitted two sets of needed process data:

- Bauxite mining life-cycle data
- Ingot casting energy-consumption data

Missing data have been worked out from currently available information, as described in Tables 4.6 and 4.7. Life-cycle data thus selected were considered quite representative.

Data Interpretation Items

As previously noted, the life-cycle inventory does not include data on nonelementary unit processes such as energy production, transport, petroleum coke and pitch production, caustic soda production, and so on. Other databases or reports can be considered for data on additional unit processes. Examples of other data sources include:

- The global breakdown by source of electricity used at primary aluminum smelters in the IAI energy survey for 2000 was as follows: hydro, 52.5%; coal, 31.6%; oil, 0.8%; natural gas, 9.0%; and nuclear, 6.1%.
- Environmental aspects from transport can be illustrated with a case from Ref 4.4, which yields the following air emission levels from transport in proportion of those generated from primary aluminum production: particulates, 1.1%; hydrocarbons, 29%; NO_x, 18.5%; and SO_2, 6.7%.

Other important items on data interpretation are summarized as follows. When constructing a life-cycle assessment related to aluminum (which includes aluminum production, fabrication, product usage, and product end-of-life issues), a methodology in accordance with internationally accepted practice (ISO 14040 series standards) should be used.

Missing data for bauxite mining and ingot casting energy have been worked out, as discussed previously in the section "Missing Process Data Supplemented" in this chapter.

Production-weighted mean values, which can be derived from the effective responses to the survey, are normally reported. In several situations, however, this led to inconsistent results. Data reporting then took place as industry-weighted use, that is, the ratio of total consumption or emission reported from the survey (total of all responses) to the total corresponding industry production. This was used for the following:

- Petroleum coke and pitch consumption for anodes (see Table 4.11 (a) is industry weighted, because mean values from anode production tables are altered due to anode butt recycling.
- For fresh water and sea water use (inputs and outputs), this correction avoided the apparent input-output imbalance visible when using weighted mean values.
- For fuel and electricity consumption for alumina and anode production, the rationale was that the mix of fuel use is typically company-specific; that is, using weighted mean values led to overestimated results.

Some data on air emissions (particulates, SO_2 and NO_x emissions) from fuel combustion (an energy unit process not documented in the present work) are included along with process emissions in the results reported from plants. This arises from the fact that the two emission types (from fuel combustion and from process) do occur together. An improved reliability on actual emission levels can correspond to fuel combustion as compared to general fuel combustion emission data. In particular, the actual sulfur content of fuel oil used, for example, in alumina production, is thus most accurately accounted for in SO_2 emissions. Accordingly, it is recommended that life-cycle assessment practitioners using data sets about emissions from fuel combustion remove collected data on particulates and SO_2 and NO_x emissions in order to avoid double counting. (Note: This applies only to fuel combustion and not to precombustion data sets.)

Other Exceptions. A review of the results could raise remarks for apparent inconsistency or weakness of evidence. Noted examples are the steel input-output imbalance (likely to be attributed to a used steel output from general maintenance work), the ingot casting mass balance (probably linked to the cold metal contribution), and a low response rate for dioxin emissions in ingot casting. These cases can probably be considered as relatively nonsignificant, given the overall purpose of the present inventory. It is also clear that the present survey produced the best currently available knowledge for the worldwide primary aluminum industry.

Dioxin emissions from ingot casting have not been included in the final inventory calculation, because they are related to aluminum scrap

Table 4.6 Supplemental bauxite mining energy data

	Selected	Two Australian mines	Nine mines (Aachen)	North American life-cycle inventory study 1998*
Inputs				
Diesel	2 kg/t(a)	0.90 kg/t	0.67–1.8 kg/t	4.37 kg/t
Other oil	. . .	(1.16 kg/t medium fuel oil, 0.27 kg/t gasoline)		1.43 kg/t
Electricity	. . .	0.002 kWh/t	. . .	0.4 kWh/t
Fresh water	. . .	0.031 m^3/t	. . .	. . .
Outputs				
Particulates	2.35 kg/t(a)	. . .	0.002–0.005 kg/t	2.35 kg/t
Solid waste	136 kg/t(a)	. . .	. . .	136 kg/t

(a) Per tonne (10^3 kg) of bauxite output
*Note: See chapter 3 for more information on the North American life-cycle inventory study 1998.

Table 4.7 Supplemental ingot casting energy data

Input	Selected	European Aluminum Assoc. (25 cast houses. 3.3 Mio t, typical primary)	Alcoa US. (K. Martchek)	Two Australian (J. Pullen)	North American life-cycle inventory 1998*
Fuel oil	10.0 kg/t(a)	10.9 kg/t	approx. 10 kg/t	. . .	17.4 kg/t
Diesel	0.1 kg/t(a)	0.1 kg/t	. . .	0.9 L/t	< 0.2 kg/t
Gas	52 m^3/t(a)	17.6 m^3/t	52 m^3/t	84 m^3/t	52 m^3/t
Electricity	81 kWh/t(a)	16 kWh/t	111 kWh/t	81 kWh/t	211 kWh/t

(a) Per tonne of cast metal output
*Note: See chapter 3 for more information on the North American life-cycle inventory study 1998.

remelting, which was outside the scope of the survey. Chlorine present in aluminum scrap (from painting or coating residues) is the origin element for dioxin emissions during aluminum remelting, and there is no chlorine in primary aluminum. The relation with aluminum scrap remelting has been confirmed from the survey results, with higher dioxin emissions results reasonably correlated with higher scrap use, despite the limited number of answers received (6 over 72 cast houses) and a high scatter in values.

Unit Processes and Results by Process

Results from the IAI Aluminum Life-Cycle Survey 2000 are presented for the following unit processes:

- Alumina production
- Anode production (prebake)
- Paste production (Soderberg)
- Reduction (electrolysis)
- Ingot casting

The life-cycle data are reported either as production-weighted mean values, which is the basic situation, or as industry-weighted use, as discussed in the section "Data Interpretation Items" in this chapter. The applicable definitions are as follows:

- *Production-weighted mean (wt. mean)*: Total consumption or emission reported, divided by total corresponding industry production of those plants that have reported data
- *Industry-weighted use (industry wt. use):* Total consumption or emission reported, divided by total corresponding industry production

For each unit process, the reference flow is 1 metric tonne. The interrelationships of these unit processes are shown in Fig. 4.1, which provides an overview of material flows in the primary aluminum production. As previously noted, no specific additional unit processes, in particular about energy production, transport, petroleum coke and pitch production, caustic soda production, ansd so on, have been added to the process in order to avoid nonelementary flows.

Inventory data for the worldwide primary aluminum are consolidated in Tables 4.1 and 4.2. The flow diagram of Fig. 4.2 combines processes with an input of 5168 kg of bauxite for a 1000 kg aluminum output. The main components of this mass distribution of 1000 kg

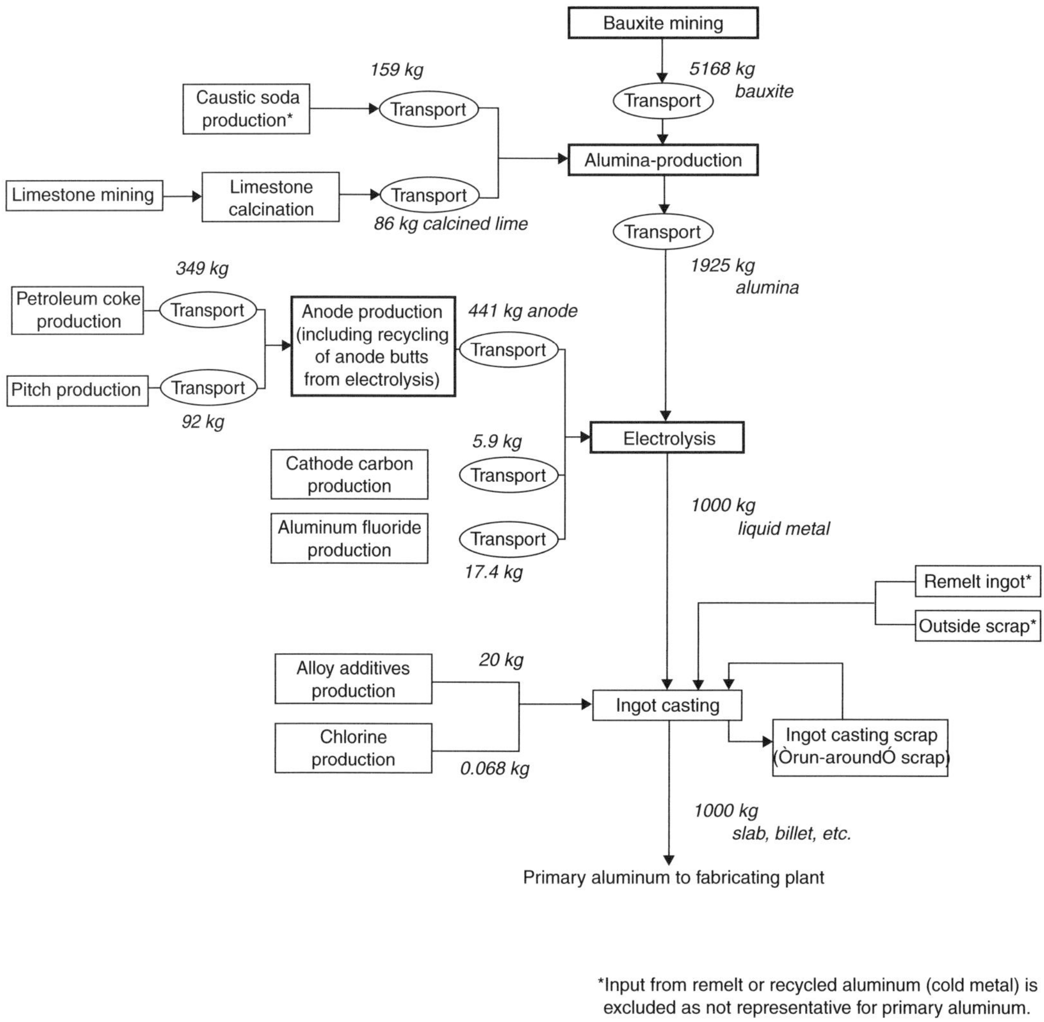

Fig 4.1 Interrelationships and materials flow of unit processes in aluminum primary production. Unit material processes not documented in the survey data are also shown. The production flow is as follows: aluminum is extracted from bauxite as aluminum oxide (alumina); this oxide is then broken down through an electrolysis process into oxygen, emitted as CO_2 by reaction with a carbon anode, and aluminum as liquid metal; next, aluminum is cast into an ingot, the usual form suitable for further fabrication of semifinished aluminum products.

aluminum output from 5168 kg of bauxite input (Fig. 4.2) need some explanation, because this is not an exact calculation. For example, there is always a significant water component in the bauxite (5168 kg), typically approximately 20% (1034 kg). The non-aluminum-containing part of the bauxite is disposed of as bauxite residue (red mud, 1905 kg). The mass balance out of the alumina production process would be approximately 2200 kg, after deduction of water component and bauxite residue.

Aluminum oxide (alumina) is converted in the electrolysis process (primary aluminum smelting) by the following reaction:

$$2Al_2O_3 + 3C = 4Al + 3CO_2$$

with a stoichiometric minimum requirement of 1890 kg Al_2O_3 for 1000 kg of primary aluminum. The actual production process could be described as an alumina breakdown by electrolysis, producing 1000 kg of aluminum and releasing

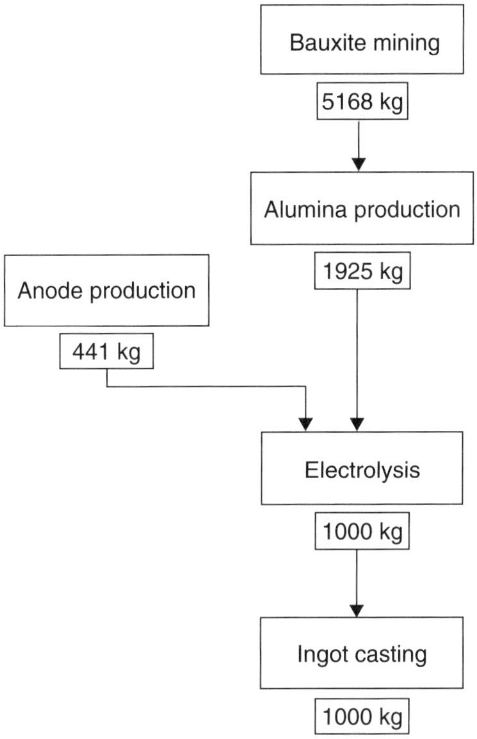

Fig 4.2 The flow diagram of processes from 5168 kg of bauxite input to produce 1000 kg primary aluminum ingot

oxygen on the carbon anode as CO_2 (where 441 kg coke and pitch input, considered as carbon weight, yield 1176 kg oxygen by difference, with 1617 kg CO_2 corresponding output).

Bauxite Mining

This unit process begins with the removal of overburden from a bauxite-rich mining site. Reusable topsoil is normally stored for later mine site restoration. The operations associated with this unit process include:

- Extraction of bauxite-rich minerals from the site
- Beneficiation activities, such as washing, screening, or drying
- Treatment of mining site residues and waste
- Site restoration activities, such as grading, dressing, and replanting

The output of this unit process is the bauxite that is transported to an alumina refinery.

Bauxite mining activities mainly take place in tropical and subtropical areas of the Earth. Almost all bauxite is mined in an open-pit mine. The known reserves of alumina-containing ore will sustain the present rate of mining for 300 to 400 years. Commercial bauxite can be separated into bauxite composed of mostly alumina trihydrates and those composed of alumina monohydrates. The trihydrate aluminas contain approximately 50% alumina by weight, while monohydrates are approximately 30%. Monohydrates are normally found close to the surface (e.g., Australia), while trihydrates tend to be at deeper levels (e.g., Brazil).

The only significant processing difference in bauxite mining is the need for beneficiation. Beneficiation occurs with ores from forested areas, while the grassland type typically does not require washing. The wastewater from washing is normally retained in a settling pond and recycled for continual reuse.

Alumina Production

This unit process begins with the unloading of process materials to their storage areas on site. The operations associated with this unit process include:

- Bauxite grinding; digestion and processing of liquors
- Alumina precipitation and calcination
- Maintenance and repair of plant and equipment
- Treatment of process air, liquids, and solids

The output of this unit process is smelter-grade alumina transported to an electrolysis plant (primary aluminum smelter).

In alumina production, also commonly called alumina refining, bauxite is converted to aluminum oxide using the Bayer process, which uses caustic soda and calcined lime (limestone) as input reactants. Bauxite is ground and blended into a liquor containing sodium carbonate and sodium hydroxide. The slurry is heated and pumped to digesters, which are heated pressure tanks. In digestion, iron and silicon impurities form insoluble oxides called bauxite residue. The bauxite residue settles out, and a rich concentration of sodium aluminate is filtered and seeded to form hydrate alumina crystals in precipitators. These crystals are then heated in a calcining process. The heat in the calciners drives off combined water, leaving alumina. Fresh water (input taken conservatively whether the water used is from fresh, underground, mine wastewater, or other sources) or sea water is used as the cooling agent.

The major differences in processing are at the calcination stage. Two types of kilns are used: rotary and fluid bed. The fluid bed or stationary kiln is newer and significantly more energy efficient. Energy requirements (coal, diesel oil, heavy oil, natural gas, electricity) have almost been halved over the last 15 years with the introduction of higher-pressure digesters and fluid flash calciners.

Air emissions primarily arise from the calcination stage (particulates; NO_X, as NO_2 and SO_2 from fuel combustion; mercury found in bauxite ores), while water emissions come from cooling use (fresh water, sea water, oil/grease) or are linked with the digestion stage (suspended solids, mercury found in bauxite ores).

Most of the bauxite residue currently turns out as solid waste, while a small but growing fraction is reused. Other by-products for external recycling are reaction chemicals. Other landfill wastes are typically inert components from bauxite, such as sand, or waste chemicals.

Survey results of inputs and outputs for alumina production are summarized, respectively, in Tables 4.8(a) and (b) based on production numbers of:

- 30,786,116 tonnes (metric tons) with life-cycle survey data from 23 refineries (which is 59% of the estimated survey coverage of the total world production)
- 36,911,495 tonnes with the IAI energy survey from 31 refineries (which is 70% of the estimated survey coverage of the total world production)

Anode Production

Survey results for anode production are summarized for prebake (Tables 4.9a–c) and for paste (Soderberg) anode production (Tables 10a–c). This unit process begins with the unloading of process materials to their storage areas on site. The operations associated with this unit process include:

- Recovery of spent anode materials
- Anode mix preparation; anode block or briquette forming and baking
- Rodding of baked anodes
- Maintenance and repair of plant and equipment
- Treatment of process air, liquids, and solids

The output of this unit process is rodded anodes or briquettes transported to an electrolysis plant (primary aluminum smelter).

There are two types of aluminum smelting technologies that are distinguished by the type of anode that is used in the reduction process: Soderberg and prebake. Soderberg design uses a single anode that covers most of the top surface of a reduction cell (pot). Anode paste (briquettes) is fed to the top of the anode, and as the anode is consumed in the process, the paste feeds downward by gravity. Heat from the pot bakes the paste into a monolithic mass before it reaches to the electrolytic bath interface. The prebake design uses prefired blocks of solid carbon suspended from steel axial busbars. The busbars both hold the anodes in place and carry the current for electrolysis.

The process for making the aggregate for briquettes or prebake blocks is identical. Petroleum coke is calcined, ground, and blended with pitch to form a paste that is subsequently formed into blocks or briquettes and allowed to cool. While the briquettes are sent directly to the pots for consumption, the blocks are sent to a separate baking furnace.

Baking furnace technology has evolved from simple pits that discharged volatiles to the

Table 4.8a Alumina production inputs to produce 1000 kg of alumina

Inputs	Industry-weighted use	Wt. mean	Standard deviation	Unit(a)	Response rate, %	Min	Max	DQI average(b)	DQI range(b)
Raw Materials									
Bauxite	...	2685	513	kg/t	100	2102	3737	1.1	1–2
Caustic soda	...	82	30	kg/t	100	29	181	...	...
Calcined lime	...	45	44	kg/t	100	5	221	...	...
Fresh water	3.3	3.5	4.4	m^3/t	91	0.003	14	1.4	1–3
Sea water	3.4	11.4	17	m^3/t	22	1.2	42	1.2	1–2
Fuels and electricity									
Heavy oil	115	154	137	kg/t	81	0.5	437	...	...
Diesel oil	0.6	1.7	2.4	kg/t	35	0.1	8	...	...
Gas	121	222	118	m^3/t	35	2	384	...	...
Coal	96	374	266	kg/t	23	127	843	...	...
Electricity	106	170	188	kWh/t	65	8	749	...	...

(a) Per tonne of alumina. (b) DQI, data quality indicator

Table 4.8b Alumina production outputs to produce 1000 kg of alumina

Outputs	Industry-weighted use	Wt. mean	Standard deviation	Unit(a)	Response rate, %	Min	Max	DQI average(b)	DQI range(b)
Air emissions									
Particulates	...	0.63	1.9	kg/t	91	0.12	7.2	1.6	1–3
SO_2	...	5.3	8.9	kg/t	87	0.000003	24	1.6	1–2
NO_x (as NO_2)	...	1.17	0.68	kg/t	74	0.39	2.3	1.8	1–3
Mercury	...	0.10	0.06	g/t	26	0.004	0.16	2.1	1–3
Water emissions									
Fresh water	3.3	3.7	7.7	m^3/t	87	0.58	33	1.9	1–3
Sea water	3.4	11.4	17	m^3t	22	1.2	42	1.3	1–2
Suspended solids	...	0.74	1.1	kg/t	52	0.0002	3.7	1.6	1–3
Oil and grease/ total hydrocarbons	...	0.069	0.10	kg/t	48	0.000001	0.27	1.9	1–3
Mercury	...	0.00094	0.0035	g/t	30	0.00003	0.0095	1.9	1–3
By-products (for external recycling)									
Bauxite residue	...	1.2	0.5	kg/t	13	0.33	1.3	1.5	1–3
Other	...	1.8	3.2	kg/t	43	0.07	10.6	1.8	1–3
Solid waste									
Bauxite residues (red mud)	...	990	407	kg/t	96	204	1916	1.7	1–3
Other landfill wastes	...	24.7	37	kg/t	87	0.18	161	1.9	1–3

(a) Per tonne of alumina. (b) DQI, data quality indicator

Table 4.9a Inputs to produce 1000 kg of baked anode

Inputs	Industry-weighted use	Wt. mean	Standard deviation	Unit(a)	Response rate, %	Min	Max	DQI average(b)	DQI range(b)
Raw materials									
Petroleum coke	...	683	57	kg/t	100	599	888	...	...
pitch	...	160	15	kg/t	100	135	200	...	...
Total	...	843	...	kg/t	...	...	...	...	...
Fuels and electricity									
Coal	2.0	66	65	kg/t	4	4.1	96	...	...
Heavy oil	17.0	75	32	kg/t	23	14	160	...	...
Diesel oil	3.9	24	30	kg/t	13	0.02	72	...	...
Gas	63	90	33	m^3/t	70	13	217	...	...
Electricity	158	185	96	kWh/t	86	0.19	450	...	...
Other inputs									
Fresh water	1.2	2.2	3.4	m^3/t	56	0.004	12	1.8	1–3
Sea water	0.0022	0.22	NA	m^3/t	2	0.22	0.22	1	1
Refractory material	...	12.5	18	kg/t	72	0.16	98	1.9	1–3
Steel	...	3.1	3.0	kg/t	44	0.04	10	1.9	1–3

From life cycle survey data with total production (baked) of 6,443,997 tonnes from 54 anode plants; and IAI energy survey data with total production (baked) of 6,694,481 tonnes from 56 anode plants. (a) Per tonne anode. Note: Recycled anode butts account for the raw material mass balance. (b) DQI, data quality indicator

Table 4.9b Outputs when producing 1000 kg of baked anode

Outputs	Average	Standard deviation	Unit(a)	Response rate, %	Min	Max	DQI average(b)	DQI range(b)
By-products for external recycling								
Refractory	6.9	6.2	kg/t	39	0.5	22.4	1.4	1–3
Steel	3.9	2.4	kg/t	37	0.3	9.3	1.8	1–3
Other	6.5	9.4	kg/t	31	0.0006	33.2	1.6	1–3
Solid waste not recycled								
Waste carbon or mix	6.0	4.9	kg/t	57	0.1	16.6	1.5	1–3
Scrubber sludges	1.9	2.4	kg/t	15	0.04	6.5	1.4	1–3
Refractory (excluding spent potliner)	5.7	9.5	kg/t	46	0.1	33.6	1.9	1–3
Other landfilled waste	5.3	6.4	kg/t	43	0.4	23.5	1.6	1–3

From life-cycle survey data with total production of 6,443,997 tonnes from 54 anode plants. (a) per tonne anode. (b)DQI, date quality indicator

Table 4.9c Air emissions when producing 1000 kg of baked anode

Outputs	Wt. mean	Standard deviation	Unit(a)	Response rate, %	Min	Max	DQI average(b)	DQI range(b)
Air emissions								
Particulates	0.33	0.39	kg/t	89	0.02	1.8	1.4	1–3
SO_2	1.8	1.6	kg/t	72	0.001	6.2	1.6	1–3
NO_x (as NO_2)	0.31	0.23	kg/t	69	0.02	1.3	1.7	1–3
Particulate fluoride (as F)	0.010	0.02	kg/t	57	0.000002	0.06	1.4	1–3
Gaseous fluoride (as F)	0.046	0.16	kg/t	76	0.00001	0.9	1.2	1–3
Total polycyclic aromatic hydrocarbons	0.060	0.11	kg/t	76	0.00003	0.5	1.5	1–3
Benzo-a-pyrene	0.27	0.68	g/t	54	0.00003	3.4	1.7	1–3

From, life-cycle survey data with total production of 6,443,997 tonnes from 54 anode plants. (a) Per tonne anode. (b) DQI, data quality indicator

Table 4.10a Inputs to produce 1000 kg of anode paste

Inputs	Industry-weighted use	Wt. mean	Standard deviation	Unit(a)	Response rate, %	Min	Max
Raw materials							
Petroleum coke		713	26	kg/t	100	669	760
Pitch		284	25	kg/t	100	240	331
Total		997	...	kg/t	...	...	...
Fuels and electricity							
Coal	2.6	65	NA	kg/t	4	65	65
Heavy oil	0.40	6.4	83	kg/t	11	0.6	145
Diesel oil	...	...	...	kg/t	...	...	...
Gas	6.6	22	12	m^3/t	19	5	37
Electricity	65	106	69	kWh/t	63	33	264
Other inputs							
Fresh water	1.5	2.0	2.6	m^3/t	53	0.0004	8.2
Sea water	...	...	...	m^3/t	...	...	...
Refractory material	...	...	...	kg/t	...	...	...

From life-cycle survey data with total production of 948,457 tonnes from 17 paste plants; and IAI energy survey data with total production of 1,466,841 tonnes from 27 paste plants. (a) Per tonne anode

Table 4.10b Outputs when producing 1000 kg of anode paste

Outputs	Average	Standard deviation	Unit(a)	Response rate,%	Min	Max	DQI average(b)	DQI range(b)
By-products for external recycling								
Refractory	...	...	kg/t	...	...	...	...	...
Steel	...	...	kg/t	...	...	...	...	...
Other	0.63	NA	kg/t	6	0.63	0.63	...	...
Solid waste not recycled								
Waste carbon or mix	0.11	0.01	kg/t	24	0.13	0.17	1.7	1–3
Scrubber sludges	...	...	kg/t	...	...	...	...	...
Refractory (excluding spent potliner)	...	...	kg/t	...	...	...	...	...
Other landfilled waste	14.7	30	kg/t	18	0.4	53	1.7	1–3

From life-cycle survey data with total production of 948,457 tonnes from 17 paste plants. (a) Per tonne anode. (b) DQI, data quality indicator By-products for external recycling

Table 4.10c Air emissions when producing 1000 kg of anode paste

Air emissions	Wt. mean	Standard deviation	Unit(a)	Response rate, %	Min	Max	DQI average(b)	DQI range(b)
Particulates	0.11	0.20	kg/t	71	0.01	0.72	1.3	1–3
SO_2	1.0	1.1	kg/t	41	0.002	2.57	1.6	1–2
NO_x (as NO_2)	0.11	0.09	kg/t	29	0.05	0.26	2.4	2–3
Particulate fluoride (as F)	...	...	kg/t	...	...	...	...	...
Gaseous fluoride (as F)	...	...	kg/t	...	...	...	...	...
Total polycyclic aromatic hydrocarbons	0.0092	0.010	kg/t	53	0.001	0.03	1.1	1–2
Benzo-a-pyrene	0.079	0.084	g/t	41	0.0002	0.23	1.6	1–3

From life-cycle survey data with total production of 948,457 tonnes from 17 paste plants. (a) Per tonne anode. (b) DQI, data quality indicator

atmosphere during the baking cycle to closed-loop-type designs that convert the caloric heat of the volatile into a process fuel that reduces energy consumption for the process. Baking furnaces use refractory materials for linings, and fresh water (input taken conservatively whether the water used is from fresh, underground, mine wastewater, or other sources) (or possibly sea water) as the cooling agent. Baking furnaces account for most of the energy consumption (coal, diesel oil, heavy oil, natural gas, electricity).

Air emissions—fluoride gaseous (as F), or fluoride particulate (as F)—arise from recovered spent anode materials (unused anode ends, or anode butts) from electrolysis (see the following section) that are recycled within prebake anode production. Particulates—NO_X (as NO_2) on SO_2 — typically come from fuel combustion. Total polycyclic aromatic hydrocarbons (PAH), which includes benzo-a-pyrene (BaP), are air emissions generated from the basic anode production process. A common practice for their prevention and monitoring is water scrubbing, a process using fresh water (input taken conservatively whether the water used is from fresh, underground, mine wastewater, or other sources) or sea water as input and resulting in corresponding fresh water or sea water discharges (the latter accounted for in the inventory together with the electrolysis water discharges from scrubbing; see the following section).

By-products for external recycling include the recovery of used steel from anode bars or used refractory material from baking furnaces. Various other by-products are also recovered, for example, carbon recovered for reuse.

Solid waste that is not recycled (landfill) includes waste carbon or a mixed residue from anode production, scrubber sludges arising from water scrubbing used for control of air emissions mentioned previously, refractory waste from baking furnaces, and other landfill wastes that arise as various residues, for example, carbon fines.

Electrolysis

Survey results for electrolysis (smelting) are data that came from all existing major technology types (Tables 4.11a–c). Approximately 15% of the total capacity surveyed was from Soderberg facilities, and the remaining 85% was produced in prebake facilities. The supplied alumina production data came from facilities currently in operation.

The unit process of electrolysis begins with the unloading of process materials to their storage areas on site. The operations associated with this unit process include:

- Recovery, preparation, and handling of process materials
- Manufacture of major process equipment (e.g., cathodes)
- Process control activities (metal, bath, heat)
- Maintenance and repair of plant and equipment
- Treatment of process air, liquids, and solids

Table 4.11a Inputs to produce 1000 kg of liquid aluminum from reduction (electrolysis)

Inputs	Industry-weighted use	Wt. mean	Standard deviation	Unit(a)	Response rate, %	Min	Max	DQI average(b)	DQI range(b)
Raw materials									
Alumina (dry)	...	1925	27	kg/t	96	1871	2033	1.3	1–3
Anode prebake (net)	...	426	25	kg/t	65	388	546	...	...
Soderberg paste	...	510	32	kg/t	29	440	584	...	...
Anodes (net)/Soderberg paste	441	...	...	kg/t	95	...	...	...	...
Petroleum coke	349	...	...	kg/t	...	...	...	...	...
Pitch	92	...	...	kg/t	...	...	...	...	...
Electricity consumption									
Electricity		15,365	1179	kWh/t	100	13,405	19,446	...	...
Other inputs									
Fresh water	2.9	3.9	7.8	m^3/t	73	0.002	32	1.6	1–3
Sea water	20.7	163	203	m^3/t	16	0.5	594	1.3	1–3
Cathode carbon	...	6.1	3.4	kg/t	93	1.1	16	1.6	1–3
Refractory material	...	6.0	4.2	kg/t	88	0.2	18	1.8	1–3
Steel	...	5.5	5.0	kg/t	85	0.1	35	1.8	1–3
AlF_3	...	17.4	5.4	kg/t	94	6.9	32	1.1	1–2

From life-cycle survey data with total production of 14,692,748 tonnes from 82 smelters representing approximately 60% total world production, of which Soderberg was used for 2,001,454 tonnes (29 smelters), with 68 smelters using baked anodes; and from IAI energy survey data with total (production) of 16,822,420 tonnes from 112 smelters representing approximately 69% coverage of total world production. (a) Per tonne liquid aluminum. (b) DQI, data quality indicator

Table 4.11b Outputs when producing 1000 kg of liquid aluminum with electrolysis

Outputs	Industry-Weighted use	Wt. mean	Standard deviation	Unit(a)	Response rate, %	Min	Max	DQI average(b)	DQI range(b)
Water discharges									
Fresh water	3.1	4.7	8.0	m_3/t	71	0.02	31.8	1.7	1–3
Sea water	20.9	192	206	m_3/t	15	0.1	594	1.4	1–3
Suspended solids	...	0.21	0.56	kg/t	61	0.0002	2.7	1.4	1–3
Oil and grease/ total hydrocarbons	...	0.0078	0.014	kg/t	41	0.00002	0.05	1.6	1–3
Fluorides (as F)	...	0.20	0.7	kg/t	70	0.00001	3.9	1.4	1–3
Polycycylic aromatic hydrocarbons (six Borneff components)	...	3.8	9.3	g/t	32	0.000002	32	1.5	1–3
By-products for external recycling									
Spent potlinen (SPL) carbon	...	9.9	11	kg/t	35	0.25	41	1.2	1–3
SPL refractory	...	5.5	6.1	kg/t	26	0.60	19	1.2	1–3
Refractory (other)	...	0.53	0.8	kg/t	12	0.11	2.6	1.3	1–3
Steel	...	6.9	4.8	kg/t	74	0.13	20	1.6	1–3
Other	...	5.1	7.7	kg/t	49	0.13	26	1.5	1–3
Solid waste not recycled									
SPL	...	17.3	12	kg/t	79	0.09	53	1.4	1–3
Waste alumina	...	4.7	7.3	kg/t	43	0.06	30	1.5	1–3
Waste carbon or mix	...	4.6	5.4	kg/t	40	0.01	20	1.4	1–3
Scrubber sludges	...	13.7	20	kg/t	16	0.04	50	1.3	1–3
Refractory (excluding SPL)	...	1.2	1.6	kg/t	40	0.05	6	1.6	1–3
Other landfilled waste	...	7.3	9.1	kg/t	71	0.06	33	1.6	1–3

From life-cycle survey data with total production of 14,692,748 tonnes from 82 smelters representing approximately 60% of total world production, of which Soderberg had 2,001,454 tonnes (with 29 smelters) and 68 smelters with prebaked anodes. (a) Per tonne liquid aluminum. (b) DQI, data quality indicator

Table 4.11c Air emission when producing 1000 kg of liquid aluminum from electrolysis

Air emissions	Wt. mean	Standard deviation	Unit(a)	Response rate, %	Min	Max	DQI average(b)	DQI range(b)
Particulates	3.3	5.0	kg/t	89	0.04	26	1.3	1–3
SO_2	13.4	6.6	kg/t	89	0.5	25	1.7	1–3
NO_x (as NO_2)	0.35	0.8	kg/t	52	0.000004	3.9	1.8	1–3
Particulate fluoride (as F)	0.50	0.8	kg/t	88	0.005	3.7	1.3	1–3
Gaseous fluoride (as F)	0.55	0.9	kg/t	91	0.02	4.7	1.2	1–3
Total polycyclic aromatic hydrocarbons	0.13	0.4	kg/t	44	0.0001	1.3	1.8	1–3
Benzo-a-pyrene	5.0	16	g/t	35	0.0001	59.4	1.8	1–3
CF_4	0.22	0.40	kg/t	100	0.007	1.8	...	...
C_2F_6	0.021	0.040	kg/t	100	0.001	0.18	...	...

From life-cycle survey data with total production of 14,692,748 tonnes representing 60% of total world production with 82 smelters, of which soderberg smelters (with 2,001,454 tonnes production) and 68 smelters with prebaked anodes; and from IAI per fluorocarbon survey data with total production of 16,079,065 tonnes representing 66% of total world production from 115 smelteres. (a) Per tonne liquid aluminum. (b) DQI, data quality indicator

The output of this unit process is hot metal transported to an ingot casting facility.

Molten aluminum is produced from alumina (aluminum oxide) by the Hall-Heroult electrolytic process that dissolves the alumina in a molten cryolite bath (requiring aluminum fluoride input) and passes current through this solution, thereby decomposing the alumina into aluminum and oxygen. Aluminum is tapped out of the reduction cell (pot) at daily intervals, and the oxygen combines with the carbon of the anode to form carbon dioxide.

The pot consists of a steel (for cathodes) shell lined with refractory materials insulation and with a hearth of carbon (cathode carbon for electrolysis). This is known as the cathode. The cathode is filled with a cryolite bath and alumina, and an anode is suspended in the bath to complete the circuit for the pot. When started, a pot will run continuously for the life of the cathode, which may last in excess of 10 years. At the end of its life, each pot is completely refurbished. Steel from used cathodes is recovered for recycling. Refractory materials are either recycled as by-products or landfilled (refractory waste—landfill). Spent potlinings (SPLs), which include a carbon-base (SPL carbon) and a refractory-based part (SPL refractory bricks), are either recycled as by-products (SPL carbon fuel—reuse, SPL refractory bricks—reuse) or landfilled (SPL—landfill).

The current in a pot varies from 60,000 to over 300,000 A at a voltage drop of 4.2 to 5.0 V. Pots produce approximately 16.2 +/– 0.6 lb per day of aluminum for each kiloampere, at an operating efficiency of 91 +/– 4%. Electricity consumption is the major energy aspect of electrolysis.

Aluminum smelters typically use an air pollution control system to reduce emissions. The primary system is usually a scrubber. Some plants use dry scrubbers, with alumina as the absorbent that is subsequently fed to the pots and allows for the recovery of scrubbed materials. Other plants use wet scrubbers, which recirculate an alkaline solution to absorb emissions; the wet scrubbing process uses fresh water (input taken conservatively whether the water used is from fresh, underground, mine wastewater, or other sources) or sea water as input and results in corresponding fresh water or sea water discharges. Unlike dry scrubbers, wet scrubbers absorb carbon dioxide, nitrogen oxide, and sulfur dioxide that are entrained in the wastewater liquor (which is subsequently treated prior to final discharge). Scrubber sludges are landfilled.

Air Emissions. Specific aluminum electrolysis process emissions are fluoride gaseous (as F) and fluoride particulate (as F), which arise from the molten bath; total PAH, including BaP, which arises from anode consumption; and CF_4 and C_2F_6, commonly reported as perfluorocarbons, are gases generated with an uncontrolled anode overvoltage situation called anode effect. Particulates, NO_X (as NO_2), and SO_2, typically come from fuel combustion.

Water Emissions. Fluoride (as F) and PAH (Six Borneff components that are monitored because of their particular environmental effect) arise from the same origin as their aforementioned air emission equivalents. Suspended solids and oil and grease (or total hydrocarbons) are monitored in water discharges from wet scrubbing.

Solid Waste. Other landfill wastes typically consist of approximately 60% environmental abatement waste (such as dry scrubber filter bags) and 40% municipal waste (Ref 4.5).

Ingot Casting

Survey results for ingot casting are summarized in Tables 4.12(a) to (c). With the focus on primary production, secondary ingots are not included, and contributions of remelt or recycled aluminum are excluded. Survey results for the process ingot casting yielded a higher weight output (1000 kg) than the corresponding electrolysis metal input (874 kg), due to a "cold metal" input contribution from remelt aluminum (133 kg remelt ingot) and recycled aluminum (101 kg outside scrap). Because the scope of this inventory report is primary aluminum and not remelt or recycled aluminum, data for the unit process ingot casting were calculated excluding the contribution from "cold metal," that is, all inputs and outputs from the survey average were adjusted by a factor of 0.79 (input ratio [electrolysis metal + alloy additives = 892 kg] / [total metal input = 1126 kg]—see Table 4.12a). According to the ISO standards on life-cycle assessment, this can be described as a situation of joint process where a mass allocation approach is applied.

Table 4.12a Inputs for casting 1000 kg of aluminum ingot

Inputs	Industry-weighted use	Wt. mean	Standard deviation	Unit(a)	Response rate, %	Min	Max	DQI average(b)	DQI range(b)
Electrolysis metal(c)	...	874(d)	165	kg/t	96	373	1229	1.03	1–2
Remelt ingot	...	133	101	kg/t	67	0.09	451	1.2	1–3
Outside scrap	...	101	107	kg/t	47	0.5	388	1.1	1–3
Alloy additives(c)	...	18(e)	14	kg/t	90	0.004	67	1.06	1–3
Fresh water	3.9	4.6	10.7	m^3/t	89	0.001	48	1.6	1–3
Sea water	0.23	10.5	12.7	m^3/t	3	0.8	19	1.2	1–3
Chlorine	...	0.086	0.11	kg/t	61	0.001	0.42	1.3	1–3

From life-cycle survey data with total production of 14,016,405 tonnes representing 57% of total world production from 72 cast houses.
(a) Per tonne of aluminum.
(b) DQI, data quality indicator.
(c) Metal input adjusted to exclude contribution from cold metal for 1000kg ingot output.
(d) Electrolysis metal adjusted to 1000 kg/t for share of ingot casting inputs and outputs.
(e) Alloy additives adjusted to 20 kg/t for primary aluminum life-cycle calculation

Table 4.12b Outputs per 1000 kg of cast aluminum ingot

Outputs	Industry-weighted use	Wt. mean	Standard deviation	Unit(a)	Response rate, %	Min	Max	DQI average(b)	DQI range(b)
Water discharges									
Fresh water	4.8	7.6	13	m^3/t	72	0.001	43	1.7	1–3
Suspended solids	...	0.027	0.047	kg/t	47	0.0002	0.20	1.8	1–3
Oil and grease/ total hydrocarbons	...	0.011	0.026	kg/t	46	0.0000004	0.10	1.9	1–3
By-products for external recycling									
Dross	...	16	7.4	kg/t	92	3.2	36	1.04	1–3
Filter dust	...	0.72	0.5	kg/t	10	0.2	1.4	1.2	1–3
Scrap sold	...	2.8	3.3	kg/t	29	0.08	10	1.1	1–3
Refractory	...	0.61	0.41	kg/t	11	0.014	1.2	1.4	1–3
Solid waste not recycled									
Dross	...	9.7	10	kg/t	18	2.0	30	1.3	1–3
Filter dust	...	0.50	0.5	kg/t	22	0.001	1.6	1.6	1–3
Refractory	...	0.81	0.7	kg/t	49	0.04	4.3	1.7	1–3
Other landfilled waste	...	1.6	2	kg/t	49	0.01	7.8	1.7	1–3

From life-cycle survey data with total production of 14,016,405 tonnes representing 57% of total world production with 72 cast houses. (a) Per tonne aluminum. (b) DQI, data quality indicator

Table 4.12c Air emissions per 1000 kg of cast aluminum ingot

Air emissions	Wt. mean	Standard deviation	Unit(a)	Response rate, %	Min	Max	DQI average(b)	DQI range(b)
Particulates	0.10	0.13	kg/t	71	0.001	0.53	1.5	1–3
SO_2	0.29	0.89	kg/t	51	0.00005	3.2	1.9	1–3
NO_x (as NO_2)	0.16	0.14	kg/t	78	0.001	0.72	1.9	1–3
HCl	0.085	0.085	kg/t	39	0.0003	0.33	1.7	1–3
Dioxin/furans	0.0061	0.014	mg/t	8	0.0000003	0.035	2.5	1–3

From life-cycle survey data with total production of 14,016,405 tonnes representing 57% of total world production with 72 cast houses. (a) Per tonne aluminum. (b) DQI, data quality indicator

This unit process begins with the unloading of process materials to their storage areas on site. The operations associated with this unit process include:

- Pretreatment of hot metal (cleaning and auxiliary heating)
- Recovery and handling of internal process scrap
- Batching, metal treatment, and casting operations
- Homogenizing, sawing, and packaging activities
- Maintenance and repair of plant and equipment
- Treatment of process air, liquids, and solids

The output of this unit process is packaged aluminum ingots or alloyed hot metal transported to an aluminum fabricating facility.

Molten metal siphoned from the pots (electrolysis metal) is sent to a resident casting complex found in each smelter. In some cases, due to proximity, molten metal is transported directly to a shape casting foundry. Remelt ingot and outside scrap may also be used as metal input. Molten metal is transferred to a holding furnace, and the composition is adjusted, by use of alloy additives, to the specific alloy requested by a customer. In some instances, depending on the application and on the bath composition in the pots, some initial hot metal treatment to remove impurities may be done.

When the alloying is complete, the melt is fluxed to remove impurities and reduce gas content. The fluxing consists of slowly bubbling a combination of nitrogen and chlorine or carbon monoxide, argon, and chlorine through the metal (chlorine use results in hydrogen chloride, or HCl, air emissions). Fluxing may also be accomplished with an in-line degassing technology, which performs the same function in a specialized degassing unit.

Fluxing removes entrained gases and inorganic particulates by flotation to the metal surface. These impurities (typically called dross) are skimmed off. The skimming process also takes some aluminum, and as such, drosses are normally further processed to recover the aluminum content and to make products used in the abrasives and insulation industries.

Depending on the application, metal is then processed through an in-line filter to remove any oxides that may have formed. Metal is then cast into ingots in a variety of methods: open molds (typically for remelt ingot), direct chill molds for various fabrication shapes, electromagnetic molds for some sheet ingots, and continuous casters for aluminum coils. Fresh water (input taken conservatively whether the water used is from fresh, underground, mine wastewater, or other sources), seldom sea water, is used for cooling (often with recirculation through a cooling tower and water treatment plant) and is subsequently discharged, where suspended solids and oil/grease (or total hydrocarbons) are monitored.

Energy used for ingot casting is electricity, natural gas, or heavy oil. Diesel oil is normally used for internal plant transport.

While recovery and handling of internal process scrap is usually included in the ingot casting operation, as mentioned previously, some prefer to sell it out (scrap sold as by-product for external recycling). Dross, filter dust from melting furnace air filtration, and refractory material from furnace internal linings are either recovered as by-products for external recycling or landfilled (dross—landfill, filter dust—landfill, refractory waste—landfill).

Solid Waste. Other landfill wastes typically consist of approximately 80% of environmental abatement waste (such as metal filter box and baghouse) and 20% municipal waste (Ref 4.5).

Aluminum Life-Cycle Assessment with Regard to Recycling Issues

This section illustrates the ISO rules and focuses on the major crucial aspects of aluminum life-cycle assessment (LCA) in terms of energy aspects, the high recycling rates, and the high value of recycled aluminum. In the past, LCA studies varied to a high degree because of different methodological approaches, which caused market risks when different materials were compared. The ISO 14040 series of standards now have set common methodological rules, which should be applied to all LCA studies, including those dealing with aluminum.

The recommended method in assessing life-cycle factors with recycling is the substitution method, which is explained and illustrated by examples. Because all metals are characterized by their ability to maintain their inherent properties after recycling, contrary to wood, paper, concrete, or plastics, this method can be applied for all metals, including aluminum. In addition to this substitution method, the so-called value-corrected substitution method is also presented. This method tries to take into account the loss of substitution ability of the recycled material in special cases where the value of the material is not maintained by recycling. More detailed guidance and scientific background, particularly on recycling issues, is given by Werner (Ref 4.6).

General

Any LCA study, especially for comparative purposes, should be based on methodologies within the framework of the following international standards:

- ISO 14040 Life-Cycle Assessment—Principles and Framework
- ISO 14041 Life-Cycle Assessment—Goal and Scope Definition and Inventory Analysis
- ISO 14042 Life-Cycle Assessment—Life-Cycle Impact Assessment
- ISO 14043 Life-Cycle Assessment—Life-Cycle Interpretation

In these standards, LCA is considered as a technique to assess the environmental aspects and potential impacts associated with a product or a service on a life-cycle basis. The LCA study includes four different phases:

- Goal and scope definition
- Life-cycle inventory analysis
- Life-cycle impact assessment
- Interpretation

The LCA deals with product systems that comprise the full life cycle of a product, including raw material acquisition, fabrication, transportation, use, recycling/disposal, and the operations of energy supply, ancillary material supply, and transports. Ideally, such a product system should only have input and output, which are elementary flows, that is, material or energy that is drawn from the environment or is discarded to the environment without subsequent human transformation.

In LCA studies where aluminum is compared to other materials, this comparison should be based on the same functional unit; for example, 1 kg of aluminum in a car may, in a specific case, correspond to 1.8 kg of conventional steel, in order to fulfill the same function.

Quantitative aggregation differs over impact categories; for example, the calculation of a single score is not permitted by ISO 14042 for studies to be used for comparative assertions that are made available to the public, where the overall environmental superiority or equivalence of one product versus a competing product that performs the same function is claimed.

A thorough interpretation of the results of an LCA study is required. This may include the need of additional information about the data of this report, that is, data quality or further information about the data, for example, a temporal or spatial differentiation of the potential environmental impact.

Energy Flows

Inventory data should not be added up if they represent different types of potential environmental impact. It is not appropriate to add up all emissions, for example, on per kilogram basis or all energy flows in megajoules. A cumulative energy, which is understood as the sum of renewable (hydro) energy and nonrenewable fossil energies, has no ecological sense.

In the impact-assessment phase, data representing elementary flows are aggregated to so-called indicator results if they belong to the same impact category, possibly after having been multiplied with a characterization factor. The elementary flow data related to energy consumption can be determined as the mass or the energy content of the relevant energy resource.

Electricity supply data cannot be considered as elementary flow data, because electricity is not directly extracted from nature. The elementary flows of the relevant power plant, including extraction of fossil energy resources, must be considered instead.

An example of an indicator result for the impact category "extraction of energy resources" is the low calorific value of the different fossil fuels extracted for the supply of a certain quantity of energy (see ISO/TR 14047). If, for example, 100 MJ of energy from natural gas are consumed in a plant, this may indicate an extraction of gas resources of 110 MJ, because 10 MJ may be used for the gas supply system or lost by leakage.

It is not appropriate to assign solar radiation or dam water, the elementary flow input of power plants based on renewable energy, to the impact category "extraction of energy resources." Renewable energy flows have a different associated type of impact (if considered) than fossil or nuclear energy flows. Generally, it is difficult to justify an impact category that comprises elementary flows from both renewable and nonrenewable energy in accordance with the criteria as formulated in ISO 14042. On the other hand, other environmental impacts of power plants based on renewable energy, for example, the land use of hydroelectric power plants, must be considered.

Recycling

For most aluminum products, aluminum is not completely consumed but rather used. Therefore, a life cycle of an aluminum product is not cradle-to grave but rather "cradle-to-cradle. This means that the life cycle of an aluminum product usually ends when the recycled aluminum is rendered in a form usable for a new aluminum product, for example, as an ingot.

According to ISO 14041, allocation principles and procedures where input and output of specific processes must be shared by more than one product system also apply to recycling situations. In such cases, the environmental burdens related to extraction and processing of raw materials and final disposal of products may have to be shared with subsequent product life cycles. This can also be addressed by using the substitution method, based on parameters such as recycling rates and related metal yields. If a change in inherent properties is considered, the value of recycled material may play a role.

System Expansion and Substitution. System expansion, an ISO 14041 recommended procedure, means expanding the system under study to include the end-of-life recycling, resulting in substitution of recycled aluminum to primary aluminum. The system boundaries must be defined in the goal and the scope definition phase of an LCA study. For the design for environment of aluminum products, it is appropriate to define the system boundaries in a way that they include all processes that can be influenced by the design. For aluminum products, material losses from recovery operations and metal yield by remelting depend on the form of the product, for example, wall thickness and the way aluminum is joined with other materials that can be significantly influenced by the design. This means that the end-of-life recycling processes of the product under study should be included, and the recycled aluminum should leave the system in the form of the recycled ingot.

A closed-loop allocation procedure is not only applicable for really closed-loop product systems, but also applies to open-loop product systems, for example, when aluminum, after use, is recycled into a raw material that has the same inherent properties as primary aluminum. In this case, the system expansion and substitution method can be applied. Unlike materials such as wood, paper, plastics, or concrete, aluminum has the ability to readily maintain its inherent properties through recycling. The inherent properties of pure aluminum are not changed by remelting.

In the case of the open-loop recycling approach, the substitution method can only be applied if the recycled raw material is similar enough to the primary raw material that it substitutes. In many instances, however, there will be a difference between a recycled material and the primary material it may substitute. Strict interpretation of this rule thus limits the application of the substitution method to only special cases.

In practice, aluminum products follow generally an open-loop recycling scheme. Recycling operations that include collecting, sorting, remelting, and refining produce recycled aluminum that fulfills the requirements for primary aluminum. Within these markets, recycled aluminum perfectly substitutes primary aluminum. As a result, in most cases, the substitution method is fully applicable to the LCA of aluminum products. However, in some particular cases, the recycling operations can lead to a significant change of the inherent properties of the recycled material compared with primary aluminum, for example, by the presence of metallic impurities that are entrapped during the remelting operation of poorly sorted or contaminated scrap. In this case, an allocation procedure with a value-corrected approach is appropriate (see the sections "Value-Corrected Substitution" and "Substitution Method with Recycled Aluminium as Input" in this chapter.

For example, if 100 kg of primary aluminum is required for a product system, then 80 kg of recycled aluminum ingots (with the same inherent properties as primary aluminum) are obtained after recovery of the end-of-life product and scrap remelting. The 20 kg of aluminum difference is due to lost metal, for example, littered or landfilled. In this case, only the environmental burdens of the production of the lost aluminum (20 kg of primary metal) must be charged to the product system under study, together with the burdens of the recovery operations. The environmental burdens of the production of 80 kg of primary aluminum must be charged to the next user(s) of the 80 kg of recycled aluminum.

Value-Corrected Substitution. The so-called value-corrected substitution method considers that the recycled metal is not able to fully substitute primary metal. This method assumes that the substitution ability is reflected by the ratio between the market prices of the recycled and primary material. As a result, if the price of the recycled material is 90% of the price of the primary material, 1 kg of recycled material will substitute only 900 g of primary material. In the case of aluminum, this method assumes a proportion of the environmental loads caused by primary aluminum production and final disposal of aluminum and the value change of the recycled metal. This procedure is in line with ISO 14041, whereby allocation must be considered when one or more unit processes are shared by different product systems, including the product system under study. In this case, it is required to identify these multifunctional processes and to select the appropriate allocation factors. The option to use economic parameters, for example, market prices, for the calculation of allocation factors, is permitted by the standard, provided that:

- No option to avoid allocation can be justified.
- No physical factor can be justified as the allocation factor.

In the case of recycling, the ISO 14041 standard offers the option to share, between subsequent product systems, inputs and outputs related only to the processes of raw material acquisition and final disposal. This means that only specific unit processes within the product system are considered to be shared with other future product systems, which can be unknown at the LCA study stage. This allocation principle avoids double-counting of inputs and outputs and ensures that the product system is charged according to its actual material consumption. Moreover, guidance is given in this standard to use economic value proportions (e.g., scrap value versus primary value). This option makes sense as long as reasonably free market prices of the primary and recycled material exist. Nevertheless, other methodologies, such as how to treat open-loop recycling, may be justified as alternatives. In this case, a comparison through a sensitivity analysis is required. If aluminum is compared with other materials, then it must be clarified that the value-corrected substitution method can be applied to the other materials as well.

For example if from 100 kg of primary aluminum in a product system, 80 kg of recycled aluminum ingots have 90% of the value of primary aluminum, then:

- 20 kg of aluminum are lost, for example, littered or landfilled.
- An additional loss by value correction is 10% of 80 kg (= 8 kg).

The total value-corrected loss is thus 28 kg. After value correction, the recycled aluminum ingots (which have the value level of 72 kg of primary ingots) substitute 72 kg of primary aluminum ingots. The environmental burdens of the production of only the lost aluminum, that is, 28 kg of primary metal, must be charged to the product system under study, together with the burdens of the recovery operations. The environmental burdens of the production of 72 kg of primary aluminum must be charged to the next user(s) of the 80 kg of recycled aluminum.

In many cases, scrap from different products and different alloys is melted together in one furnace batch, and alloying elements may be added. If, for example, pure alloy scrap is melted together with $AlMg_3$ (EN-AW-5754) scrap, then this melt may be cast to $AlMg_{1.5}$ (EN-AW 5050) rolling ingots. In this case, one could argue that the input material and the output material have different inherent properties. However, not considering the value of the alloying elements, it can be shown that the $AlMg_{1.5}$ rolling ingots have the same market value as the unalloyed ingots and the $AlMg_3$ ingot that were the origin of the scrap.

If the market value analysis shows that the market value of the aluminum ingots obtained from recycling of the end-of-life product is the same as the market value of primary aluminum, then a value correction is not necessary. In this case, the substitution can take place as in the case of identical inherent properties, effectively treating the product system as a closed-loop one.

Substitution Method with Recycled Aluminum as Input. ISO 14041 requires that allocation procedures must be uniformly applied to similar inputs and outputs of the system under consideration. The rules on how to treat incoming recycled aluminum must correspond with the methods for treating recycled metal that leaves the system. If the substitution method is applied, there is no need to consider the incoming portion of recycled aluminum, because only the metal loss during the complete life cycle of the product is considered. This means that if 100 kg of aluminum ingots enter the system, and 100 kg of recycled aluminum with the same inherent properties leave the system, then the environmental loads associated with the input metal and those associated with the output metal should be considered to be the same, even if the products from which the input metal is recycled are not known.

For a product, if 100 kg of a secondary raw material, usually recycled from a mixture of production scrap and postconsumer scrap, with the same value of primary aluminum is used and no recycling of this product happens, then the environmental loads of the production of 100 kg of primary aluminum must be charged to the product. For example, with 100 kg of aluminum used for a product system, it may consist of 50 kg of primary aluminum and 50 kg of recycled aluminum with the same inherent properties as the primary aluminum. If 80 kg of recycled aluminum ingots (with the same inherent properties as the primary aluminum) result from recycling, including remelting (20 kg of aluminum are lost), the environmental burdens of only the production of the lost aluminum, that is, 20 kg of primary metal, must be charged to the product system under study, together with the burdens of the recovery operations. These environmental burdens are valid no matter the recycled content of the product system, as long as no value correction is necessary. The environmental burdens of the production of 80 kg of primary aluminum must be charged to the next user(s) of the 80 kg of recycled aluminum.

Value-Corrected Substitution Method with Recycled Aluminum as Input. If the value-corrected substitution method is applied for recycled metal at the output side, then the value of the incoming recycled metal must be considered as well. For a product, if 100 kg of a secondary raw material, usually recycled from a mixture of production scrap and old scrap, with a value of 90% of the value of primary aluminum is used and no recycling of this product happens, then the environmental loads of the production of 90 kg of primary aluminum must be charged to the product.

If end-of-life recycling can be considered, then a (possibly value-corrected) substitution can occur, as described previously. In this case, the mass of the lost aluminum on the value level of primary aluminum, that is, the value-corrected mass of the input aluminum minus the value-corrected mass of the output aluminum, must be calculated, and the environmental burdens of the production of this quantity of primary aluminum must be charged to the product system under study.

For example, 100 kg of aluminum required for a product system consist of 40 kg of primary aluminum and 60 kg of recycled aluminum ingots with 90% of the value of primary aluminum. Of the primary aluminum (100 kg primary) used as the source for the recycled aluminum, 20 kg are lost (for example, littered or landfilled), while 80 kg of recycled aluminum ingots have 90% of the value of primary aluminum as a result from recycling, including remelting. The net aluminum loss, based on the value level of primary aluminum, is calculated as:

- Value-corrected mass of input aluminum minus value-corrected mass of output aluminum, both on the value level of primary aluminum
- Value-corrected mass of input aluminum is 40 kg + (60 kg × 0.9) = 94 kg aluminum ingots of the value level of primary ingots
- Value-corrected mass of output aluminum is 80 kg × 0.9 = 72 kg aluminum ingots of the value level of primary ingots

The net aluminum loss, based on the value level of primary aluminum, is 94 kg – 72 kg = 22 kg.

The environmental burdens of the production of the lost aluminum, that is, 22 kg of primary metal, must be charged to the product system under study, together with the burdens of the recovery operations. The environmental loads of the production of 6 kg of primary aluminum had already been charged to the producer of the 60 kg of recycled aluminum. The environmental burdens of the production of 72 kg of primary aluminum must be charged to the next user(s) of the 80 kg of recycled aluminum.

Long Lifetime Products. The substitution method for recycling may apply for any lifetime of a product, not only for aluminum cans but also for aluminum in buildings. Aluminum products often have extended lifetimes because of their high corrosion resistance, often further enhanced by specific measures of corrosion protection. Such products should not be mistreated in LCA studies by omitting recycling credits, as described earlier in the section "System Expansion and Substitution" in this chapter.

Any uncertainty about recycling rates and recycling techniques for long-life aluminum products is not sufficient to refuse recycling credits. Rather, it must be dealt with by applying different recycling scenarios in the form of sensitivity analyses, which should include the state-of-the art recycling technique for the product under study.

Moreover, it may be necessary in the interpretation phase of comparative studies to consider the temporal aspects of the environmental impacts of the different materials, for example, current greenhouse gas (GHG) emissions compared with GHG emissions in 50 years.

The Recycled Material Content Approach. There have been cases of LCA studies in which the recycling of a product is disregarded and no recycling credits are given on the output side, even in the case of closed-loop recycling, that is, when the recycled aluminum is used for the same product from which it has been recovered. Recycling credits are only given on the input side if the aluminum product contains a certain amount of recycled aluminum (recycled material content approach).

There have also been cases of LCA studies in which a different form of the recycled material content approach is applied. Here, recycling credits according to the substitution principle are given only in the case where the closed-loop recycling approach can be applied. In the case of open-loop recycling, no or only limited recycling credits are given. In addition, recycling credits are given on the input side if the aluminum product contains a certain amount of recycled aluminum.

According to ISO 14041, allocation principles and procedures also apply to recycling situations. The recycled material content approach is not mentioned in this standard as a method to avoid allocation. This approach can only be used in specific cases where the omission of specific outputs of recycled material can be justified according to well-defined cut-off criteria, for example, if recycled concrete is given free of charge for road construction.

The high value of aluminum recycled from end-of-life products can be demonstrated by the market price of recycled aluminum ingots, which is identical or close to the price of primary aluminum. In this context, methods such as the recycled material content approach that do not consider the end-of-life recycling or even the value of recycled material should be avoided. If, under certain conditions, the recycled content method can be justified, the LCA study should at least include a sensitivity analysis with other methods that better address the end-of-life recycling of aluminum products.

For example, an LCA study of different window frame materials (aluminum, aluminum/wood, steel, copper, polyvinyl chloride) was performed by K. Richter et al. (Ref 4.7) in 1996. The study used the recycled material content approach and was based on a window measuring 1650 by 1300 mm (5.4 by 4.3 ft) of two wings and a use time of 30 years. For the aluminum frame that had a mass of 40 kg, two different recycling scenarios were assumed:

- A realistic scenario, assuming a recycled aluminum content of 35% (alternative 1)
- A future scenario, assuming a recycled aluminum content of 85% (alternative 2)

The study did not show general disadvantages of the two aluminum alternatives compared to other window frame materials. Nevertheless, the greenhouse potential of alternative 1 was significantly higher (385 kg CO_2 equivalents); only alternative 2 showed results equivalent to the other materials (172 kg CO_2 equivalents).

From this study, the conclusion could be drawn that the aluminum industry should increase the recycled aluminum content, for example, by buying recycled aluminum on the market. Nevertheless, because recycled aluminum is an expensive commodity available in a relatively constant quantity, any increase of the recycled aluminum content in a window frame would lead to a reduction of the recycled aluminum content in other products, which, accordingly, need more primary metal. Finally, new scrap and old scrap often have the same inherent properties and therefore are mixed together, which makes the determination of the recycled metal content impossible.

After the publication of the comparative window frame study, ISO 14041 was published. Therefore, an additional study on aluminum window frames was performed by K. Richter and F. Werner, where the ISO rules were properly applied according to the guidance given in this paper (Ref 4.8). Because the recycling processes of window frames had been cut off in the first study, a main part of the second study was to identify the unit processes for the end-of-life recycling of window frames and to define the value of recycled aluminum when leaving the system boundaries. This work included visits to various recycling plants and the acquisition of data from such locations.

The recycling studies showed that two different types of aluminum window frames existed:

- *Type 1 window frames*: Contain components of zinc and brass or small steel parts that could not be separated from the aluminum fraction
- *Type 2 window frames*: Contain only aluminum, steel, and plastic/rubber parts that can be easily separated by shredding, magnetic sorting, and eddy-current sorting

The greenhouse potential of these window frames has been calculated as:

- 235 kg CO_2 equivalents for type 1 windows
- 179 kg CO_2 equivalents for type 2 windows

One recommendation of the new study was to apply design for recycling in such a way that type 2 window frames should be built in a manner where aluminum can be easily separated from other materials and contamination of the metal with impurities can be avoided.

REFERENCES

4.1. Board Resolution, International Aluminum Institute, 1998

4.2. "Life Cycle Assessment of Aluminum: Inventory Data for the Worldwide Primary Aluminum Industry," International Aluminum Institute, March 2003

4.3. "Aluminum Applications and Society: Life Cycle Inventory of the Worldwide Aluminum Industry with Regard to Energy Consumption and Emissions of Greenhouse Gases, Paper 1—Automotive," International Aluminum Institute, May 2000

4.4 "Environmental Profile Report for the European Aluminium Industry," April 2000

4.5 "North American Aluminum Association LCI Report," 1998

4.6 F. Werner, "Treatment of Aluminum Recycling in LCA; Development and Evaluation of the Value-Corrected Substitution

Procedure Applied to Window Frames," Research and Work Report 115/47, Swiss Federal Laboratories for Materials Testing and Research (EMPA), Duebendorf, 2002
4.7 K. Richter, T. Künninger, and K. Brunner, "Ökologische Bewertung von Fensterkonstruktionen," Swiss Central Office for Windows on Facade Construction, 1996
4.8 F. Werner and K. Richter, Economic Allocation in LCA: A Case Study about Aluminum Window Frames, *Int. J. LCA,* Vol 5 (No. 2), 2000, p 79

CHAPTER 5

Sustainable Development for the Aluminum Industry*

ALUMINUM HAS ADVANTAGES to help take on the challenge of sustainable development and environmental impact assessments, which can only be judged in terms of a full cradle-to-grave life-cycle analysis. It is a material that can be profitably recycled without loss of quality for use by future generations, and the use of aluminum has the potential to help control CO_2 emissions from weight reduction of transportation vehicles. For industry, the concern for the environment involves reducing the consumption of resources. This includes minimizing the use of energy, materials, water, and land while enhancing recycling and product durability. It also means reducing the impact on nature by minimizing air emissions, water discharges, waste disposal, and the dispersion of toxic substances.

To achieve sustainable development goals, the aluminum industry needs to focus on the following:

- Continuing the reduction program for greenhouse gas emissions
- Increasing the recycling rate of used aluminum
- Minimizing the use of hazardous or toxic materials and finding alternatives to landfill disposal for spent potlining and bauxite residues
- Achieving further energy-efficiency improvements
- Establishing consistently understood and applied sustainability metrics that can be used as leading indicators of progress
- Developing performance indicators on the social aspects of sustainability
- Encouraging an industry-wide approach to sustainability through awareness training, benchmarking, monitoring, and reporting

With these objectives, aluminum producers continue to develop models of sustainability that are being integrated into daily company operational management practice.

The aluminum industry has a number of elements in its current environment, health, and safety program that reflect a pioneering role in the field of global industrial reporting on sustainable development. For example, the International Aluminum Institute (IAI) has been monitoring and reporting on a number of widely used sustainable development indicators, such as energy use, greenhouse gas emissions, and safety performance. Since 1997, the IAI has collected comprehensive benchmarking data on safety performance in the global aluminum industry, with shared information on accidents, near misses, and their causes. Such benchmarking data are a driving force for continual improvement. The IAI is also currently working on:

- Generating global scrap and recycling statistics to help identify the scope for increased recycling
- Publishing annual global surveys of the aluminum industry's energy consumption
- Publishing an annual global survey of the industry's perfluorocarbon (PFC) emissions and PFC reduction performance, backed up by benchmarking graphs and training seminars, to encourage plants to match the best performers
- Conducting regular surveys of land use for bauxite mining and rehabilitation, and reporting the results every 5 years

*Adapted from Ref 5.1 and 5.2

- Collecting annual accident statistics globally and issuing global benchmarking reports to encourage participating companies to match the best performance of their colleagues
- Collecting data on the industry's global economic contribution

This chapter summarizes some IAI reports on sustainability progress from surveys through the year 2000 (Ref 5.1) and a 2005 update (Ref 5.2). Survey data on sustainability performance are collected from IAI member companies, collectively responsible for over 70% of global primary aluminum production and approximately 20% of recycled metal production. The 2005 survey data were collected from:

- Seventy-five smelters representing almost 18 million tonnes of primary aluminum production, a 21% increase on 2004 coverage and equivalent to 55% of total global aluminum production
- Twenty-four refineries representing over 36 million tonnes of smelter-grade alumina, a 41% increase on 2004 coverage and equivalent to approximately 60% of total global alumina production
- Eleven mines representing over 81 million tonnes of bauxite, a 13% increase on 2004 coverage and equivalent to approximately 50% of total global bauxite production

Recycling

Aluminum is a metal that can be recycled and reused with as little as 5% of the energy that would be required to produce it from raw materials. In effect, the world's increasing stock of aluminum acts like an energy resource bank, over time delivering more and more practical use and value from the energy embodied in the primary production for future generations by conserving energy and other natural resources. For example, recycling of postconsumer aluminum now saves an estimated 84 million tonnes of greenhouse gas emissions per year, equivalent to the annual emissions from 15 million cars. Since its inception, the recycling of postconsumer aluminum scrap has already avoided over 1 billion metric tonnes of CO_2 emissions.

The contribution of scrap metal recovery has steadily increased in terms of global output of aluminum metal (Fig. 5.1). The percentage of global output of aluminum metal from scrap recovery has also increased from 17% in 1960 to 33% today (2007) and is projected to rise to almost 40% by 2020 (Fig. 5.2). Of an estimated total of over 700 million tonnes of aluminum produced in the world since commercial manufacture began, approximately three-quarters (between 400 and 500 million tonnes) is still in productive use. As an "energy bank," the aluminum in use accounts for almost 50,000 petajoules of

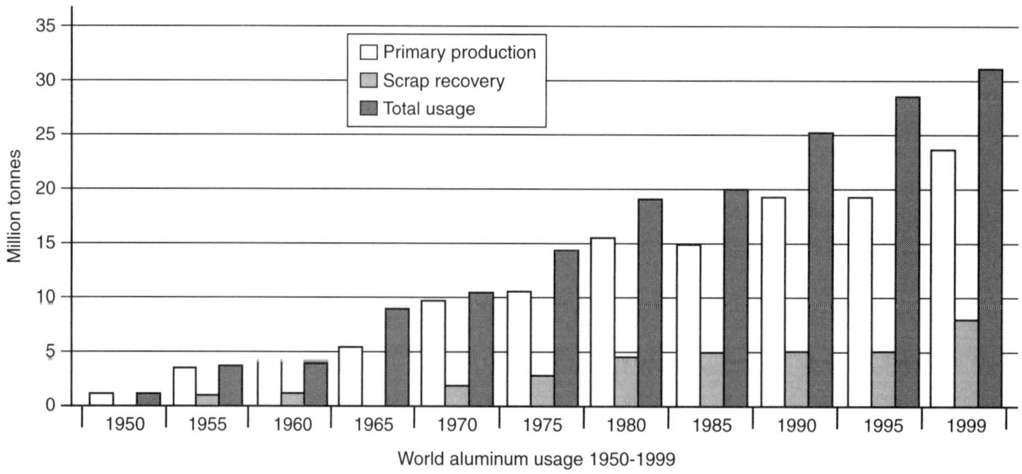

Fig. 5.1 Scrap and primary components of aluminum usage. Source: Ref 5.1

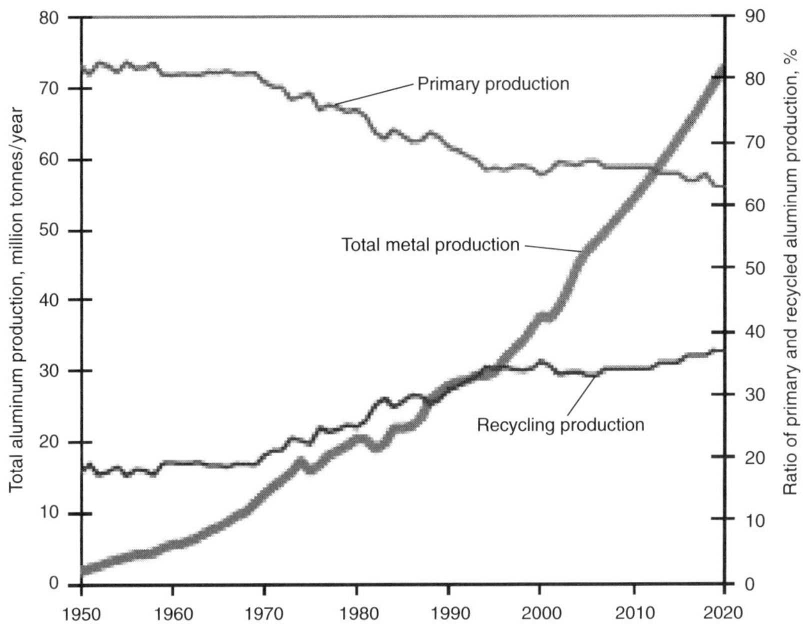

Fig. 5.2 Percentage of recycling component in aluminum production. Source: Ref 5.2

energy. This is higher than the current combined annual energy demand of Africa and Latin America and is equivalent to the annual total electrical energy generated globally from coal. If this metal is recycled, the banked energy is a resource not just once but repeatedly, and over future generations. If landfilled, this valuable resource is potentially lost forever.

Global aluminum recycling rates are high, approximately 90% for transport and approximately 60% for beverage cans. In Europe, aluminum has high recycling rates, ranging from 41% in beverage cans to 85% in construction and 95% in transportation. In Japan, the recycling rate for cans was 79%, and Brazil achieved 78%. A kilo of used cans is worth more to collectors than 15 kilos of plastic or 10 kilos of paper. Americans recycled 62.6 billion aluminum cans in 2000, for a beverage can recycling rate of 62%. For many groups, used aluminum cans turn into new-found money, thanks to aluminum can industry initiatives. Initiatives include the American Aluminum Association's partnership with Habitat for Humanity, called "Aluminum Cans Build Habitat for Humanity Homes," designed to boost public interest in aluminum can recycling while helping volunteers and families build homes.

The recycling rate for end-of-life aluminum depends on the product sector, the lifetime of each product, and society's commitment to collect aluminum. Each application requires its own recycling solutions. Just over 15 million tonnes of recycled aluminum were produced worldwide in 2004, which met 33% of the global demand for aluminum. Of the almost 7 million tonnes of aluminum recycled from end-of-life products, 28% came from packaging, 44% from transport, 7% from building, and 21% from other products.

Aluminum enjoys a high recycling rate of 85% in the building industry. The global industry is keen to increase collection rates and is working with producers of building applications to enable even more efficient collection of scrap from demolished buildings. In 2004, Delft University of Technology conducted a study into the aluminum content of, and collection rates from, demolished buildings in six European countries and found that the average collection rate for aluminum was close to 96%. The transport sector also has high rates of recycling, currently approximately 90% globally, because dismantlers and recyclers recognize the high intrinsic value of end-of-life aluminum products.

The aluminum industry is working with manufacturers to enable easier dismantling of aluminum automotive components to improve the sorting and recovery of scrap aluminum. The Japanese Aluminum Association, for example, is currently undertaking a study of advanced separation techniques to identify ways in which aluminum components from Shinkansen ("Bullet Train") carriages can be efficiently separated from other materials, thus increasing the recycling rate and the quality of scrap collected. Electromagnetic sorting is a powerful tool, currently used in many regions of the world, that separates even the smallest aluminum shards from waste material. Applying heat to end-of-life vehicle parts to separate the lacquer from the aluminum also facilitates recycling.

Can collection is approximately 60% globally, and in some countries, the collection rate is already above 80%. Sweden and Switzerland collect 86 and 88% of their aluminum beverage cans, respectively. Sweden's success lies in a deposit/refund system, whereas in Switzerland, a voluntary prepaid recycling charge covers the costs of collection. In Japan, a collection rate of 92% for used beverage cans is achieved with a voluntary system. In Brazil, recycling is not mandatory, but every region has a recycling market that facilitates the collection and transportation of end-of-life products. This has encouraged communities, supermarkets, condominiums, shopping centers, and clubs to collect. As a result, in 2005, 96% of cans were recycled in Brazil, and this is considered the world's highest recycling rate for used beverage cans.

As a tool, the IAI has designed a mass flow model (critically reviewed by Yale and Delft Technical Universities) to track aluminum throughout its life cycle, from mining to product use to recycling. The main objective for creating the model is to identify present and future recycling flows and the scope for further recycling. The model traces the flow of aluminum from 1888 to the present (2007) along the complete value chain of aluminum processing: bauxite mining, alumina refining, aluminum and aluminum ingot production, fabrication (rolling, extrusion, and casting), manufacturing (production and assembly of finished products), use, and recycling. New scrap is generated immediately during the production and processing stages, not having yet reached the use phase. Old scrap is generated when an aluminum-containing product reaches its end of life and is collected for recycling. To calculate the amount of aluminum still in productive use and leaving the use stage, a product residence time model is applied. Here, the average lifetime of each of the main products in which aluminum is used and the historical tonnage of aluminum in those products are considered. The results for 2004 are shown in a mass flow diagram (Fig. 5.3).

At present, approximately one-third of world demand is met from recycled metal. Much better global scrap statistics are required to improve the use of available scrap material and ensure that valuable metal is not going to the landfill. The IAI and associations from the recycling industry have established a Global Aluminum Recycling Committee to monitor scrap flows and recycling rates. A study is also being conducted to evaluate the contents in landfills of various regions.

Perfluorocarbon Emissions

Two PFC compounds (CF_4 and C_2F_6) are produced during the electrolysis process from brief upset conditions known as anode effects. These PFCs have long atmospheric lifetimes and are potent greenhouse gases, with a greenhouse gas warming potential of 6500 and 9200 times that of CO_2 for CF_4 and C_2F_6, respectively. From annual IAI surveys of PFC emissions, there has been a steady trend in the reduction of PFC emissions (Fig. 5.4a), and a voluntary industry objective is to obtain an 80% reduction in PFC greenhouse gas emissions per tonne of aluminum produced for the industry as a whole by 2010 versus 1990 levels. The trend and 2010 target of CO_2 equivalents for greenhouse gas emissions per tonne of aluminum is plotted in Fig. 5.4(b).

Data on PFC emissions from the annual IAI surveys allow benchmarking and comparisons, so individual plants can monitor their performance with other nonidentified plants using the same technology. Those companies responding to requests for anode effect data increased from 61% of the world's primary aluminum production in 1990 to 66% in 2000. In 1997, the industrial processes of the primary aluminum industry emitted 110 million tonnes of CO_2 equivalents, of which approximately 50 million tonnes (45%) originated from the two PFC compounds. Preliminary results of the IAI surveys for the years 1998, 1999, and 2000 for IAI member companies reporting anode effect data indicate a continuing declining trend with PFC emissions. The data indicate a reduction of

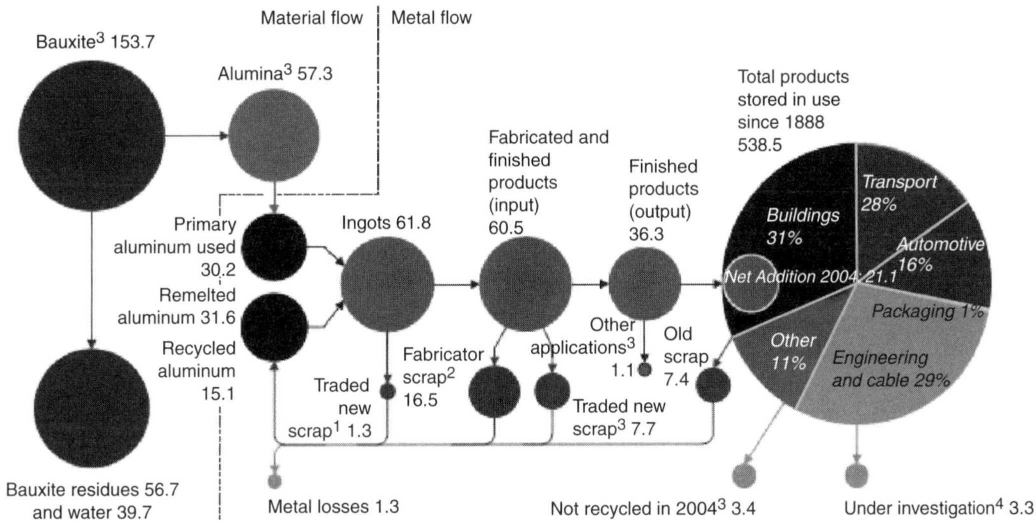

Fig. 5.3 Mass flow model of aluminum throughout its life cycle, from mining to product use and recycling. Values in millions of metric tonnes. Values may not add up, due to rounding. Production stocks not shown. 1, aluminum in skimmings; 2, scrap generated by foundries, rolling mills, and extruders. Most is internal scrap and not taken into account in statistics: 3, such as powder, paste, and deoxidation aluminum (metal property is lost); 4, area of current research to identify final aluminum destination (reuse, recycling, or landfilling); 5, calculated. Includes, depending on the ore, between 30 and 50% alumina; 6, calculated. Includes, on a global average, 52% Al; 7, scrap generated during the production of finished products from semis; 8, landfilled, dissipated into other recycling streams, incinerated, incinerated with energy recovery. Source: Ref 5.2

specific CF_4 emission rates by 60% between 1990 and 2000, even though total primary production increased from 20 to over 30 million tonnes per annum in the same period. Over the same period, the specific emission rate for C_2F_6 was reduced by 62%. Between 1990 and 2005, PFC emissions were reduced by 76% per tonne of aluminum produced, and the industry objective is to continue with this beneficial trend.

Worldwide estimates of PFC emissions have been based on an extrapolation of the IAI survey data, using knowledge of the reduction technologies at those facilities that did not report anode effect data. These results show that, while worldwide aluminum production has increased by approximately 24% since 1990, there has still been an overall reduction in the total annual emissions of PFCs. These reductions in PFC emissions to the atmosphere are estimated to amount to over 34 million tonnes as carbon dioxide equivalents, a reduction of approximately 39% from the 1990 baseline for worldwide PFC emissions. This is one of the few examples where the global emissions of a greenhouse gas from an industry sector are actually in decline.

The surveys also show that smelters in the developing world, which often use state-of-the-art technology, are performing as well as, if not better than, some plants in Europe or North America. Voluntary agreements between government and industry have played a significant role in encouraging this reduction in PFC emissions in many countries, such as Australia, Bahrain, Brazil, Canada, France, Germany, New Zealand, Norway, and the United Kingdom. Together, they represent approximately 50% of world production.

Fluoride Emissions

For decades, fluoride emissions (as gases and particulates) were considered to be the single most important pollutant from aluminum smelters. Depending on the local conditions, fluorides could have a serious environmental impact on the local flora and fauna. Fluorides accumulate in vegetation and can cause damage to coniferous trees. They also accumulate in the teeth and bones of ruminants eating fluoride-contaminated forage.

Figure 5.5(a) shows the development in fluoride reductions through three "generations" of aluminum smelters. Plants with modern control systems to remove and recycle the fluorides do not generate local concerns. Optimal fume

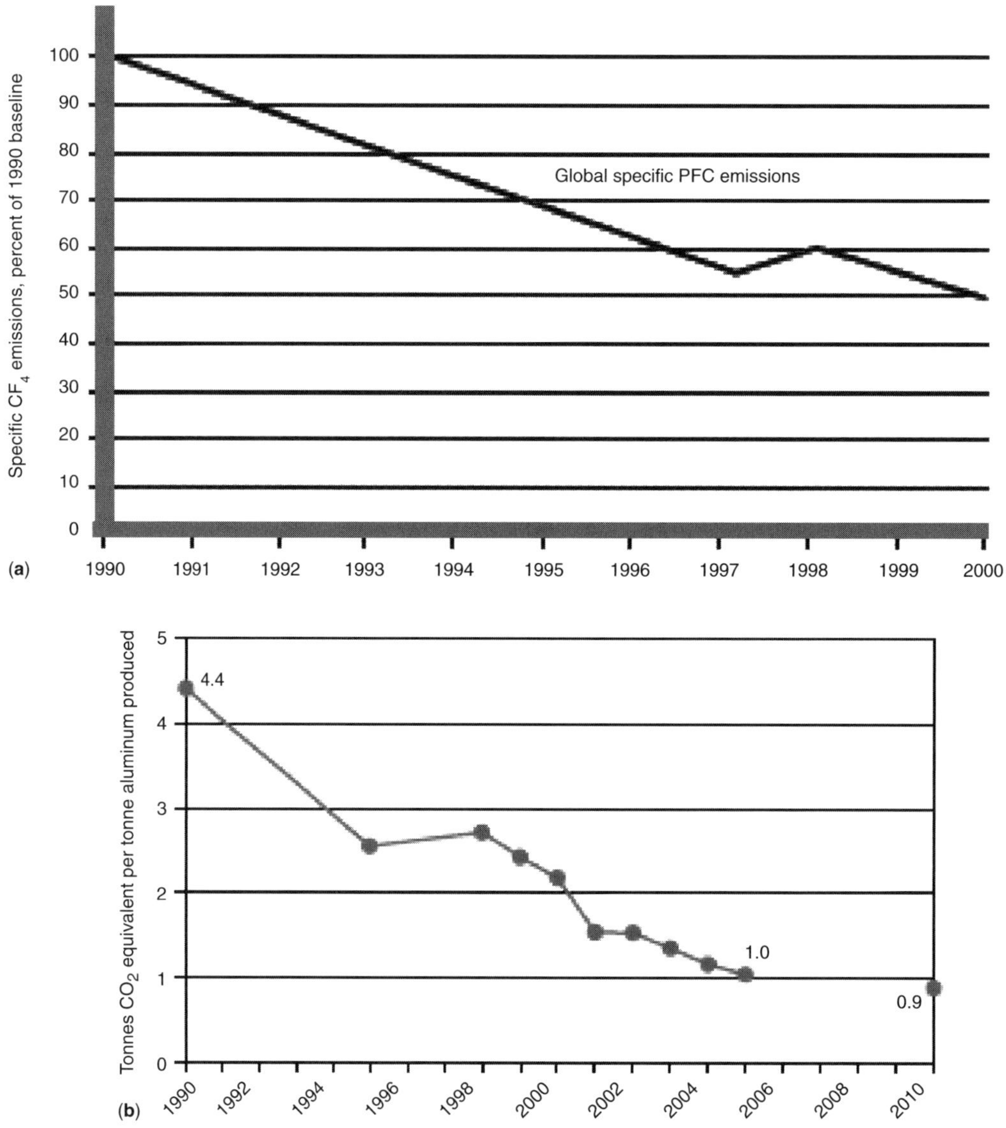

Fig. 5.4 Global emission of greenhouse gases from electrolysis reduction. (a) Percentage (from 1990 baseline) of specific perfluorocarbon (PFC) emissions. Source: Ref 5.1 (b) CO_2 equivalents including PFC components of CO_2 equivalents. Source: Ref 5.2

collection from the electrolytic cells, coupled with specific workplace-related training of the employees, should also lead to further improvement in the future. Figure 5.6 illustrates the reduction of atmospheric fluoride emissions in Norwegian smelters from 1960 to 2000, despite the increase in production from 185,000 tonnes per year in 1960 to 1,030,000 tonnes in 2000.

In Russia over the last decade, specific energy consumption decreased by 11%, and hydrogen fluoride and sulfurous anhydride emissions were reduced by 45 and 59%, respectively. In the case of the majority of plants around the world that use modern control equipment with efficient gas collection and wet/dry scrubbing systems, the damage from fluoride emissions to surrounding flora and fauna has largely been eliminated. Fluoride control systems are operated as closed-loop systems with no residual wastes.

Data collected from facilities representing 83% of IAI member company production indicate a reduction in total fluoride emissions (gaseous and particulate) of over 50% per tonne of aluminum produced between 1990 and 2005. A voluntary objective is to obtain a minimum of a 33% reduction in fluoride emissions by IAI

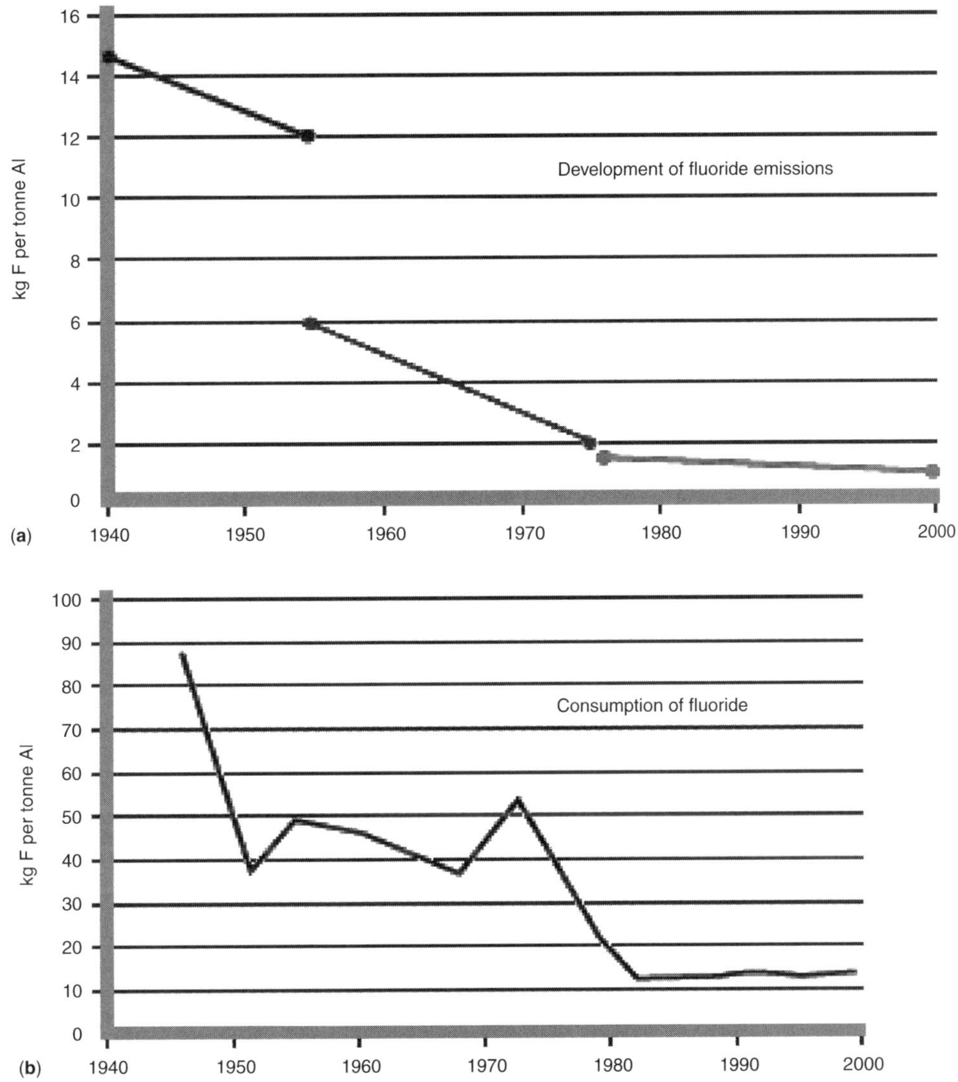

Fig. 5.5 Trends in fluoride emissions and consumption. (a) Fluoride emissions from first, second, and third generations of aluminum electrolysis (smelting) technology. (b) Fluoride consumption due to losses to air, water, spent potlinings, and waste at a representative smelter. Source: Ref 5.1

member companies per tonne of aluminum produced by 2010 versus 1990.

Energy Efficiency

Energy represents a large part (~25%) of the costs associated with primary aluminum production. The two main components of energy consumption for primary aluminum occur in the production of metallurgical alumina and electrolysis. Trends in global energy consumption rates for these two components of primary production are given in Fig 5.7 and 5.8.

The aluminum industry has a long tradition of improving energy efficiency. Figure 5.9 shows a steady reduction in electrical power used in primary aluminum production from 1899 to 1999. The average energy consumption and subsequent emissions per tonne of production fell by 70% over those one hundred years, due to research and continuing process developments. In the 1950s, it took, on average, approximately 21 kWh (kilowatt-hours) to make a kilogram of aluminum from alumina.

The three main energy sources in alumina refining are fuel combustion, electricity production, and energy use in lime production. These

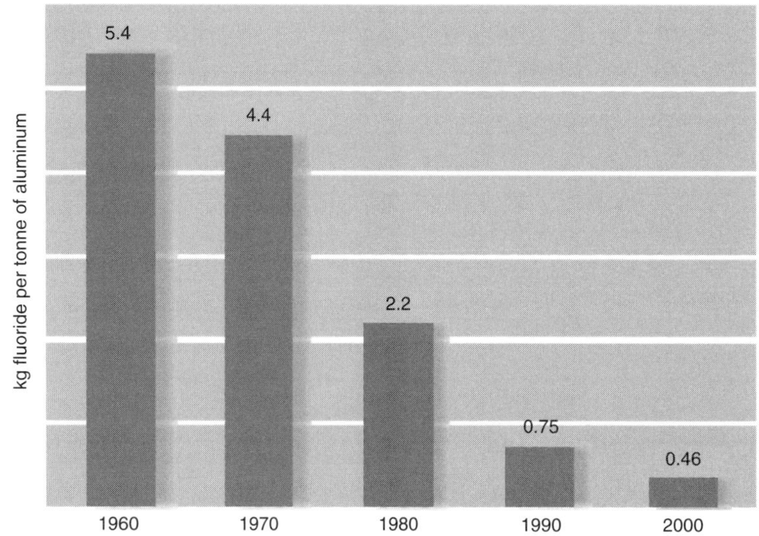

(a)

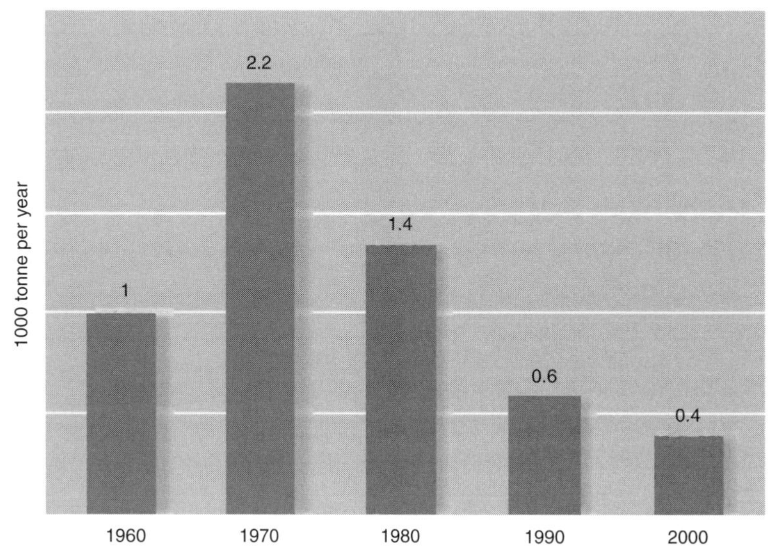

(b)

Fig. 5.6 Reduction of atmospheric fluoride emissions in Norwegian smelters from 1960 to 2000. (a) Specific emissions. (b) Total emissions. Source: Ref 5.1

three sources are monitored together as the total energy used in alumina production. The average energy used to produce 1 tonne of metallurgical alumina decreased by 5% between 1990 and 2004 (Fig. 5.10). The IAI is developing a quantitative voluntary objective for alumina refining energy efficiency. The objective of IAI member companies is to seek reduction in greenhouse gas emissions from the production of alumina per tonne of alumina produced.

Smelters in the 1990s used one-third less electricity per tonne than equivalent plants in the 1950s. In 1999, it took one of the newest smelters just 13 kWh. The average electrical energy required to smelt 1 tonne of aluminum from alumina was cut by 5% between 1990 and 2005, mainly through investment in modern, more efficient technologies. New smelters generally use the best available technologies, no matter where in the world. The trend of improving energy

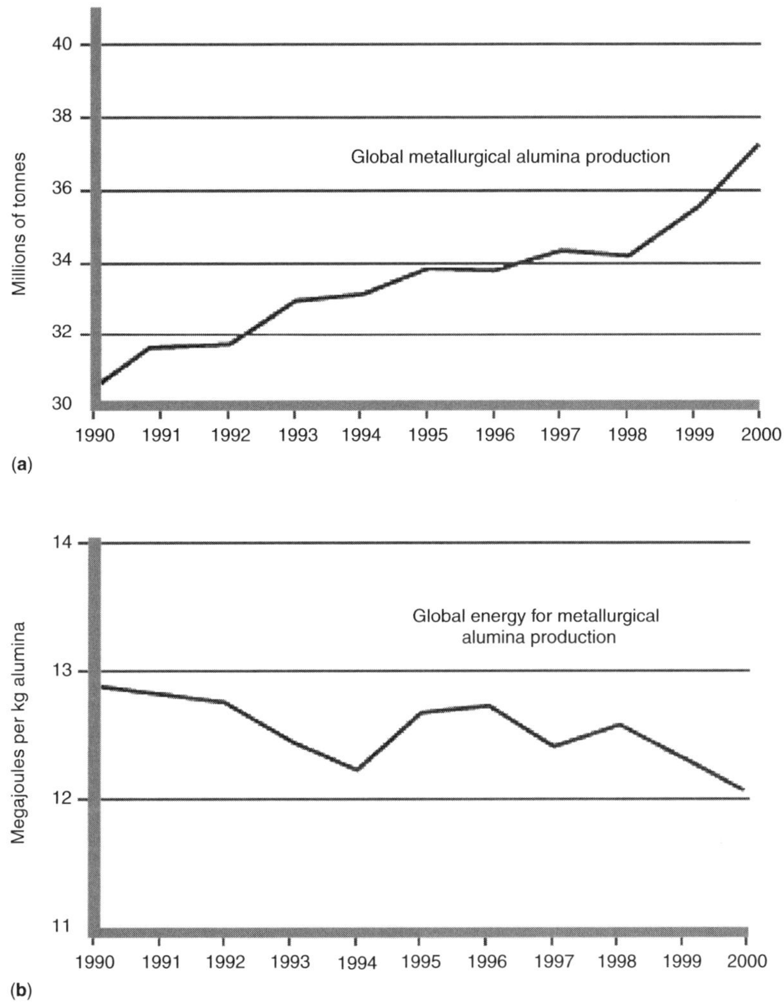

Fig. 5.7 Metallurgical alumina production. (a) Global tonnage. (b) Global energy consumption rates

efficiency continues today (2007), and the industry objective is a 10% reduction in average smelting energy use by IAI member companies per tonne of aluminum produced by 2010 versus 1990. For benchmarking to encourage improved performance, the IAI carries out an annual energy consumption survey covering 70% of the world's primary production facilities.

Aluminum in Transportation

The global use of aluminum in the automotive sector increased from 2.5 million tonnes in 1991 to nearly 4.5 million tonnes in 1999. For the first decade of this millennium, the use of aluminum in cars is predicted to double because of more cars worldwide and more aluminum in cars. The U.S. automobile fleet contained 116 kg of aluminum per vehicle in 2000 and is expected to contain 159 kg by 2010. The cumulative impact of using aluminum in this way will yield greenhouse gas savings of 180 million tonnes per year by 2010.

In seeking to reduce the environmental impacts caused by transport, aluminum can contribute to improved performance from weight reduction. Current estimates show that globally there will be a 35% increase in CO_2 emissions from all vehicles by the year 2020. An increased use of aluminum could reduce this figure to 28% and thus help toward making

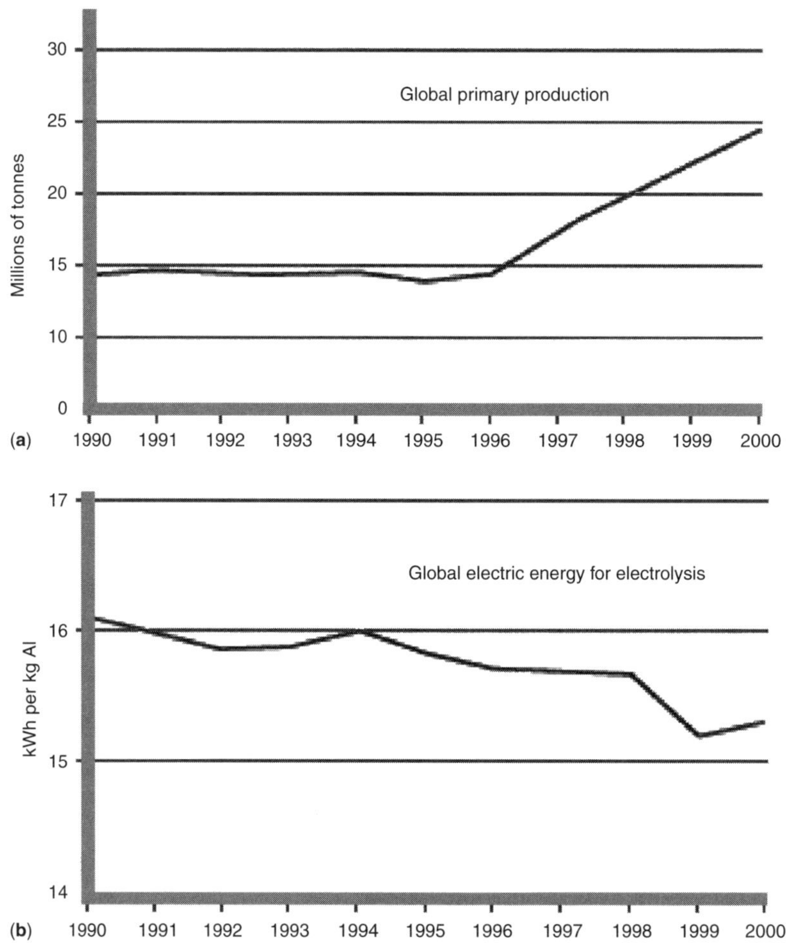

Fig. 5.8 Primary aluminum production. (a) Global tonnage. (b) Global electric energy rates for electrolysis. Source: Ref 5.1

the transportation sector more sustainable. This use of aluminum for automobiles in 1999 alone had the potential, over the lifespan of the vehicles, to reduce overall greenhouse gas (GHG) emissions by 90 million tonnes, assuming that all this aluminum was used to replace denser materials.

Annually, aluminum shipments for use in transport are being monitored by industry in order to track aluminum use for reducing GHG emissions from road, rail, and sea transport. Different life-cycle assessments have shown that 1 kg of aluminum in a car body, replacing 2 kg of steel in a conventional car body, saves, during the lifetime of the car, approximately 20 kg of GHG emissions (in CO_2 equivalents). In addition, if, for example, 1000 kg of GHG is saved by the use of less gasoline in a lightweight vehicle, this also means a reduction of other potential environmental impacts, including:

- 15,800 MJ of crude oil resources; 933 kg of water
- 1.8 kg ethylene equivalents of ozone-forming hydrocarbons
- A savings of 2.1 kg SO_2 equivalents of acidification potential

Greater opportunities for reduction of GHG emissions also are possible for vehicles such as buses and long-haul trucks (which operate over much longer distances during their lifetimes, often five times longer than that of a typical passenger car). For example, the use of 1 kg of aluminum replacing 1.5 kg of steel in a typical bus or truck reduces the GHG emission by approximately 40 kg over its lifetime (i.e., twice

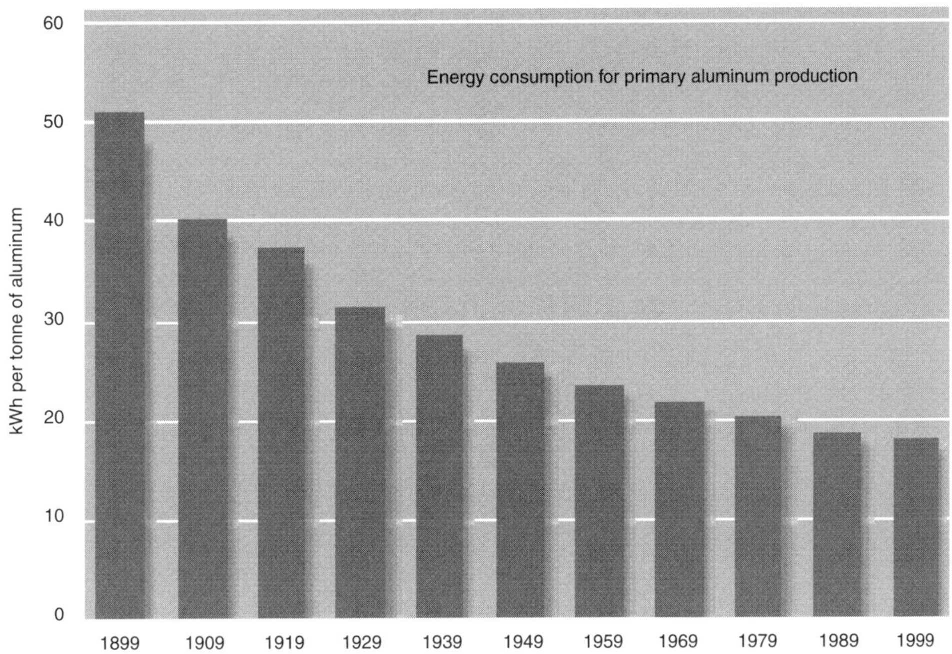

Fig. 5.9 Energy consumption of primary aluminum from 1899 to 1999. Source: Ref 5.1

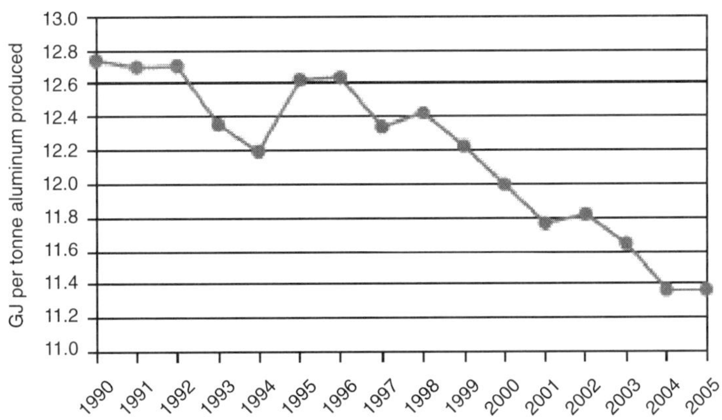

Fig. 5.10 Energy use in alumina production. Source: Ref 5.2

as much as for a car). Finally, a railroad train, during its lifetime, runs a distance of between 30 and 50 times the distance of a typical car. The use of 1 kg of aluminum replacing 1.6 kg of steel in a railway car reduces the GHG emissions by more than 200 kg (i.e., ten times as much as for a car). Reduction in the "dead weight" of ships through the use of lighter-weight aluminum can also result in energy savings and GHG reductions.

Natural Resources

Land Use. The use of mineral resources for the production of aluminum is quite modest as compared with many other materials. Between 4 and 5 tonnes of bauxite are required to produce 1 tonne of alumina, and 2 tonnes of alumina are required to produce 1 tonne of aluminum metal. Globally, bauxite mining disturbs approximately 25 km^2 a year, an area equivalent in size to one-third of Manhattan Island, NY. Every year, approximately 20 km^2 is rehabilitated. The area of land rehabilitated as a percentage of land mined since operations began, in currently operating mines, is 70%.

There are numerous bauxite deposits, mainly in tropical and subtropical regions, that are generally extracted by open cast mining from strata, typically some 4 to 6 m (13 to 20 ft) thick under a shallow covering of topsoil and vegetation. In most cases, the topsoil is removed and stored. Of the four main climate groups, the distribution of bauxite mining locations in 1998 was: temperate, 0.5%; Mediterranean, 39%; tropical 48%; and subtropical, 13%. The vegetation types disturbed each year are mainly 76% forests, 19% agriculture and pasture, and 2% shrubland. Postmining land use shows 70% being returned to native forest, 3% to commercial forest, 17% to pasture and agriculture, and 7% used for urban and industrial development, housing, and recreational purposes.

The IAI's first Bauxite Mine Rehabilitation Survey was prepared in 1991, covering 65% of total world bauxite production. It described the overall impact of bauxite mining on the environment and the rehabilitation programs in place at 18 mining locations. A second survey was carried out in 1998 and covered 27 mining locations covering 72% of the total world bauxite production. The results of both of these surveys indicate that the bauxite mining industry is broadly in harmony with the approach set out in the United Nations Environment Programme (UNEP) "Guidelines for the Environmental Management of Alumina Production."

The area disturbed for mine development has increased by some 14% over the past decade. In 1991, it was approximately 1400 hectares, and by 1998, it had risen to 1591 hectares. The output of bauxite ore per hectare had increased from 52 thousand tonnes per hectare to 56.5 thousand tonnes per hectare in 1998. Of the total area mined per year in 1998, 80% was identified as wildlife habitat, and 175 hectares were tropical rainforest. The majority of mines have introduced measures such as wildlife reserves and corridors, wildlife research, and monitoring programs. In Jamaica, the industry has created orchid sanctuaries to help preserve the local orchid species. The 577 hectares of mined area were identified as possessing important fauna species; of these, 527 hectares, or 91%, will be restored to wildlife habitat by the planting of suitable vegetation. Companies representing 60% of global bauxite production had their own plant nurseries for the purposes of reforestation and revegetation.

Rehabilitation plans are in place at most mines (25 out of 27, or 90%, as compared to 10 out of 18, or 55%, in 1991). A very high proportion of mine managements have long-term plans for the mine areas that will leave a self-sustaining ecosystem in place when all mining operations have been completed. The industry continues to seek increases in the proportion of bauxite mining land rehabilitated annually.

Water Use. Besides energy production from hydroelectric facilities, water in the aluminum industry is used for cooling in certain processes and sanitary purposes. The local programs to reduce water consumption vary from location to location, depending on the ease of access to water and on water consumption costs. A voluntary objective of IAI member companies is to reduce their fresh water consumption per tonne of aluminum and alumina produced. The IAI member companies will concentrate efforts to minimize fresh water consumption where there are limited available fresh water resources.

REFERENCES

5.1. "The Aluminum Industry's Sustainability Development Report," International Aluminum Institute, 2003

5.2. "Aluminum for Future Generations—Sustainability Update 2006," International Aluminum Institute, 2006

CHAPTER 6

Material Flow Modeling of Aluminum for Sustainability*

Kenneth J. Martchek, Alcoa Inc., Alcoa Corporate Center

STATISTICS ON THE RESOURCE FLOWS required to produce primary and recycled aluminum are incomplete on a global basis. In addition, there has not been a consensus or quantitative estimate of future resource flows related to aluminum production or the potential availability of less resource-intense end-of-life aluminum (recycled) metal to meet ever-increasing consumer and developing nation needs.

A model was developed by Alcoa Inc. (Ref 6.1) to provide a quantitative understanding of historic and today's (year 1950 through year 2003) worldwide aluminum mass flows and systems losses. In addition, current and future resource requirements and life-cycle inventory flows were estimated by coupling these global aluminum mass flows with global, average life-cycle inventory intensity data developed from a majority of producers in a report (Ref 6.2) by the International Aluminum Institute (IAI) (see also Chapter 4, "Life-Cycle Assessment of Aluminum: Inventory Data for the Worldwide Primary Aluminum Industry" in this book). The model was also developed to provide quantitative scenario-development capability to determine the positive impact of enhanced recycling, lower resource-intense production, and product usage scenarios. The model and key results information were developed to be shared with global aluminum industry technical experts, executives, and external stakeholders to better understand potential paths to more globally sustainable aluminum.

Modeling

Modeling of historic and current flows was built around aluminum product net shipments statistical data provided by governments, such as the U.S. Geological Survey (Ref 6.3–6.5), or regional aluminum associations, such as the European Aluminum Association, Australian Aluminum Council, the Japan Aluminum Association, or the North American Aluminum Association (Ref 6.6). The data were gathered, starting in year 1950, into a comprehensive spreadsheet model by year, by region (European Union, South America, China, etc.), and in accordance with the following customer (market) segmentation:

- Building and construction
- Transportation—auto and light truck, aerospace, and other (heavy trucks, trains, etc.)
- Packaging—aluminum containers and other packaging (foil, etc.)
- Machinery and equipment
- Electrical—cable and other electrical
- Consumer durables
- Other (such as aluminum used for propellant or steel deoxidation)

From the product net shipments, the model estimates both internal (runaround) aluminum

*Adapted with permission of *International Journal of Life Cycle Assessment* from the following article by Kenneth J. Martchek: "Modelling More Sustainable Aluminium: Case Study," *International Journal of Life Cycle Assessment*, Volume 11 (No. 4), 2006.

facility recycle flows and new (prompt, fabricator) customer recycle flow amounts, based on estimation of average use, yield, and melt loss rates identified in the literature and reviewed and agreed on by a subteam of global aluminum technical experts.

This subteam of experts was commissioned by the IAI Global Aluminum Recycling Committee (GARC). Postconsumer (end of product life) aluminum flows are estimated from product net shipments in previous years, estimated product lifetimes (worldwide by market by year), scrap recollection rates (by region by market by year), and recovery factors again based on industry statistics, published literature, and review and agreement by the IAI GARC committee. An illustration of some of these data on scrap recovery rates and melting recovery efficiency is provided in Table 6.1.

Modeling of future aluminum and resource flows is based on literature (Ref 6.7, 6.8) and expert projections of life-cycle inventory intensity rates (Ref 6.2, 6.9) and aluminum product shipments by market (currently with a weighted-average compounded annual growth rate of 2.5% per year). The availability of recycle flows to meet these market demands is based on projected use, yield, melt loss, recovery rates, postconsumer recycling rates, and anticipated future product lifetimes. Primary aluminum production is then calculated to determine the market demand for additional primary capacity and resulting resource requirements.

Validity Checks. There are two validity checks in the model:

- Comparison of estimated postconsumer and new scrap by year with published values by year. (This check is informational because published values for global recycled metal are considered to be incomplete.)
- Comparison of the estimated market demand for primary aluminum by year with published primary production by year

Figure. 6.1 shows the aluminum production that is estimated by the model to have been required for the years 1970 to 2003, based on product net shipments less postconsumer and

Table 6.1 Example of global average worldwide collection (recycle) rates and melting recoveries by market

Market	Collection, % 1990	Collection, % 2000	Melting recovery, %
Buildings	69	70	96
Autos and light trucks	75	75	96
Aerospace	76	75	96
Other transport	76	75	96
Containers	61	59	85 (net of 4 cycles/yr)
Packaging—foil	13	16	30
Machinery	40	44	96
Electrical cable	45	51	96
Electrical other	30	33	96
Consumer durables	20	21	96

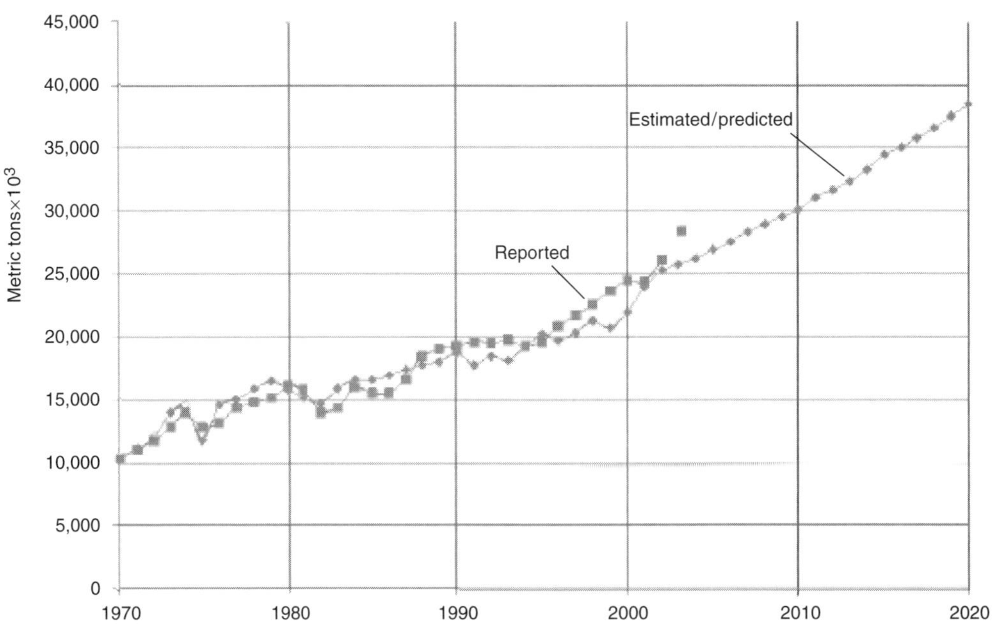

Fig. 6.1 Estimated versus reported worldwide primary production

new recycle flows and system losses. This is shown to be in fairly good agreement with reported worldwide primary aluminum production for the years 1970 to 2003. Required primary aluminum production is then projected to the year 2020, with assumptions about each market segment growth rates, postconsumer scrap collection rates, and anticipated recoveries based on latest trends.

Key Results

The model assessment of global aluminum mass flows is shown schematically in Fig. 6.2. The size (area) of the circles illustrates relative volume of flows. In year 2003, recovered postconsumer and new customer recycled metal supplied 33% of the global aluminum industry's product net shipment supply.

An estimated 516 million metric tons of aluminum, approximately 73% of all of the aluminum ever produced, is contained in current transportation, cable, and building product inventory (in service), as illustrated in Fig. 6.3. The model also projects future product inventory volumes by market segment to year 2020.

Additional key results also included system losses, such as aluminum packaging lost in

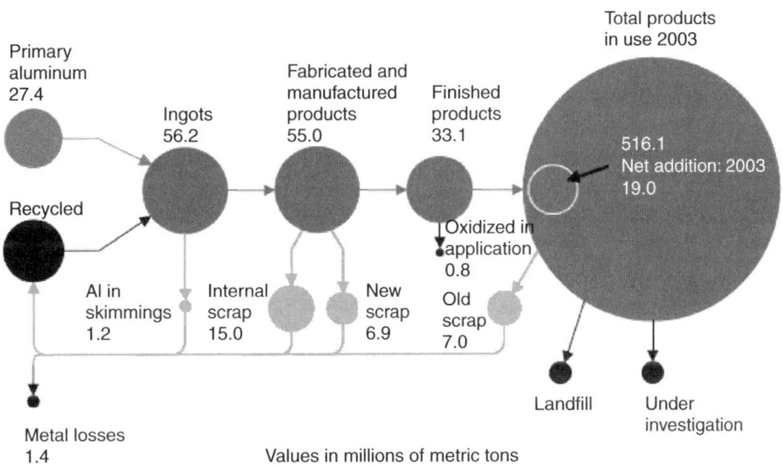

Fig. 6.2 Global aluminum mass flows for the year 2003

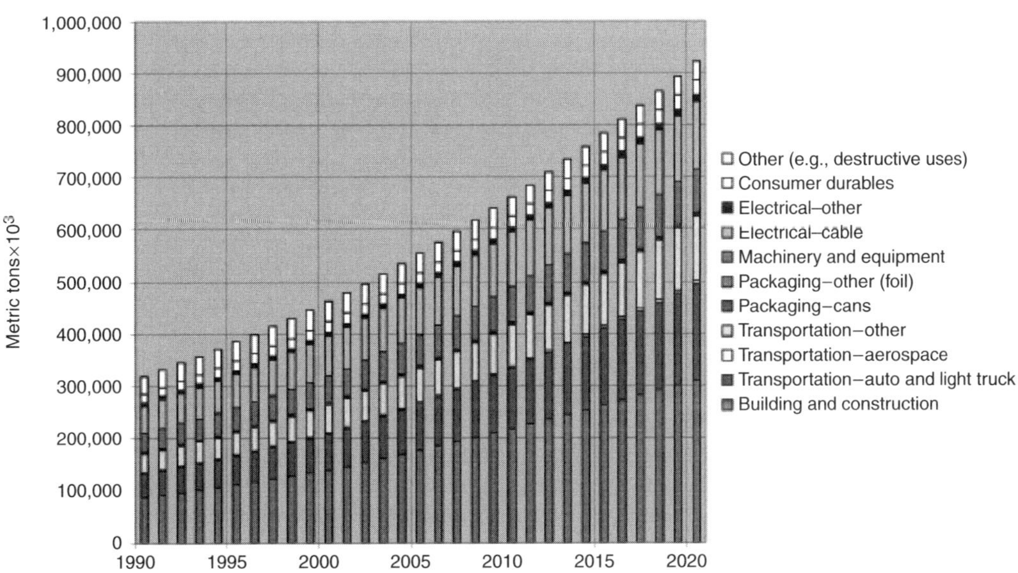

Fig. 6.3 Worldwide aluminum product inventory by market

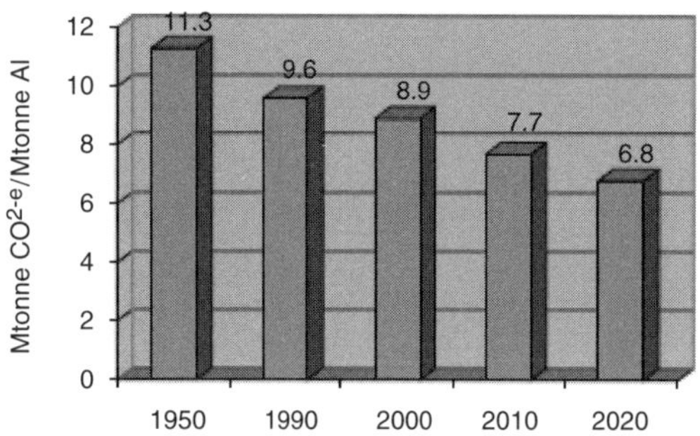

Fig. 6.4 Greenhouse gas emissions intensity of aluminum shipments

landfills or metal oxidized to aluminum oxide when used as a propellant or for deoxidizing steel melts. The model also provides an assessment of past, current, and projected energy and emissions intensity of aluminum semifabricated shipments, as illustrated in Fig. 6.4, based on the latest assessment of average greenhouse gas (GHG) emissions intensity for aluminum processes (Ref 6.9).

On average, worldwide aluminum products are becoming less GHG intense on a per-ton-shipped basis due to two reasons:

- Increase in the percent recycled metal relative to primary metal. Only 5% of the energy and GHG emissions is required to produce aluminum ingot compared to primary (bauxite/Al_2O_3/electrolysis) aluminum.
- Lower emissions from primary aluminum facilities due to reductions in energy intensity and significant reductions in perfluorocarbon emissions

The results of the model were initially shared with the IAI Board of Directors in May 2004. Model development, the GARC expert review and contribution, and results of global aluminum flow were described, including a quantitative assessment of GHG emissions from global industry aluminum facilities today (2003) and projected into the future as a guide to the industry's contribution to climate change effects. At that time, the Board requested a "what-if" case scenario that later indicated that industry factory and indirect emissions from purchased electricity could be stabilized, despite significant industry growth by 2020, based on currently projected recycled metal flows and moving all of the global industry toward today's (2003) global benchmark technologies and operating best practices. Furthermore, the model indicated that fuel efficiency and emissions savings due to additional aluminum transportation products had the potential to surpass the global industry's production emissions by 2020.

In May 2005, additional model results were shared with the IAI Board, and they added the following voluntary objective to their list of sustainable development quantitative goals in recognition of the ecological and economic value of enhanced recycling to reduce natural resource consumption and life-cycle inventory effects:

> "The IAI has developed its Sustainability Material Flow Model to identify future recycling flows. The model projects that global recycled metal supply (back to the industry) will double by 2020 from today's (2005) level of 6.4 million metric tons. The aluminum industry will report annually on its global recycling performance."

The IAI also continues to develop and improve the model, collect supporting life-cycle inventory intensity data, and use the scenario capability to quantitatively assess current and future production paths and sustainable strategies.

REFERENCES

6.1. P.R. Bruggink and K.J. Martchek, Worldwide Recycled Aluminum Supply and Environmental Impact Model, *Light Metals, Proceedings of the 2004 Annual Meeting,* March 14–18, 2004 (Charlotte, NC), A.T. Tabereaux, Ed., TMS (The Minerals, Metals and Materials Society), p 907–911

6.2. "Life Cycle Assessment of Aluminum: Inventory Data for the Worldwide Primary Aluminium Industry," International Aluminum Institute, London, U.K., March 2003; http://www.world-aluminum.org/iai/publications/documents/lca.pdf (accessed July 2007)

6.3. P.A. Plunkert, *Aluminum: Minerals Yearbook,* U.S. Geological Survey, U.S. Department of the Interior, Reston, Virginia, 2001, p 6.1–6.19; http://minerals.er.usgs.gov/minerals/pubs/commodity/aluminum/050401.pdf (accessed July 2007)

6.4. P.A. Plunkert, *Aluminum: Minerals Yearbook,* U.S. Geological Survey, U.S. Department of the Interior, Reston, Virginia, 2002, p 5.1–5.18; http://minerals.er.usgs.gov/minerals/pubs/commodity/aluminum/alumimyb02r.pdf (accessed July 2007)

6.5. P.A. Plunkert, *Aluminum: Minerals Yearbook,* U.S. Geological Survey, U.S. Department of the Interior, Reston, Virginia, 2003, p 5.1–5.16; http://minerals.er.usgs.gov/minerals/pubs/commodity/aluminum/alumimyb03.pdf (accessed July 2007)
6.6. "Aluminum Statistical Review for 2002," The Aluminum Association Inc., Washington, D.C., 2003
6.7. G. Rombach, Future Availability of Aluminum Scrap, *Light Metals, Proceedings of the 2002 Annual Meeting and Exhibition*, Feb 17–21, 2002 (Washington, D.C.), W. Schneide, Ed., TMS (The Minerals, Metals and Materials Society), p 1011–1018
6.8. *Economics of Aluminium*, 8th ed., Roskill Information Services, London, U.K., 2003
6.9. "Life Cycle Inventory of the Worldwide Aluminium Industry with Regard to Energy Consumption and Emissions of Greenhouse Gases," International Aluminum Institute, London, U.K., March 2000; http://www.world-aluminum.org/iai/publications/documents/expanded_summary.pdf (accessed July 2007)

CHAPTER 7

Recycling of Aluminum*

AMONG THE MORE IMPORTANT CHARACTERISTICS of aluminum is the instantaneous formation of a self-limiting tenacious oxide film in liquid and solid states. The protective oxide makes the metal impervious to degradation or weight loss, except in cases of galvanic corrosion or specific chemical attack. For this reason, most of the aluminum produced from the beginning remains recoverable as a vital resource in satisfying current and future industry metal supply requirements.

Aluminum recycling became a distinct manufacturing activity less than 20 years after the commercialization of the Hall-Heroult process in 1888, driven by high value and growing demand. In the early days of the developing aluminum industry, primary producers attempted to maximize new metal sales, reduce the unit price, and develop the alloys and processes that were essential to become more competitive with existing materials. They were not interested in scrap recycling, leaving that activity to others, who in time developed an independent secondary industry.

As both industries grew, their objectives changed. Primary producers began acquiring and consuming scrap in their operations, and secondary producers began producing more sophisticated end products, thus reducing the originally distinctive differences between the two industries. Similar developments occurred in the technology arena.

Secondary producers were originally low-capital salvage operators. Some still operate in that mode, whereas others have become more sophisticated enterprises using advanced technology to maximize recoveries, reduce costs, and produce competitive-quality remelt ingot and differentiated products. The latter group patterned its development after that of the primary industry, which has taken a more capital-intensive approach to recycling challenges.

With the modern demands of energy conservation and environment, the aluminum industry is a leading proponent of global sustainability and strongly advocates the use of recycled metal. Aluminum can be recycled with large energy and emission savings and essentially without any loss in potential material properties. Further, it can be recycled repeatedly without significant degradation, although there are oxidation losses and some contamination by impurity elements. This recycling advantage is so dominant for aluminum that it has become a key factor in the industry's sustainability. In addition to significant energy savings, recycling requires relatively low-capital-cost facilities when compared to the capital required for primary production.

Recycling advantages explain the aluminum industry's proactivity in promoting and exploiting the recovery of aluminum scrap in all forms. Aluminum cans and other aluminum products are the most valuable components of the municipal waste stream, often providing the only economic justification for separation processes. Legislation mandating minimum standards for end-of-life recycling promotes the increased use of highly recyclable aluminum in transportation applications. Of course, recycling is not the only relevant factor. The use of aluminum in transportation is a major factor in light-weighting of vehicles, and lighter vehicles are more fuel efficient. It has been demonstrated that a 10% reduction in vehicle weight results in a 6 to 8%

*Adapted from "Recycling of Aluminum" by Elwin L. Rooy and J.H.L. Van Linden in *Properties and Selection: Nonferrous Alloys and Special-Purpose Materials*, Volume 2, *ASM Handbook*, 1990, and "Recyclable Aluminum Rolled Products" by John Green and Michael Skillingberg in *Light Metal Age*, Aug 2006, p 33. Adapted with permission of *Light Metal Age*.

reduction in fuel use and thus in CO_2 emissions. The growth in the use of aluminum in cars and light trucks, 81 lb/vehicle in 1973 to 319 lb/vehicle in 2006, reflects recognition of these essential advantages. Studies suggest that the contribution of aluminum in improving fuel efficiency results in rates of reduction in emissions of greenhouse gases (GHG) from the transportation sector that exceed emissions from aluminum production. This has led to the claim that the aluminum industry will become GHG neutral by the year 2020 (Ref 7.1).

Recycled aluminum is a vital supply component. It minimizes the need for imported metal in the U.S. domestic market, saves ~95% of the energy, reduces ~95% of the emissions associated with the production of metal from ore, and conserves raw materials. These benefits affect virtually all aluminum products but are especially pertinent to rolled products that include sheet, plate, and foil, which constitute 45.7% of all U.S. and 39.9% of North American aluminum industry shipments (Ref 7.2). However, a dominant issue in the use of recycled metal in the production of rolled aluminum products is the control of impurities, especially iron and silicon that increase modestly with each remelting cycle. Alloys used in the production of engineered castings typically contain much higher silicon concentrations, so that mixing scrap can substantially and negatively affect the chemistry of the wrought alloy stream. Iron impurities generally increase as results of extraneous contamination and refractory reactions.

In spite of the issues with impurity control, there is absolutely no doubt that the advantages inherent in recycling aluminum are enormous. The "energy bank" concept and the opportunity to regain almost all the embedded energy in the product again and again when the metal is recycled will drive further growth in the secondary industry. In effect, the world's increasing stock of aluminum acts like an "energy resource bank," over time delivering more and more practical use and value from the energy embodied in the primary production for future generations by conserving energy and other natural resources. For example, recycling of postconsumer aluminum now saves an estimated 84 million tonnes of GHG emissions per year, equivalent to the annual emissions from 15 million cars. Since its inception, the recycling of postconsumer aluminum scrap has already avoided over 1 billion metric tonnes of CO_2 emissions. For these reasons, it is clear that recyclable rolled products will continue to be building blocks toward sustained manufacturing viability.

Industry and Recyling Trends

Several generalizations from the early 1990s concerning the U.S. and world aluminum industries continue to be relevant today:

- *The aluminum industry today is truly international:* Primary aluminum is produced in virtually every global region, and the metal produced competes in global markets. Aluminum alloy scrap is now traded internationally as well.
- *Domestic U.S. primary aluminum production will not expand beyond the capacity of existing smelting facilities*: Energy costs, labor rates, and environmental standards in the United States suggest that there will be substantial decreases in primary output in the absence of any forseeable new, more economical smelting technologies.
- *New risks and uncertainties:* The European and, to a lesser but significant extent, the U.S. aluminum industries face new risks and uncertainties caused by the growing geographic separation of primary production from major fabricating facilities and markets.
- *Aluminum is a U.S.-dollar-based commodity*: Exchange rate fluctuations represent a major complication in stabilizing prices and regulating international competition and metal supply.
- *The world aluminum production capacity will expand:* This will occur in countries with low energy costs, such as Canada, Venezuela, Brazil, Australia, and parts of the Middle East.
- *Recycling continues to have increasing importance:* For the United States, and ultimately for the rest of the aluminum-consuming world, recycling and resource recovery will play an increasingly important strategic role in ensuring a reliable and economical metal supply.
- *The United States will import aluminum:* On the basis of the best assumption, the United States will become an importing nation as aluminum requirements exceed domestic smelting capacity. Most imports will consist of unalloyed smelter ingot for remelting, casting, and fabrication in North American facilities. Some products will be imported but only

for specialty applications with unique and/or cost advantages over domestic products.

- *World competition for scrap units will intensify:* This competition will be based on the relative costs and availability of primary aluminum and scrap.

The U.S. sources of aluminum have undergone considerable change. Primary production from the reduction of alumina dominated the U.S. domestic metal supply for 112 of 115 years of commercial production since the first year of recorded statistics in 1893 (when some 91,000 metric tons of metal were produced). In recent years, the domestic metal supply reflects increasing imports and higher levels of secondary recovery. Recycled aluminum and imports have actually exceeded primary production in the United States since 2002. These changes are shown in Table 7.1 (Ref 7.2) and graphically in Fig. 7.1 (Ref 7.2).

Table 7.1 U.S. metal supply

Year	Metric tons x10^3	Primary production,%	Imports,%	Secondary recovered,%
1960	2406	78.0	7.5	16.5
1970	4950	72.9	8.8	18.3
1980	6833	68.1	8.9	23.0
1990	7863	51.5	18.1	30.4
1999	11,154	33.9	33.0	33.1
2000	10,699	34.2	33.5	32.2
2002	9579	28.5	40.8	30.7
2004	10,112	24.9	45.1	30.0

Overall, the United States reached the highest primary production at 4653 thousand metric tons in 1980; the largest U.S. metal supply on record occurred in 1999 with a combined value of 11,154 thousand metric tons. Year 2000 shows the marked reduction in primary metal supply resulting from the effects of energy costs and availability and competitively priced metal in the international market. These conditions now appear to be permanent and have resulted in the shutdown of much primary capacity, particularly in the Pacific Northwest. Restarting this capacity will require access to low-cost power, unanticipated changes in the supply of unalloyed metal from foreign sources, or government action involving national security issues.

Overall aluminum production and the use cycle are strongly linked for the United States and Canada, and the combined U.S. and Canadian production of primary and secondary aluminum is shown in Table 7.2 (Ref 7.3). Primary production is still nearly 50% of the total North American metal supply, and the portion supplied from recycled metal drops to a little over 30%. Still, the long-term trend has been an increase in the importance of recycled and imported material. Accordingly, from an energy-efficiency standpoint, it is logical to increase the

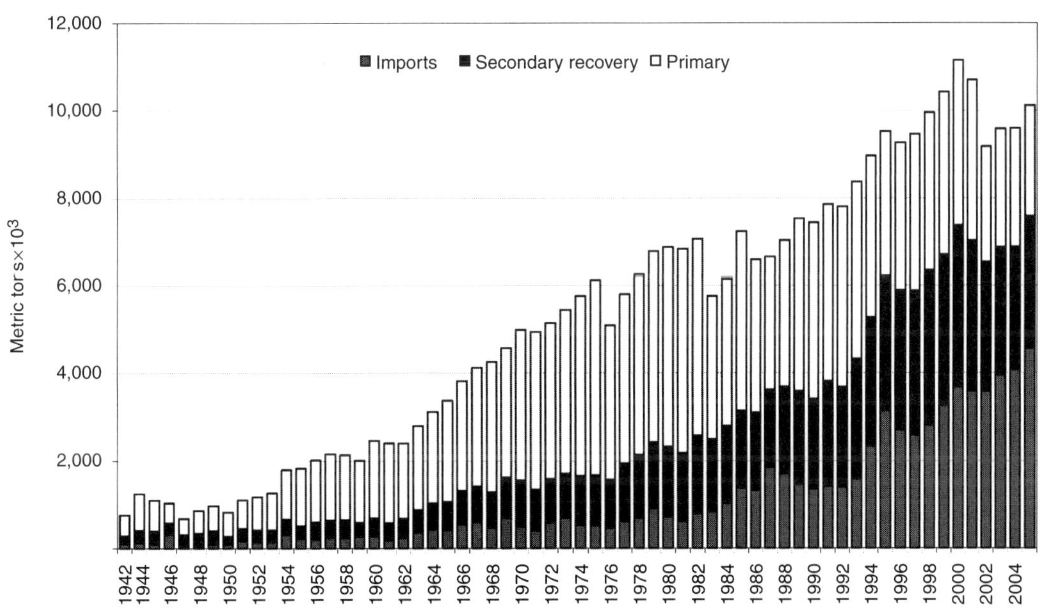

Fig. 7.1 U.S. aluminum supply. Source: Ref 7.2

Table 7.2 North American metal supply

Year	Metric tons	Primary production,%	Imports,%	Secondary recovered,%
1960	2978	84.6	1.9	13.5(a)
1970	5650	81.1	2.4	16.6(a)
1980	7511	76.2	2.0	21.9
1990	8562	65.6	5.5	28.9
1999	11,765	52.4	15.0	32.5
2000	11,438	52.8	15.7	31.5
2002	10,320	52.5	17.4	30.2
2004	10,639	48.0	21.5	30.5

(a) U.S. + estimate of Canadian secondary. Source: Ref 7.3

amount of recycled aluminum in the metal supply reservoir to the maximum extent possible.

Recycling Trends in the 1980s (Adapted from Ref 7.4). The objective of all collection activities is the conversion of scrap forms to products having the highest commercial value. One trend in the 1980s was the consumption of scrap in secondary fabricating facilities. A number of extruders, foundries, and minimill operations produce billet, castings, and common alloy sheet products directly from scrap. The die casting industry relies heavily on secondary compositions for the production of automotive and other parts, and the larger die casters have, in many cases, expanded metal supply through direct scrap purchases.

A number of industry segments are in competition for the available aluminum scrap, although not necessarily the same types of scrap. As shown in Fig. 7.2, primary producers experienced the largest increase in scrap consumption during the decades of 1970 to 1990. The sporadic data available for the period prior to 1940 suggest that recycling of aluminum grew steadily to approximately 15% of total shipments. World War II disrupted the pattern drastically, but the stockpile reduction in the years following the war reduced the recycling rate to the prewar level (Fig. 7.3).

Current Trends (Adapted from Ref 7.5). The contribution of scrap metal recovery has steadily increased in terms of global output of aluminum metal (Fig. 7.4). The percentage of global output of aluminum metal from scrap recovery has also increased from 17% in 1960 to 33% today and is projected to rise to almost 40% by 2020 (Fig. 7.5) (Ref 7.6). Of an estimated total of over 700 million tonnes of aluminum produced in the world since commercial manufacture began, approximately three-quarters (between 400 and 500 million tones) is still in productive use. As an "energy bank," the aluminum in use accounts for almost 50,000 petajoules of energy. This is higher than the current

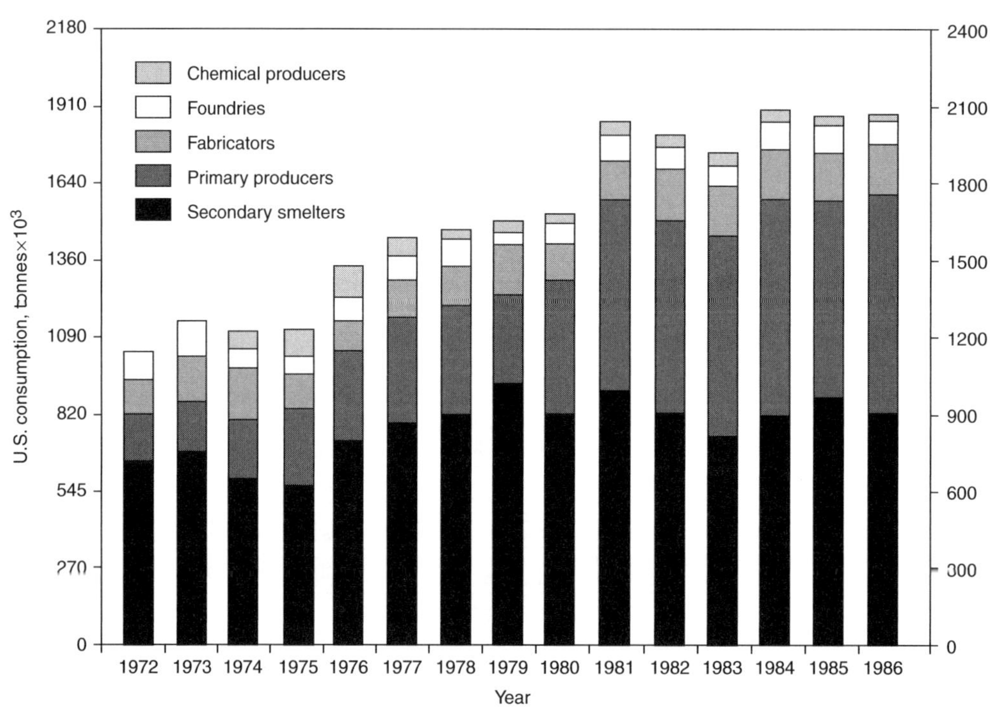

Fig. 7.2 U.S. aluminum scrap consumption by type of company for the years 1972 to 1986. Source: U.S. Bureau of Mines

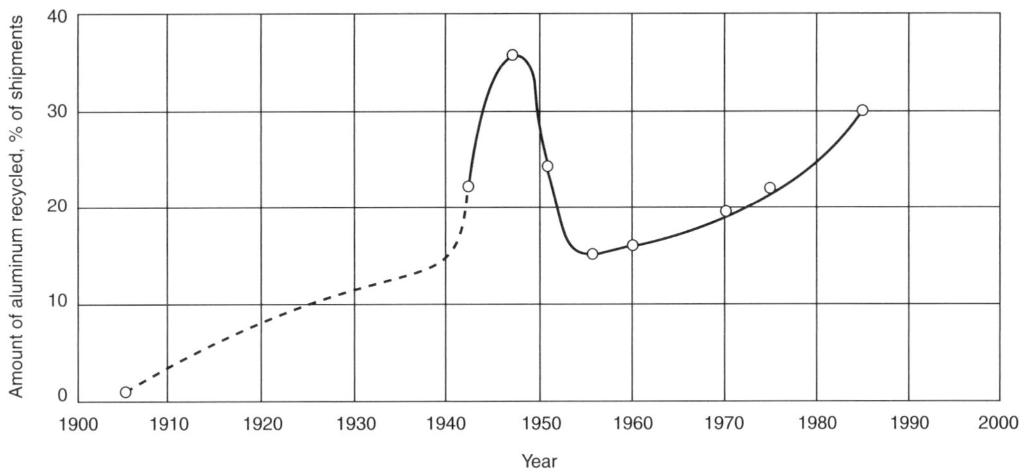

Fig. 7.3 Aluminum recycling trends in the United States. The percentage of shipments recycled is only now approaching the peak experienced during World War II

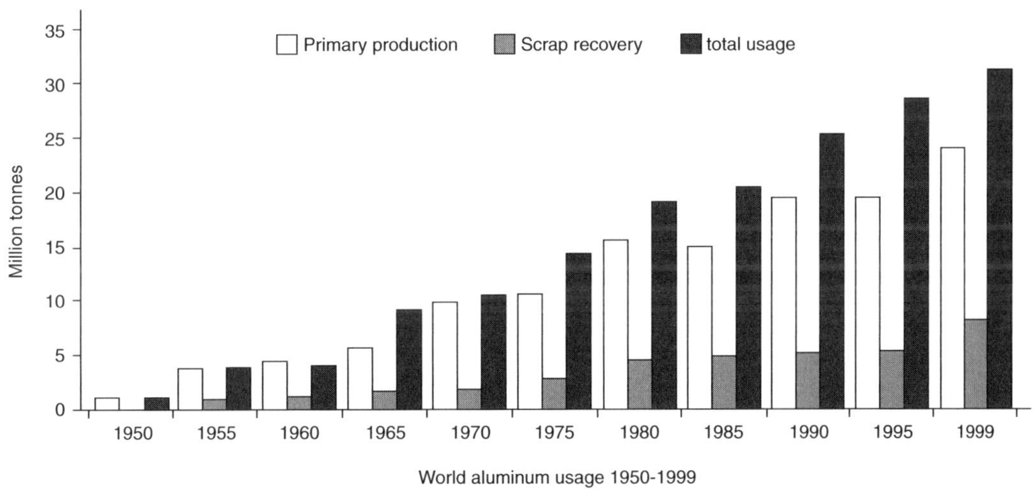

Fig. 7.4 Scrap and primary components of aluminum world usage. Source: Ref 7.5

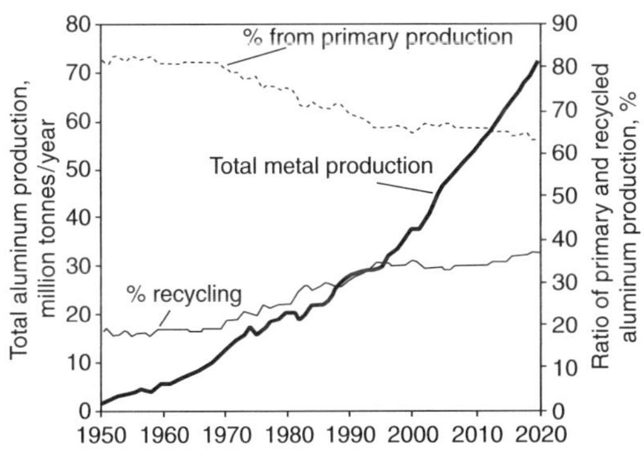

Fig. 7.5 Percentage of recycling component in aluminum production. Source: Ref 7.6

combined annual energy demand of Africa and Latin America and is equivalent to the annual total electrical energy generated globally from coal. If this metal is recycled, the banked energy is a resource not just once but repeatedly and over future generations. If landfilled, this valuable resource is potentially lost forever.

Global aluminum recycling rates are high, approximately 90% for transport and approximately 60% for beverage cans. In Europe, aluminum has high recycling rates, ranging from 41% in beverage cans to 85% in construction and 95% in transportation. In Japan, the recycling rate for cans was 79%, and Brazil achieved 78%.

A kilo of used cans is worth more to collectors than 15 kilos of plastic or 10 kilos of paper. Americans recycled 62.6 billion aluminum cans in 2000, for a beverage can recycling rate of 62%. For many groups, used aluminum cans turn into new-found money, thanks to aluminum can industry initiatives. Initiatives include the American Aluminum Association's partnership with Habitat for Humanity, called "Aluminum Cans Build Habitat for Humanity Homes," designed to boost public interest in aluminum can recycling while helping volunteers and families build homes.

The recycling rate for end-of-life aluminum depends on the product sector, the lifetime of each product, and society's commitment to collect aluminum. Each application requires its own recycling solutions. Just over 15 million tonnes of recycled aluminum were produced in 2004 worldwide, which met 33% of the global demand for aluminum. Of the almost 7 million tonnes of aluminum recycled from end-of-life products, 28% came from packaging, 44% from transport, 7% from building, and 21% from other products.

Aluminum enjoys a high recycling rate of 85% in the building industry. The global industry is keen to increase collection rates and is working with producers of building applications to enable even more efficient collection of scrap from demolished buildings. In 2004, Delft University of Technology conducted a study into the aluminum content of, and collection rates from, demolished buildings in six European countries, which found that the average collection rate for aluminum was close to 96%. The transport sector has high rates of recycling, currently approximately 90% globally, because dismantlers and recyclers recognize the high intrinsic value of end-of-life aluminum products.

Can collection is approximately 60% globally, and in some countries, the collection rate is already above 80%. Sweden and Switzerland collect 86 and 88% of their aluminum beverage cans, respectively. Sweden's success lies in a deposit/refund system, whereas in Switzerland, a voluntary prepaid recycling charge covers the costs of collection. In Japan, a collection rate for used beverage cans of 92% is achieved with a voluntary system. In Brazil, recycling is not mandatory, but every region has a recycling market that facilitates the collection and transportation of end-of-life products. This has encouraged communities, supermarkets, condominiums, shopping centers, and clubs to collect. As a result, in 2005, 96% of cans were recycled in Brazil, and this is considered the world's highest recycling rate for used beverage cans.

Recyclability of Aluminum

Once produced, aluminum can be considered a permanent resource for recycling, preferably into similar products. It is highly recyclable, due to the following characteristics:

- It is resistant to corrosion under most environment conditions and thus retains a high level of metal value after use, exposure, or storage.
- The energy required to remelt aluminum is only 5% of the energy required for its primary production.
- The versatility of aluminum alloying has resulted in a large number of commercial compositions, many of which were designed to accommodate impurity contamination.

The objective of recycling is to produce a salable commercial aluminum alloy product. Currently, more than 300 compositions covering wrought and cast alloys are registered with the Aluminum Association. Many of these alloys are designed to tolerate the variations in composition and ranges in impurity contents that may be experienced in the recovery of scrap.

In the early decades of the 20th century, the output of the secondary aluminum industry was largely tied to the consumption of castings by the automotive industry. As shown in Fig. 7.6, the recycling of aluminum has increased steadily from 1950 through the 1980s, despite recessions and energy crises. This increase is

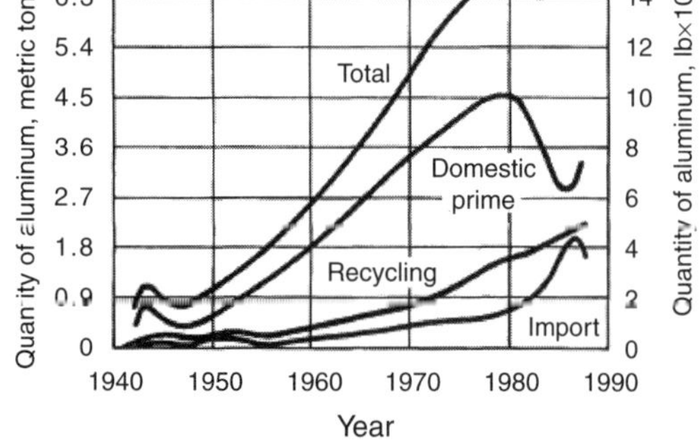

Fig. 7.6 U.S. aluminum supply distribution for the years 1940 to 1990. Source: Aluminum Association

the result of growth in the automotive market as well as the development of new and significant recycling applications, such as the consumption of used beverage cans (UBCs) in the manufacture of sheet for new cans.

In recent years, environmental concerns have contributed to an increased awareness of the importance of scrap recycling. Today (2007), the recyclability of aluminum is a major advantage to the aluminum industry in materials competitions for major product markets. The current driving forces for aluminum recycling are:

- Regulatory actions taken by government agencies to encourage resource conservation, energy conservation, and waste reduction through mandatory segregation and deposit programs
- Consumer sensitivity to environmental issues and the solid-waste crisis
- Competitive pressures from other materials
- Economic advantages based on the relative value and availability of aluminum scrap

Energy Savings. The recognized energy savings are well known for aluminum recycling when compared with the energy consumed in primary aluminum production. For example, the actual ratio of primary total energy to recycled total energy for used beverage cans is 28.5:1 (Ref 7.7). A definitive study of energy savings by recycling aluminum was obtained in the late 1990s in a comprehensive project mandated by the U.S. Council for Automotive Research under the Partnership for a New Generation of Vehicles program. At the request of the automotive companies, the aluminum industry was challenged to produce a rigorous life-cycle assessment (see Chapter 3, "Life-Cycle Inventory Analysis of the North American Aluminum Industry"). This assessment represents a comprehensive evaluation of the aluminum industry operations involving 15 separate unit processes (refining, smelting, hot rolling, shredding and decoating, and secondary ingot casting are examples of unit processes) located in 213 plants throughout North and South America, Africa, Australia, Europe, and the Caribbean.

Specifically, the life-cycle assessment demonstrated that the production of primary aluminum, when all the electrical generation, transmission losses, and transportation are accounted for, requires ~45 kWh and emits ~12 kg of CO_2 per kilogram, whereas recycled aluminum only requires ~2.8 kWh of energy (~5%) and only emits ~0.6 kg (~5%) of CO_2 for each kilogram of metal. In this sense, the intrinsic energy of the metal is recovered when the material is recycled, hence the "energy bank" concept.

The Recycling Loop

Some amount of aluminum recycling has always occurred. Producers remelted internal scrap, and foundries, the first important customers of the emerging aluminum industry, remelted gates and risers and scrap castings. In time, the recovery of new and old scrap became a function of secondary smelters, whose principal customers were the aluminum castings industry. The emergence of the aluminum beverage can in 1965 and the initiation of "buy-back" programs by primary producers, who were the source of rigid-container sheet (RCS), created the first significant public awareness of the importance and value of product recycling. The importance of aluminum recyclability was not lost on the producers of RCS nor on can makers, who realized the economic and environmental advantages over competing container materials. With the success and magnitude of the aluminum can market, recycling efforts by major producers quickly spawned a new phase in recycling based on closed-loop processing and the development of radically new technologies specifically designed to maximize recoveries from light-gage coated scrap. The lessons of can recycling were quickly adapted to other scrap forms. The primary industry that had been single-mindedly devoted for decades to the promotion and sale of primary metal and products realized the comprehensive advantages of maximizing recycling as a means of reducing costs, serving the environment, and conserving raw materials and energy.

The reclamation of aluminum scrap is a complex interactive process involving collection centers, primary producers, secondary smelters, metal processors, and consumers. Figure 7.7 depicts the flow of metal originating in primary smelting operations through various recycling activities. The initial reprocessing of scrap takes place in the facilities of primary producers. In-process scrap, generated both in casting and fabricating, is reprocessed by melting and recasting. Increasingly, primary producers are purchasing scrap to supplement primary metal supply; an example of such activity is the purchase or toll conversion of UBCs by primary

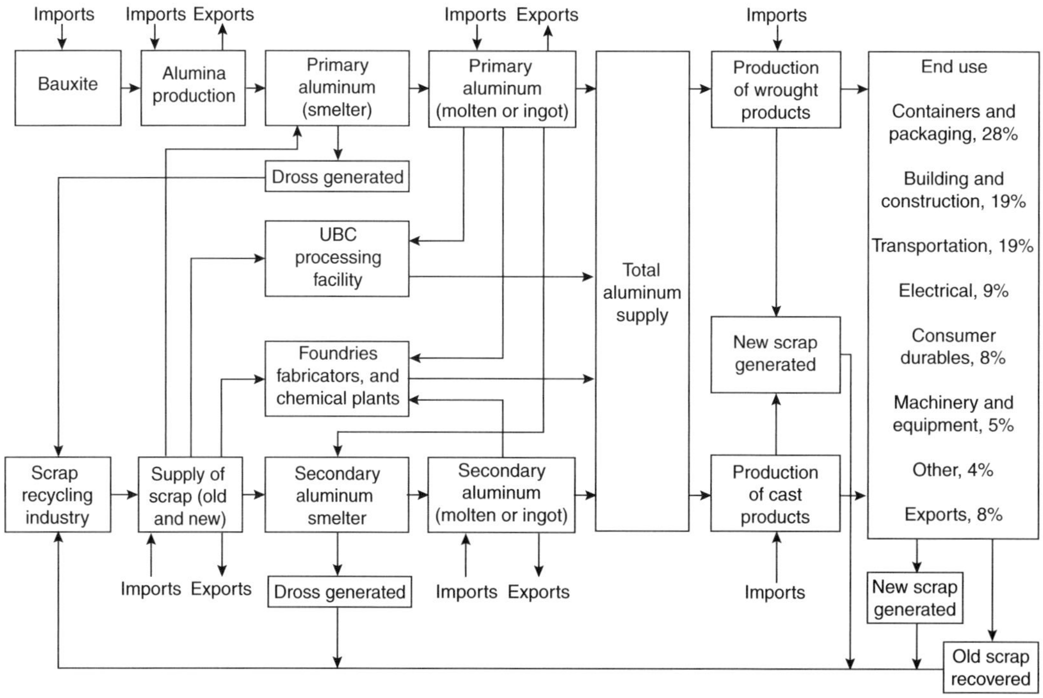

Fig. 7.7 Flow diagram for aluminum in the United States, showing the role of recycling in the industry. Scrap recycling (lower left) includes scrap collectors, processors, dealers and brokers, sweat furnace operators, and dross reclaimers. UBC, used beverage can. Source: U.S. Bureau of Mines

producers engaged in the production of rigid container stock.

Scrap incurred in the processing or fabrication of semifabricated aluminum products represents an additional source of recyclable aluminum. Traditionally, this form of new scrap has been returned to the supplier for recycling, or it has been disposed of through sale on the basis of competitive bidding by metal traders, primary producers, and secondary smelters.

Finished aluminum products, which include such items as consumer durable and nondurable goods; automotive, aerospace, and military products; machinery; miscellaneous transportation parts; and building and construction materials, have finite lives (see Chapter 6, "Material Flow Modeling of Aluminum for Sustainability," in this book for additional life-cycle discussion). In time, discarded aluminum becomes available for collection and recovery. So called old scrap, metal product that has been discarded after use, can be segregated into classifications that facilitate recycling and recovery.

Scrap specifications have been developed that allow the convenient definition of scrap types for resale and subsequent reprocessing. Those developed by the Institute of Scrap Recycling Industries are in broad use (Ref 7.8). These specifications, however, are, for the most part, physical descriptions of scrap categories useful to dealers. Chemistry specifications and limits on harmful impurities are not defined, and treatment of extraneous contaminants is inconsistent. More comprehensive scrap specifications are being developed by the industry.

Technological Aspects of Aluminum Recycling

The widely accepted range of aluminum recycling includes scrap collection and preparation, remelting, refining, and the upgrading of molten metal to a ready-to-cast condition. Multiple alloys and product forms are consumed in the generation of new product.

Molten metal treatment processes that reduce the concentration of reactive elements, such as magnesium, sodium, and calcium, to comply with specification limits, remove dissolved hydrogen, and filter entrained oxides and other nonmetallics play an extremely important role in enabling the reuse of often heavily contaminated scrap. Environmental restrictions have resulted in the development of process alternatives to reactive gas fluxing for these purposes. Various methods, including magnetic separation, are used to reduce the introduction of extraneous contamination in the solid state.

After the invention of the resealable taphole, the coal-, natural gas-, or oil-fired reverberatory, or open hearth furnace, evolved as the predominant remelting method. There are two reasons for the long-term success of the direct-firing method in which the combustion chamber and the solid charge are combined into one open hearth. First, aluminum oxidation progresses slowly due to the protective, flexible thin oxide film that forms instantaneously when a fresh aluminum surface is exposed to any oxygen-containing atmosphere. Second, the surface-to-volume ratio of scrap particles was generally small, because the scrap usually consisted of large sections or parts and heavy-gage sheet. Therefore, the penalty for direct exposure to the melting environment was generally acceptable.

The open hearth furnace has been the mainstay of the scrap-remelting business for over 50 years, but as the application of aluminum became more widespread and diverse, the need for effective and efficient recycling technology grew rapidly. The corresponding growth and diversification of the scrap-recycling industry have been driven by several factors, including increased production requirements, alloy development, scarcity of energy and resources, and increased availability of mixed scrap.

Increased Production Requirements. Large, modern casting stations may use up to 250,000 lb per production cycle. This requires not only a high melt rate in large melting furnaces but also a minimal charge time. The present rectangular open hearth furnace construction is practically limited in size to approximately a 30 m^2 (320 ft^2) bath area. Cylindrical furnace construction allows for hearth areas of up to 75 m^2 (800 ft^2) and features a removable lid. This configuration allows for very fast, well-distributed overhead charging with preloaded dump buckets.

Alloys. Magnesium, in many important compositions that include castings, extrusions, sheet, and plate, results in higher oxidation rates and melt losses. Among more important applications for alloys containing significant magnesium concentrations are light-gage sheet for automotive and container products, aerospace and other high-strength applications, and marine applications. The prevalence of this type of scrap has stimulated the development and use of methods that minimize the exposure of scrap to the furnace atmosphere; these methods include continuous melting, rapid ingestion systems, induction melting, and, more recently, isothermal melting.

Energy and Resources. The energy crisis of 1973 highlighted the enormous energy advantage of using scrap rather than primary metal and exposed the corresponding need to minimize melt losses. Primary producers began processing more-difficult-to-melt scrap using newly developed technologies, forcing secondary producers to seek new scrap sources, including less-desirable scrap types, which are often processed in rotary salt furnaces previously employed in the reclamation of skim and dross.

Concerted efforts have been made to incrementally improve the energy efficiency of reverberatory furnaces through adaptation of heat recuperation, preheating combustion air and solid charge components, mechanically or electromagnetically stirring the melt, modifying burner designs, and incorporating sensors and automated controls to optimize furnace operation. New melting processes that target energy efficiency and reduced melt loss have been developed that include flotation and isothermal melting.

Mixed Scrap. Since the early 1960s, the volume of scrap from discarded life-cycle products such as automobiles, home construction materials, trailers, and household goods, often inseparably mixed or joined with other metals, has steadily increased. Substantial amounts of otherwise lost aluminum are recovered from such scrap by sweat melting. This is a selective process that involves melting the scrap in a sloped hearth that exploits the melting temperature differences of dissimilar metals and which, during operation, melts, drains, and collects metals based on melting points.

Alloy Integrity. In addition to preservation and melt loss reduction, a key factor in successful recycling is the maintenance of alloy integrity. The ideal way is closed-loop recycling in which scrap is recycled into identical or similar products. Examples are the conversion of UBCs into new-can sheet and the conversion of auto body sheet into new auto body sheet. Mixed scrap that requires extensive treatment to chemically react and remove magnesium or that requires the addition and adjustment of alloying elements is usual but far less economically attractive. Accordingly, great efforts are made at all times to keep in-process and purchased scrap identified and segregated.

The greatest challenge for the recycling community is finding the most economical way to separate and prepare scrap for melting so that it can be used in the least-degraded form with the least number of postmelt treatments for alloy or quality adjustments. Because UBC, automotive, and municipal scrap loops are important and represent different challenges, technological developments in preparation and melting for these streams is further described.

Process Developments for Remelting

Although melt loss had become the major cost factor in ingot production, it was, in fact, the soaring cost of energy during the 1973 energy crisis that triggered the search for more efficient remelt processes. This effort also sought to develop less labor-intensive and higher-productivity solutions to scrap melting. It was recognized that the open hearth furnace was designed for remelting bulky scrap, which not only requires extensive charging time but also creates large amounts of skim. Skim, comprising oxidation products and metal, acts as an insulating blanket between radiation from superheated refractories, which is the dominant heat-transfer mechanism, and the melt, severely reducing thermal efficiency. It became clear that the commitment to recycling necessitated a fundamental review of processing methods.

Reverberatory furnaces, into which high-capacity natural gas burners introduce thermal energy, are the mainstay of primary and secondary industries and large engineered casting foundries for melting, alloying, processing, and holding. These furnaces range in capacity from 10,000 to more than 250,000 lb. Heat transfer occurs predominantly by radiation. The rate of heat transfer is a function of melt surface area and is maximized by design to enable the furnace to operate at high melt rates, typically 60 lb/h-ft^2. As a consequence, the depth of metal in the furnace hearth may approximate only 76 cm (30 in.). These design considerations impose significant process disadvantages, some of which are noted as follows:

- Aluminum in the molten state is highly reactive and readily oxidizes when exposed to the atmosphere and products of combustion. The thickness of the oxide layer depends on alloy, temperatures, and time but represents at all times an insulating barrier to the absorption of radiant energy by the melt. Charging and melting of scrap increases oxide formation as a function of surface-to-volume ratio.
- Heat concentrates at the melt surface, accelerating oxidation, increasing oxide barrier thickness, and forming more harmful oxide species, such as magnesium aluminate spinel in magnesium-containing alloys that grow at geometric rates with increasing temperature.
- Stirring of the melt to accelerate alloying, assure thermal and chemical homogeneity, and remove the oxide-metal skim layer is usually accomplished by industrial truck- or rail-mounted tools. Their use requires that the furnace design includes large access doors that, when opened, result in substantial heat losses and, even when closed, may result in pressure and heat losses that reduce energy efficiency. Too-frequent stirring exposes nascent surfaces to oxidation, resulting in greater metal losses and increased dissolved hydrogen levels.

Charges may include solid and molten aluminum, internal and purchased scrap, metallurgical metals, and master alloys. These are introduced to the hearth, after which the furnace doors or lids are closed, and firing for melting begins. Energy from combustion is absorbed by exposed refractories, from which radiant heat is reflected to the charge.

After melting, additional metal treatment steps are typically required. These metal treatments become more problematic with high proportions of scrap use and increase as unsegragated or mixed scrap are included in the furnace charge. The melt must be stirred to assure complete solution and uniform distribution of alloying elements. Adjustments in metal chemistry may be required to meet specification requirements mandating additional alloying, stirring, and sampling. In many cases, the melt is treated with solid or gaseous fluxes. Finally, the on-composition melt may be held for varying periods at controlled temperatures before transferring to other furnaces or casting stations. Additional stirring and/or skimming may be required before transfer. With each iteration, more energy is required, additional oxide losses are experienced, and dissolved hydrogen levels increase.

Because of the dramatic negative effects of shredded can and light scrap melted through conventional methods in open hearth furnaces, it was reasoned that low-bulk-density scrap must not be exposed directly to the furnace atmosphere and that scrap must be submerged quickly to optimize efficiencies and reduce melt losses.

Continuous Melting. Accordingly, continuous melting processes, distinct from typical batch operations, were developed specifically to avoid the conditions that contribute to increased melt losses. Reverberatory furnaces used extensively for scrap melting are often modified to incorporate an exposed charging or side bay into which scrap can be introduced without opening the furnace and without directly exposing the charge to the combustion atmosphere. The rapid immersion of scrap is aided by tamping devices called well walkers, or by pumping/recirculation/ingestion methods, and oxidation of the melt is reduced by introducing a nitrogen gas stream. While these simple side-bay configurations approach the objectives, further process developments provide more advanced configurations for optimizing melting efficiencies and reducing melt losses. The advantage of the latter processes is that skim can also be efficiently captured or separated and removed outside the hearth. Steady-state conditions involving heat input that balances melting energy requirements and heat losses can be established and maintained.

Three different methods (Ref 7.9–7.11) based on pump design variations have been developed. These designs create the scrap-ingesting vortex either in the pump bay itself or in an optimized adjacent charge bay (Fig. 7.8). Although these methods differ in such areas as production capacity, scrap size tolerance, and hardware simplicity, the skim generation for a specific scrap type is equally low for all three methods. Several variations of this principle have been reported by other researchers (Ref 7.12–7.14), suggesting that the method is appealing to others in the industry who are also struggling to reduce melt losses.

Reverberatory furnace energy efficiencies vary with the extent to which process modifications have been adopted. Incremental improvements result from recuperation of exhaust gas heat, cogeneration, preheating of charge components and combustion air, the use of induction or mechanical circulation/stirring, and by automated systems that monitor and control key operating parameters. All are commercially available technologies. Numerous burner variations are in use, and other fuel options, such as oxy-fuel and oxygen-enriched burner operation, have been explored. An approximation of the best energy efficiency obtained by reverberatory furnaces in good condition with the most advanced features and operated competently is 45%. More typically, energy efficiencies across the industry are in the range of 20 to 30%.

Alternatives to Reverberatory Furnaces. The challenges of efficient low-oxidation-loss melting have also resulted in the use of different furnace types.

Coreless and channel induction furnaces are routinely used to melt fine scrap charged continuously by automated conveying systems. Machining chips, delaquered UBCs, and shredded light-gage scrap are immediately drawn into the melt by convective forces. Unfortunately, this means that oxides that may otherwise contribute to and remain in the skim layer are instead entrained, requiring effective downstream melt processing for their removal.

New process developments now under industrial-scale evaluation promise quantum advances in both efficiency and melt loss control. They include flotation melting, based on the principles of cupola operation, and isothermal melting, employing high-watt-density/high-efficiency electric resistance immersion heaters. Both developments have been supported by the U.S. Department of Energy.

Developing Scrap Streams

The traditional flow of scrap through the primary and secondary aluminum industries is dominated by three major scrap streams: UBCs, automotive scrap, and municipal scrap. The recycling of cans has continued to grow and has established precedents and industrial infrastructure. Now, the more extensive recovery of aluminum from the transportation sector and the effective separation and recovery of aluminum from the municipal scrap stream are of imminent importance.

In the last two decades, some 10,000 recycling centers have been established in the United States for aluminum cans and other aluminum products. More recently, there has been consolidation in the recycling centers and also dramatic growth in municipal curbside collection of recyclables that emphasize the economic value of aluminum cans. Communities are now supplying aluminum cans to more than 300 municipal recycling facilities covering half the U.S. population. This has developed a subsidized dealer market for aluminum cans and other aluminum products that can effectively offset the costs for

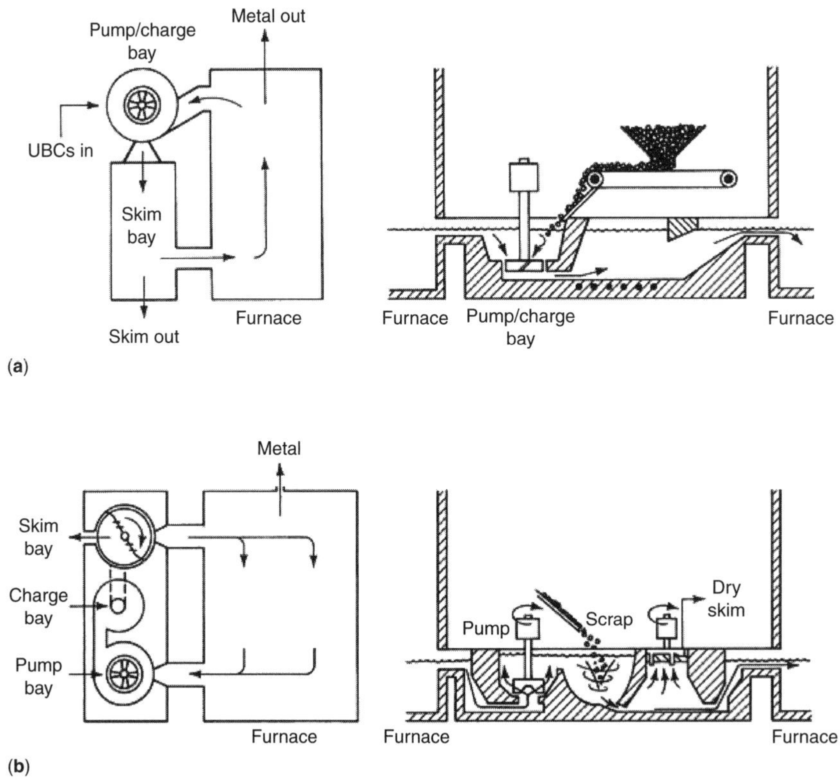

Fig. 7.8 Melting processes for used beverage can (UBC) scrap. (a) Early can scrap melter. (b) More advanced swirl scrap charge melter, which uses a continuous melting process

bins, collection, hauling, separation, aggregation, and remarketing.

The time and rate at which material becomes available for recycling is a function of the product and can be modeled (Ref 7.3). For instance, it is generally accepted that rolled aluminum components in buildings, automobiles, and beverage cans have useful lives that are respectively ~50, ~15, and ~0.2 years. The quantity of metal that is sold into these markets is known and can be forecast. The useful lifetime can be estimated and used to determine when the material becomes available as old, postconsumer scrap. In this way, a metal flow model has been constructed to estimate the availability of future scrap. This model used existing data whenever possible and projects future metal flows based on recent data trends.

Specifically, the scrap model identifies used scrap based on aluminum product markets. The percentage of total metal market and the percent of annual growth over 1990 to 2000, respectively, for these markets were assumed as follows: transportation (37%, 9.8%), containers and packaging (23%, 0.3%), building and construction (15%, 2.6%), consumer durables (8%, 4.8%), electrical (8%, 3.0%), machinery and equipment (7%, 4.8%), and others (2%, 1.7%).

The shear number of cars, sport utility vehicles, trucks and trailers produced, approximately 15 to 18 million per year, all containing an appreciable and growing amount of aluminum, causes the transportation segment to now dominate the scrap supply. Accordingly, in developing any model for recycling aluminum, it is critical to understand the changes in aluminum content in vehicles over time and to estimate when this metal may return to the recycle stream. The amount of aluminum in vehicles is shown by the annual Ducker Research Company surveys to have grown steadily. Specifically, it has grown from a value of 81 lb per vehicle in 1973 to 274 lb in 2001, and the latest number for 2006 is 319 lb per vehicle. Growth in the 1990s was higher than in the previous decades due to increased emphasis on fuel efficiency, lightweighting, safety, and recyclability. At the same time, however, vehicles are lasting longer (the previous lifespan of ~12 years is now trending toward 15 to 18 years). Data show the significant lengthening of the lifespan of vehicles from 1980 to 1990 in part is due to the application of better electrocoat paint systems and body design improvements, including the increased use of aluminum, that significantly reduce body rusting.

Figure 7.9 shows the model results for transportation metal flows and scrap material from 1960 through the year 2030. When the results of the transportation metal flows are evaluated (Ref 7.3), it indicates that at some stage during 2005, the total volume of scrap from the transportation sector vehicles exceeded that from the beverage cans and all other scrap sources. Scrap from the transportation sector is expected be the dominant form of aluminum scrap for the foreseeable future.

Beverage Cans. The recycling of UBCs is a remarkable success story (Fig. 7.10). It is not an exaggeration to suggest that recycling has played a major role in the market growth of aluminum beverage cans and their penetration into a market previously dominated by competing materials. In 1976, the aluminum can accounted for 21% of the beverage container market; 46.4 billion containers were produced, of which 4.9 billion were recycled. In 1986, 72.9 billion cans were produced, and 33.3 billion cans were recycled. In 1988, 94% of all beverage can bodies were produced from aluminum, and virtually 100% of all cans featured the aluminum easy-open end.

Aluminum producers, can makers, and the public have invested in the return system. Large-scale consumer advertising campaigns have emphasized energy savings and resource conservation. Under deposit legislation, cans have generally proven most convenient to handle by consumers, retailers, bottlers, and wholesalers. Collection activities include reverse vending machines, mobile return centers, and public information and educational campaigns. Recycling centers are active in the development of thematic programs to promote the concept of recycling. These programs often associate recycling with civic causes and medical programs, and encourage volunteer, service, and community groups and individuals to maximize the return of UBCs for charitable benefits.

Recycling of UBCs is based on the inherently high scrap value of aluminum and its convertibility into new-can stock. For the most part, UBCs are consumed by primary producers of rigid-container sheet and are employed in the regeneration of can stock. The growth in markets such as Western Europe, coupled with high energy rates in some countries (Japan, for example), has also created an active market involving the export of UBCs and can scrap. The alloy compatibility of the components of the can makes the UBC uniquely suitable for the closed-loop recycling concept, and it is responsible for the consistent high value of UBCs as well as the ever-increasing volume of UBCs in new-can sheet.

Automotive Scrap. Increased activity in the recycling of spent automobiles is based largely on new steel technologies that make scrap conversion economically attractive. However, the exceptional value of the nonferrous metals that are being separated through the efforts of metal recyclers and the secondary industry also contributes significantly to this recycling activity.

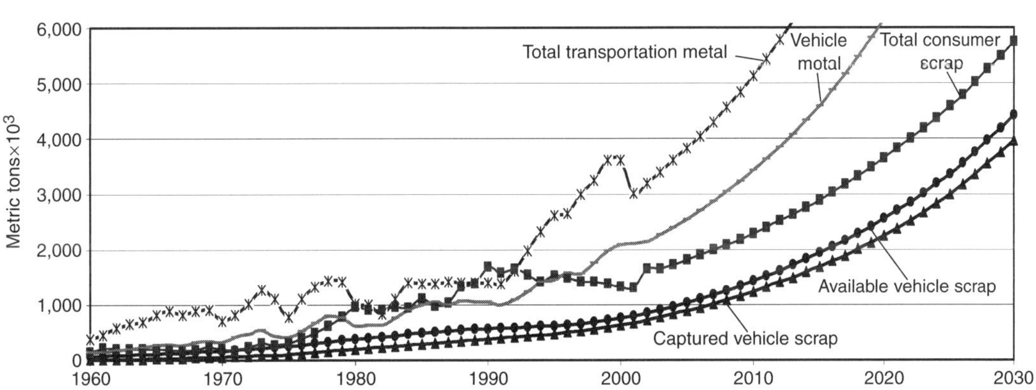

Fig. 7.9 Transportation metal flows

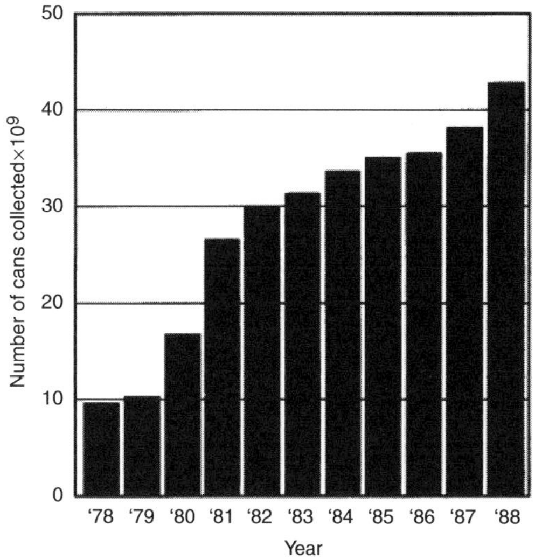

Fig. 7.10 Increase in recycling of aluminum used beverage cans from 1978 to 1988. The calculation for the number of cans collected is based on a can weight survey conducted by the Aluminum Association

Because the use of aluminum in automobiles is increasing (Fig. 7.11), the aluminum value recovered from shredded automobiles is of special importance to the metal castings industry. The potential for aluminum use in automobiles is limited only by how aggressively material advantages are pursued by the manufacturers. Growth to the present level of aluminum use in the automotive industry has been largely associated with construction concepts emphasizing castings. Aluminum alloy forgings, sheet, and extrusions are presently being developed by international automobile manufacturers in cooperation with primary aluminum producers to enhance fuel efficiency.

In 1990, 10 million retired automobiles yielded as much as 500,000 metric tons (550,000 tons) of aluminum; this figure is based on a recovery of 55 kg (120 lb) per vehicle, approximately the 1980 average. In 1995, recoverable aluminum increased to 820,000 metric tons (900,000 tons). Trucks and buses, in which aluminum is used more intensively, have not been included.

Municipal Scrap. The separation of metals from municipal refuse has not grown as originally expected. The best available information indicates that more than 12 million metric tons (13.5 million tons) of metals are lost annually through refuse disposal in the United States.

Metals make up only 9% of total refuse. Of the total metals available in metal municipal refuse, less than 1% is recovered. Because the cost of separating all components of the refuse stream must be based on the value of reusable materials and energy content, a logical conclusion is that municipal refuse processing has not yet become economically viable. Future expansion of such processing will depend either on government subsidies, changes in energy costs, or the availability of raw materials.

The solid-waste crisis has exposed the need to increase all forms of recycling activity. In 1988, the United States generated 145 million metric tons (160 million tons) of waste. In 1979, the nation had 18,500 landfills in operation. This number was reduced to 6500 by 1988, and the U.S. Environmental Protection Agency (EPA) estimates that 80% of existing landfills will close in the next 20 years. The EPA and other agencies have argued for responsible reductions in waste, increased emphasis on recycling, and the use of incineration to generate energy and greatly reduce the reliance on landfill capacity. At present, only 10% of waste in the United States is recycled. The value of aluminum scrap in the municipal waste stream can be expected to significantly affect the success of plans to increase the recycling rate.

Clearly, the recyclability of any material enhances its attractiveness in commercial applications relative to materials that are not recyclable or that can be recycled only at excessive cost. The inherent recyclability of aluminum and its value after recycling supports and is fully consistent with national environmental and waste-reduction goals.

Can Recycling Technology

Beverage can recycling has succeeded because of economic and environmental incentives. It can be conducted by individuals, by charitable fundraising groups, by municipalities, and by state mandates. Individuals can earn approximately 1 cent per can for themselves, or they can choose to donate or aggregate via an array of groups, from Boy Scouts and other organizations to the largest of corporations and charities. All channels can make money for individuals, municipalities, businesses, and charities while at the same time saving energy, preserving the environment, and recycling the metal.

Large amounts of UBCs are either toll converted for primary producers or remelted by secondary operators for resale as remelt ingot or for use in casting alloys. However, the majority

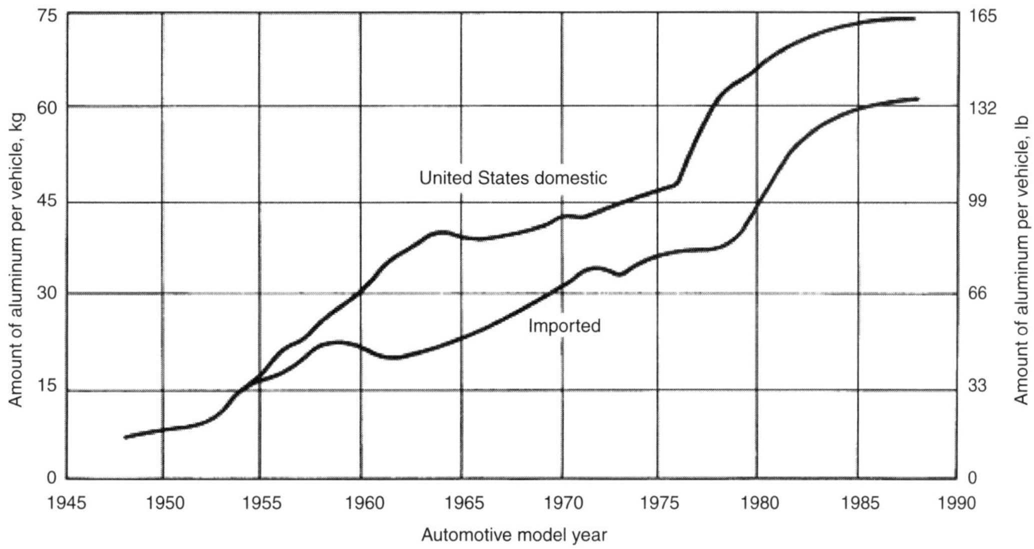

Fig. 7.11 Average use of aluminum in automobiles from 1946 to 1988. U.S. automakers have used more aluminum than world producers since the mid-1950s

(approximately 80%) of UBCs are returned directly to the primary industry, and UBC recycling is thus a prime example of closed-loop product recycling. The flow diagram in Fig. 7.12 (Ref 7.15) shows the captive nature of the can manufacturing and recycling loop. The scrap preparation and melt technologies described in this section are considered the most advanced and should not be viewed as industry standards. However, many UBC converters are using similar, rather sophisticated technologies.

In the recent past, the most successful beverage can recycling year was 1997, when 66.5% of all cans produced were bought back by the industry and some 2.05 billion 1b of metal were recovered. The rate of can recycling declined until 2004, when the trend reversed and slightly more than 50% of all cans were collected.

To encourage higher percentages of can collection, the industry has further strengthened the recycling infrastructure and embarked on extensive marketing and publicity efforts. For example, November 15 each year has been declared "America Recycles Day," and localities and communities have been challenged to recycle more. Some nations, notably Brazil, Japan, Sweden, and Switzerland, have already achieved can collection rates at the level of +90% due to a combination of economic, social, and regulatory forces.

In conjunction with the Can Manufacturers Institute, the Aluminum Association has formed the Aluminum Can Council (ACC) to develop educational programs and encourage collection through school, community, and municipal recycling promotions. The ACC formed the Curbside Value Partnership to enhance curbside collection, public awareness, and increased household bin use. The Aluminum Association has also established "Cans for Habitat" in conjunction with Habitat for Humanity, and since 1997, some $4 million dollars has been collected and 92 homes have been built for needy families (Ref 7.16).

These activities, along with a strong municipal curbside recycling network established by the paper and fiber industry, have started to drive up the aluminum recycling rates and have resulted in several marketing slogans to encourage the individual collector. Some examples are that "recycling each beverage can is equivalent to saving the energy content of 6 oz of gasoline," or that "recycling a can saves enough electricity to run an average television set for 3 hours," or "recycling buys a nail for a Habitat for Humanity house." All of this has encouraged and simplified the rationale for recycling.

When collected by industry, the cans are shredded, remelted, cast, and rolled again to form new rigid-container sheet from which new cans are made. These new cans are back on the supermarket shelves in as little as 60 days. Over 95% of recycled UBCs go back into can sheet, forming a closed-loop recycled material product. Cans in the United States currently contain an average of 49% recycled material

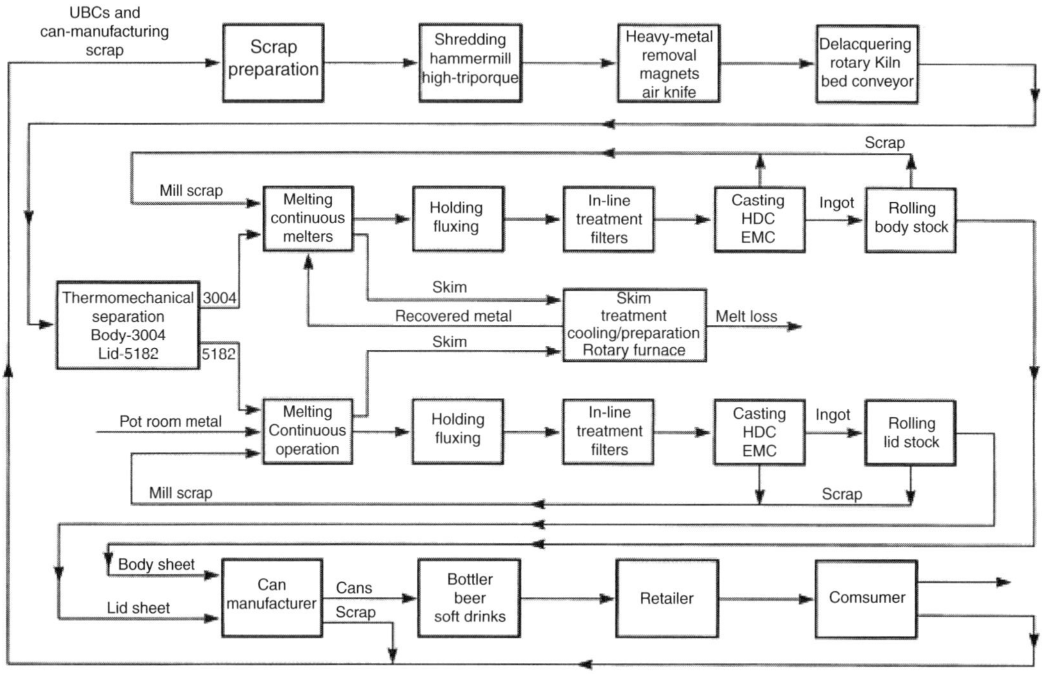

Fig. 7.12 Flow diagram of closed-loop used beverage can (UBC) recycling and manufacturing. HDC, horizontal direct chill; EMC, electromagnetic casting. Source: Ref 7.15

from this closed-loop process. Other recycled material products, such as paper coffee cups and polyethylene terephthalate soda bottles, boast approximately 10% recycled material. Alternatively, the recycled material can be rolled into other alloys to form additional building blocks of material for use in transportation, packaging, building and construction, or other applications.

Collection. The UBCs are received from collection centers as bales weighing 400 kg (880 lb) or as briquettes with a maximum density of 500 kg/m^3 (31 lb/ft^3). These bales and briquettes are broken apart and the cans shredded to ensure that no trapped liquid or extraneous material will reach the melters and cause serious damage or injuries. From the shredder, the material passes via a magnetic separator that removes ferrous contaminants and through an air knife that separates more dense nonferrous and nonmagnetic ferrous materials, such as lead, zinc, stainless steel, and high-nickel iron. The shredded aluminum cans are then typically delacquered.

Delacquering. There are two basic approaches to thermal delacquering. One is based on a relatively long exposure time at a safe temperature, and the other is based on staged temperature increases to just below melting for as short an exposure time as possible. The first approach uses a pan conveyor on which a bed of precrushed and shredded UBCs approximately 200 mm (8 in.) deep moves through a chamber held at approximately 520 °C (970 °F). The chamber contains products of combustion (POC) that are diluted with air to provide the proper atmosphere and temperature for the pyrolysis and combustion. The second approach employs a rotary kiln with a sophisticated recirculating system for POC gases at various entry points. The temperature in the last stage is near 615 °C (1140 °F), which is very close to the temperature at which incipient melting occurs in the aluminum-magnesium (5*xxx*-series) alloys typically used in can lids and tabs.

Precise practices and controls are vital for these delacquering units, which treat approximately 18 metric tons (20 tons) of scrap (~1.25 million UBCs) per hour.

Alloy Separation. Hot, delacquered UBCs then move into the thermomechanical separation chamber, which is held at a specific temperature and contains a nonoxidizing atmosphere. In this chamber, a gentle mechanical action breaks up the alloy 5182 lids into small fragments along grain boundaries, which have been weakened by the onset of incipient melting. An integrated screening action removes the fragments as soon as they can pass the screen, to avoid overfragmentation. This process requires a very narrow operating range to avoid melting and assure maximum segregation. The screened-out alloy 5182 particles are transported to the lid stock melters, and the larger alloy 3004 particles continue directly into the body stock melters.

Melting, Preparation, and Casting. At present, most melting facilities for UBCs throughout the industry are dedicated units designed to handle the enormous volumes and to minimize the melt losses inherent in melting thin-walled material. Larger companies have developed their own often-proprietary processes. Significant amounts of skim—the mixture of metal, oxides, other contaminants, and trapped gas that floats on top of the melt—are removed and treated for metal recovery. The metal from these dedicated melters is often transferred to on-line melting furnaces, where additional bulky scrap is remelted and primary unalloyed metal is charged to create the desired volume of the proper alloy composition. From these furnaces, the metal is transferred to the holding furnaces, where minor composition adjustments are made and metal quality treatments are performed (for example, gas fluxing to remove hydrogen). Some metal treatment, for example, inclusion removal, can be done in so-called in-line treatment units; again, most major companies have developed their own preferred methods and technology. The clean and on-composition metal is cast into essentially rectangular cross-sectional ingots that are typically scalped to remove nonuniform surface and subsurface structure and thermomechanically processed to finished sheet dimensions. Process scrap in the form of scalpings, edge and end trim, and discrepant product is in-house, or runaround, scrap, which is returned to the melters.

Sheet is shipped to can manufacturers, who will generate approximately 20% skeleton and other forms of scrap that are most typically returned to the aluminum supplier. On a global basis, 55% of a melt consists of new (production-related) scrap. If all cans were returned as UBCs and total melt losses were 7% of the melt, this 7% would be the only makeup metal required from primary smelters to close the loop (provided the market remained constant). Figure 7.13 illustrates this interrelationship.

As the recycling rate continues to increase, composition control and the corresponding contamination avoidance become technical challenges as important as melt loss reduction. These developments were predicted in a mathematical model of the recycling system (Ref 7.17). It appears that prudent use of salt fluxes may hold the key to improvements in these areas.

Automobile Scrap Recycling Technology

In contrast with the modern aspects of closed-loop can recycling, the recycling of aluminum from automobiles has existed for more than a century. It has traditionally been a multifaceted scavenging activity carried out by independent scrap yards. Materials selection by car manufacturers has been based on cost and performance, with little consideration given to the economic or environmental value of recyclability.

Transportation is currently the largest (33.9% in 2004) and fastest-growing market of the aluminum industry (Ref 7.2) and achieves greater than 90% recovery of recycled material back to the industry. Market growth rates for aluminum in transportation were 9.8% over the decade 1990 to 2000 and 6.1% from 2001 to 2004. This growth is driven by the need for light-weighting, increased fuel efficiency, and safety. It has been estimated that the replacement of heavier components in vehicles by aluminum saves ~16 kWh annually for each kilogram of aluminum used. Incidentally, this number implies that, on a life-cycle basis, all the energy required to produce primary aluminum is recovered in vehicle fuel savings within three years, and in less than three months when recycled metal is used. This large light-weighting advantage is expected to continue to drive the high growth of aluminum in the transportation sector. In essence, this growing quantity of metal is a very large and growing mobile "urban mine" of metal available for recycling (Ref 7.3).

The time and rate at which the urban mine material becomes available for recycling is a function of the product and can be modeled, as previously noted in the section "Developing Scrap Streams" in this chapter. The useful lifetime can be estimated and used to determine when the material becomes available as old or postconsumer scrap. In this way, an urban mine metal flow model has been constructed to estimate the availability of future scrap. The model considers the number of vehicles produced, the amount of aluminum per vehicle, vehicle life, growth rate of aluminum use, impact of dismantlers, the role of automotive shredders, and so on.

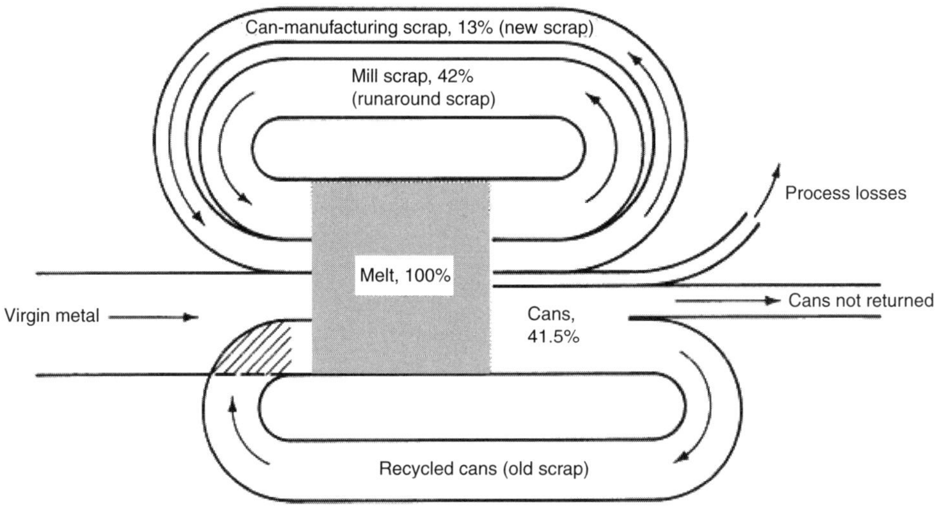

Fig. 7.13 Process by which recycled cans replace virgin metal in the beverage container market

As mentioned before, the amount of aluminum in cars has steadily increased since 1946 (Fig. 7.11). Substantial amounts of aluminum had been applied in early car manufacturing, but between 1925 and 1946, its use was minimal. Until 1975, most growth was in castings, but as a consequence of the energy crisis, more wrought alloy hang-on parts began to be used for weight reduction, and sheet metal panels and space frame constructions are now under development. An average 1980-model car contains approximately 70 kg (155 lb) of aluminum, which accounts for only 5% of the total weight of the average car. Yet, it is economically an important fraction.

Figure 7.14 is a schematic showing one of several possible paths for the recovery of the majority of the materials presently used in cars. The aluminum fraction is usually sold to an automotive cast shop, which uses open hearth as well as induction furnaces and occasionally rotary salt kilns. The fraction of inseparable multimetallic particles must be treated in a sweat furnace. In some schemes, without heavy-medium or eddy current separation, the entire nonferrous fraction goes to the sweat melter.

The can recycling loop is totally dedicated to a single product of two compatible aluminum alloys. Automotive recyclers, on the other hand, must deal with a number of fractions with different destinations and relatively low values. For automobile recycling to become as effective as can recycling, a cooperative effort is required by the scrap collectors, handlers, and manipulators to introduce advanced scrap separation and upgrading technology. A number of alternative scrap recovery methods have been developed for other mixed-aluminum scrap sources, and these methods could be adapted for use in car recycling. They include improved preprocessing methods to concentrate aluminum fractions and mechanical, physical, and chemical separation processes for upgrading scrap mixtures and recovering aluminum from low-grade sources.

One major factor that has strongly impacted vehicle recycling is the advent of automotive shredders for the industrialized collection of automotive scrap. Nowadays, most vehicles are first processed by dismantlers and scrap processors, and then the residual hulk is shredded. After shredding, followed by air separation to remove fluff, magnetic separation of ferrous material, sink-float density separation, and eddy current separation, it is possible to obtain a nonferrous metal stream containing some 90 to 95% of the available metal. In other words, shredders have industrialized and automated the scrap collection process and made it more efficient. Two additional developments by Huron Valley Steel Corporation (Ref 7.18, 7.19) have further advanced this technology. By use of color sorting, it is now possible to sort cast aluminum alloys from wrought material, and by using laser-induced breakdown spectroscopy (LIBS), it is possible to analyze the emitted plasma light from a sample and determine its chemical composition. This analysis of individual pieces of scrap has now

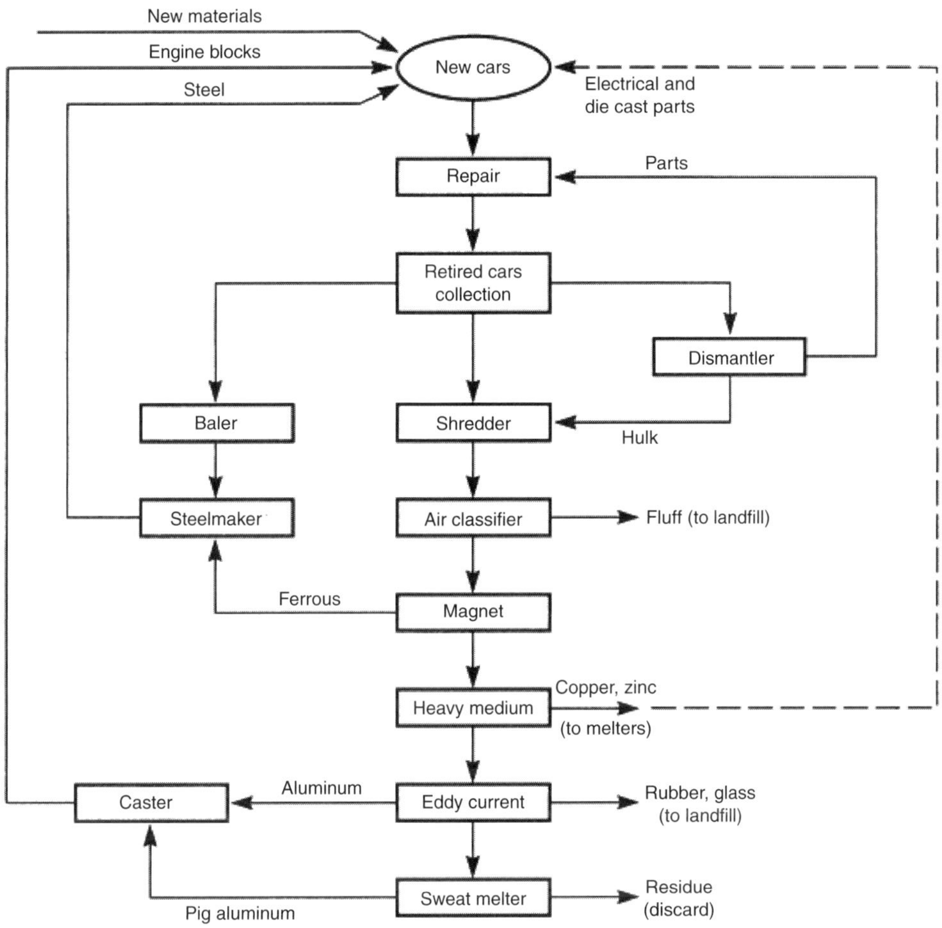

Fig. 7.14 Recycling loop for aluminum automotive components. Castings make up the bulk of aluminum automotive scrap

been demonstrated at a production scale and is shown schematically in Fig. 7.15. Although not yet economically feasible, it is technically possible to separate and sort aluminum on an alloy-by-alloy basis by virtue of these developments.

High-Temperature Process. Besides the delacquering processes for can scrap described earlier, other high-temperature processes have been considered for separating aluminum scrap. However, these high-temperature methods were not effective or cost too much energy. Methods included a fluid-bed rotary furnace to remove paint, plastic, and other combustibles from aluminum (Ref 7.20). Another was the so-called Granges box used in Sweden, which is an oven consisting of two chambers, one for containing scrap, the other for combustion of gases and fumes released from the scrap. These high-temperature batch processes are no longer used.

Other Separation Technologies. Some other separation technologies that have been used in Europe include low-temperature separation and gravity separation methods.

Cryogenic separation has been performed commercially in Belgium on shredded automotive scrap. The method is based on the difference in ductility of nonferrous and ferrous metals at extremely low temperatures. Below –65 °C (–85 °F), ferrous metals become very brittle and can be fragmentized easily, whereas nonferrous metals remain ductile. Simple screening achieves separation. This method is expensive and should be used only for separation of mixed shredder fragments (of steel attached to aluminum, for example).

Gravity seperation methods include the common heavy-media/sink-float process mentioned earlier and a process developed in the Netherlands (Ref 7.21). The latter uses the same heavy medium (ferrite/water suspension),

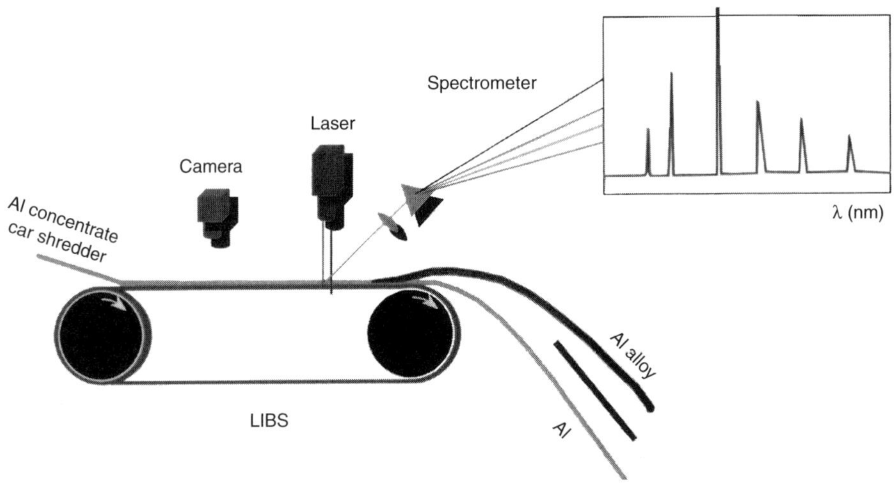

Fig. 7.15 Schematic of scrap-sorting laser-induced breakdown spectroscopy (LIBS)

but it is not a passive sink-float method. Instead, the scrap is charged in a cyclone through which the heavy medium is pumped. In this manner, the sensitivity of the process is increased, and inaccuracies due to shape differences of the particles are decreased.

Building and Construction Recycling

The recycling of rolled products from the building and construction area has long been something of an unknown quantity of recycled material, mostly because buildings have a long life cycle (~50 years) and the data are sparse. In an attempt to provide some more specific information for this activity, the European Aluminum Association recently commissioned a study by the University of Delft to provide information on the content and collection rate in several European buildings (Ref 7.22). Buildings in France, Germany, Italy, The Netherlands, Spain, and England were closely monitored, and comprehensive data were gathered. It was found that the collection rates of aluminum in this market sector varied between 92 and 98%, thereby demonstrating enormous potential for recycling and highlighting the role of aluminum in striving for full sustainability. With the enforced reduction of landfill areas and the introduction of increased landfill fees, it is anticipated that the recycling rates of building materials will tend to increase.

A survey of U.S. aluminum producers in late 2003 indicated that the total recycled content of domestically produced, flat rolled products for the building and construction market was approximately 80 to 85%. A subsequent survey of the producers indicated that, on average, approximately half of the recycled content (or 40 to 42% of the total content) is from postconsumer sources (Ref 7.23).

Aluminum Foil Recycling

While aluminum foil is fully recyclable, the amount of foil that has actually been recycled has been somewhat limited. With its thin cross section and high surface area, foil tended to be rapidly oxidized into dross, and product recovery was poor. The recycling of the combined plastic-foil aseptic packaging especially had been an issue. However, in May 2005, a partnership involving Alcoa Alumino, Tetra Pak, Klabin, and TSL Ambiental announced a solution to the recycling of aseptic packaging. Using a high-temperature plasma jet to heat the plastic and aluminum mixture under an inert atmosphere, the plastic is converted to paraffin, and the aluminum is recovered in the form of high-purity ingot, which can then be recycled to manufacture new foil. The emission of pollutants during the recycling of materials is reported to be minimal, and the energy efficiency rate is close to 90%.

Impurity Control

As noted earlier, a dominant issue in the recycling of rolled aluminum products is the control of impurities, especially iron and silicon. Both of these elements tend to increase modestly the more frequently the metal has been recycled. Casting alloys, with a generally much higher silicon content, can markedly increase impurity levels of a wrought alloy recycle stream. Iron impurities generally build up due to wear and tear of process equipment.

These impurities can adversely impact physical, chemical, and mechanical properties, and, as a result, definite alloy compositional limits are set for aluminum alloys. This is especially true in the aerospace market, where there are critical requirements for fatigue, toughness, creep, and stress corrosion, and the impurity levels must be maintained at relatively low levels. High-performance rolled, automotive alloys generally restrict both silicon and iron maximum limits but at levels considerably above those of the aerospace alloys (Ref 7.24). Better control of impurities or improved sorting of recycled materials is necessary for the widespread application of recycled alloys, and the industry is exploring several options. At present, the recycled material is "sweetened" with more pure, primary ingot to attain the required impurity limits for specific alloys.

The better the impurities can be controlled, the more economical the whole recycling process will become. Of course, most recycled wrought alloys tend to fit well within the compositional limits of casting alloys, which typically are significantly higher, especially in regards to silicon. However, this approach does not satisfy the need to recycle wrought alloys back into wrought material. Several approaches for coping with the wrought alloy impurity issues come to mind.

It may be possible to modify the compositions of certain alloys with recycling in mind. Conceivably, these alloys may be used as painted stock where corrosion issues are less critical (soffits, downspouts, etc.) and where mechanical property needs are less demanding. Already, some alloys such as 3105 and 6061 have higher compositional limits for iron (0.7% maximum in each case) (Ref 7.25) and are good alloys for potential scrap additions. Perhaps the composition limits of such alloys can be modified without detriment to their physical or mechanical properties.

Another approach is to encourage the design and assembly community to design with eventual recycling in mind. For example, it is important to minimize the use of dissimilar metals in joining and assembly. Some early automotive designs employed steel rivets or clinching to attach an aluminum roof to the aluminum body-in-white, a design that can be detrimental to the quality of the recycled feedstock after the eventual shredding of the vehicle.

With the successful development of the LIBS technology, it is quite feasible technically, although not yet economical, to conduct alloy-by-alloy sorting. Before this stage is reached, it would be possible to set the computer limits to sort and discard pieces of scrap with excessively high iron content, thereby producing a much purer melt composition subsequently. The high-iron material could then be targeted for the preparation of "de-ox" alloys and sold to the steel industry, thereby eliminating the material from the aluminum process stream.

Recently, the corrosion performance of modern zirconium-refined magnesium alloys has been markedly improved and is now close to that of aluminum alloys. This has been achieved by controlling the level of iron impurities to be < 0.005% in the alloy, and this in turn has improved the corrosion performance of the oxide. Zirconium effectively controls iron levels by forming heavy iron-zirconium intermetallic particles that sink to the bottom of the crucible during processing (Ref 7.26). Can such a similar "density separation" concept be made to work for the removal of iron from aluminum melts?

Another option may be to "learn to live with the impurities" by modifying production practices. For example, with partial support from the Department of Energy, a multipartner team involving the University of California at Irvine/Davis, Colorado School of Mines, Idaho National Laboratory, Alcoa Technical Center, Pechiney Rolled Products (now part of Alcan), Inductotherm Corp., and Metals Technology, Inc., is currently developing a spray rolling aluminum strip process (Ref 7.27). This spray rolling process is an innovative strip-casting technology that combines the benefits of twin roll casting with conventional spray forming (Fig 7.16). The project description says:

> "Compared to conventional ingot processing, spray rolling significantly reduces energy consumption and cost by eliminating direct chill casting, homogenization, and hot rolling unit

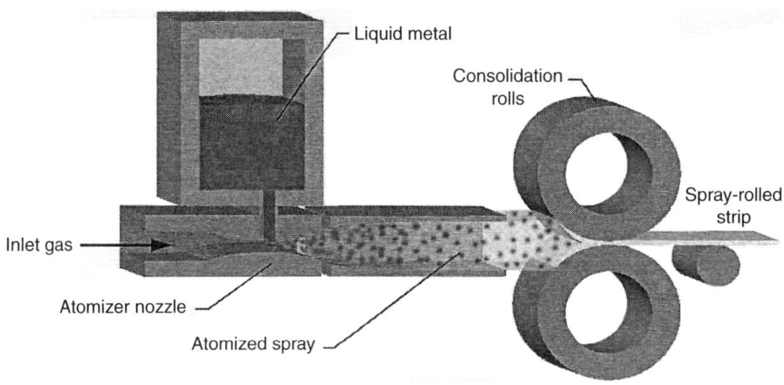

Fig. 7.16 Schematic of spray rolling process

operations. Compared to twin-roll casting, spray rolling significantly increases production rate and is applicable to a broader range of alloys. Tensile properties of strip processed by spray rolling have met or exceeded those of ingot-processed material."

Currently, the process has been demonstrated at a strip width of 20 cm (8 in.) and is to be scaled up to 51 cm (20 in.) in the next stage. More significantly, the project description goes on to say:

> "The resultant improvements in fuel efficiency and *anticipated improvements in recyclability and tramp element tolerance of spray-rolled sheet* would result in even greater energy savings."

In general, rapid-solidified materials such as those generated by spray forming have been shown to tolerate higher-impurity contents because the tramp elements are frozen in the structure and do not have as much opportunity to segregate to the grain-boundary regions, where they can form intermetallic precipitates and cause localized corrosion or embrittlement sites. It will be important to demonstrate whether this speculation about spray rolling having greater tolerance for impurities is in fact true.

Molten Metal Handling and Safety*

The major safety concerns specific to the aluminum industry involve molten metal and pot bath and finely divided aluminum particles, which, under certain circumstances, can give rise to explosions, with serious consequences. The hazards of producing and handling these materials are well recognized, and the industry has undertaken a number of programs and data collection efforts to better understand and control the risks of burns, explosions, injuries, and particular hazards, depending on the processes and/or products involved.

Molten Metal Explosions

The single greatest safety concern in the aluminum industry is a molten metal explosion. While millions of pounds of molten aluminum are produced and handled without incident every day, explosions do infrequently occur. Most are the result of the molten metal trapping and reacting with water, but other contaminants can also bring about catastrophic incidents.

In general, whenever two liquids at widely different temperatures are brought into intimate contact, an explosion can result, particularly if the colder liquid has a relatively low boiling point. Examples of these potential explosions can be found in many different industries. The paper industry has long had concerns about smelt-water reactions. The nuclear power industry has had serious concerns and sponsored a considerable amount of research about steam explosions that could arise from meltdown of core and/or cladding reacting with cooling water in a nuclear reactor. And, of course, the metals industry is concerned about reactions between molten metal and water.

In most instances of explosions involving molten metals and water, the water is trapped

*Adapted from the paper "World-Wide Efforts to Prevent Explosions in the Aluminum Industry" by Seymour G. Epstein, Technical Consultant, The Aluminum Association

between the molten metal and a solid surface. It quickly turns to steam and, in so doing, expands more than a thousand times. This rapid expansion, accompanied by a rise in temperature, is an explosion that can throw metal a large distance, injure employees, and damage equipment.

With aluminum, there is an additional factor to be recognized. Aluminum is a very reactive chemical element with a great affinity for oxygen, with which it is almost always combined in nature. Just as it requires a large amount of energy to break the aluminum-oxygen bonds and produce metallic aluminum in a reduction cell, that energy will be released if the aluminum is able to recombine with the oxygen from either water or air. From thermodynamic calculations, it has been estimated that the energy released from 1 lb of aluminum fully reacting with oxygen is equivalent to detonating 3 lb of trinitrotoluene (TNT).

Research on Causes and Prevention of Explosions. For more than 50 years, the industry has studied molten aluminum-water explosions relevant to what may happen if a bleed-out of molten metal were to occur during the direct chill (DC) casting process. These studies have been conducted at company laboratories (Ref 7.28–7.30), sponsored by the industry at independent research laboratories (Ref 7.31), and conducted at government laboratories with industry cooperation (Ref 7.32) to gain an understanding of these phenomena, how they are initiated, and how they may be prevented. The results of these studies have been published in the open literature and presented at workshops and international conferences (Ref 7.33).

These early studies revealed that certain organic coatings applied to the exposed surfaces in a DC casting pit will prevent explosions resulting from bleed-outs. Recently, a series of studies were undertaken to gain a better understanding of how the coatings afforded that protection and to identify new coatings to replace the long-used Tarset Standard, whose production had been discontinued (Ref 7.34–7.36). The studies were conducted at the Alcoa Technical Center, with a supporting effort at the Oak Ridge National Laboratory, sponsored by an international consortium arranged through the Aluminum Association and including aluminum producers, coating suppliers and an equipment manufacturer.

Based on the results of these efforts, safety procedures were developed and implemented, guidelines and training aids were prepared and distributed, and a number of workshops and conferences were held in this country and abroad to provide pertinent information to those in need of it. The efforts are continuing, and, while explosions have not been eliminated, the industry has made considerable progress in better protecting its employees.

Molten Metal Incident Reporting

For more than 20 years, the Aluminum Association has conducted an industry-wide molten metal incident reporting program to glean as much information as possible from events. Currently, the program has 230 participants reporting for approximately 300 plants located in 20 countries (Ref 7.37).

Scrap-Charging Safety. One important finding emerged from the incident reporting program. Many of the most severe explosions resulted from charging scrap into a remelt furnace, no doubt a reflection of the greatly increased worldwide recycling of aluminum since the program was initiated. In particular, contamination of the charge was suspected in many of the events. As a result, the Association began working closely with the Institute of Scrap Recycling Industries (ISRI) in the United States to increase awareness of the potential hazards of scrap contamination (Ref 7.38). Several workshops were conducted for the ISRI membership, guidelines for scrap receiving and inspection were prepared and distributed along with other training aids (Ref 7.39), a scrap rejection notification program was implemented (Ref 7.40), and information on scrap melting safety continues to be disseminated (Ref 7.41).

Sow Casting and Charging. The reporting program also identified sow charging as the second leading cause of furnace explosions. A sow is an ingot weighing from 700 to more than 2000 lb that has been cast in a steel mold. Often, solidification of the ingot leaves a shrinkage cavity below the surface. If water enters the cavity during storage and is not completely driven off by preheating and/or sectioning, an explosion can occur when the sow is charged into the furnace heel of molten metal. Reports were also received of sows exploding on a dry furnace hearth or during preheating, because the trapped moisture could not conveniently escape. Needless to say, the violence is greatest when the sow explodes under molten metal.

The Association sponsored efforts at the Kaiser Center for Technology and the Alcoa Technical

Center to develop and validate a computer model for metal solidification in a sow mold to hopefully lead to improved sow design (Ref 7.42, 7.43). Guidelines for sow casting and charging, based on these efforts and company practices, were published by the Association in 1998 (Ref 7.44).

Guidelines and Training Aids

Based on the research findings and on company experiences, comprehensive guidelines for handling molten aluminum were first published and distributed by the Aluminum Association in 1980. They were updated, and an expanded second edition was published by the Association in 1990. Through the efforts of an editorial board with considerable industry experience and expertise, the third edition of the guidelines was published in July 2002 by the Association (Ref 7.45).

Awareness of potential hazards is a key objective. The Association publishes and distributes information, guidelines, and audiovisual training aids. Presentations and short courses on safe handling of molten aluminum are arranged with professional societies, such as TMS, and workshops are conducted at members' plants. Casthouse safety workshops are held at regular intervals in the United States and have also been conducted in Europe and Asia as a joint activity with the European Aluminum Association and the International Aluminum Institute. In the United States, these are held regionally for the convenience of the plants located in those areas. In addition, potroom personal protective equipment workshops, sponsored by the Association's Primary Division, were held at members' plants.

Finely Divided Aluminum

Because of its chemical activity, affinity for oxygen, and the large amount of surface area available for reaction, finely divided aluminum represents a hazard that must be recognized. As with other reactive materials, such as flour, starch, and coal dust, fine particles of aluminum dispersed in the air as dust clouds can ignite or explode when triggered by a spark or other energy source (Ref 7.46).

Finely divided aluminum is encountered primarily in one of several forms: powder atomized from molten metal; flake produced from ball-milling operations and made into a paste for paints; and fines generated by various fabricating or finishing operations such as grinding, sawing, cutting, sanding, brushing, and so on. Generally, aluminum powders with a particle size of 420 μm (40 mesh) or smaller present a fire and explosion hazard; the finer the particles, the greater the explosion hazard.

Approximately 25 years ago, the Safety and Property Protection Committee of the Aluminum Association's Pigments and Powder Division developed *Recommendations for Storage and Handling of Aluminum Powders and Paste* (Ref 7.47). Now in its fourth edition, the publication is continually updated and made available to producers and users. The Committee was also involved in the preparation of *Guidelines for Handling Aluminum Fines Generated during Various Aluminum Finishing Operations* (Ref 7.48) and has representation on appropriate panels of the National Fire Protection Association that publishes standards for handling these products. In addition, workshops are held to help exchange information on programs and experiences.

REFERENCES

7.1. R. Overbey, "Sustainability—What More Should Companies Do?" presentation to Alcoa Conference Board, June 2005

7.2. *Aluminum Statistical Review, 2004,* The Aluminum Association, Arlington, VA, 2005, p 5

7.3. W.T. Choate and J.A.S. Green, "Modeling the Impact of Secondary Recovery (Recycling) on the U.S. Aluminum Supply and Nominal Energy Requirements," paper presented at TMS 2004 (Charlotte, NC), 2004

7.4. E.L. Rooy and J.H.L. Van Linden, Recycling of Aluminum, *Properties and Selection: Non Ferrous Alloys and Special-Purpose Materials, Vol 2, ASM Handbook,* ASM International, 1990

7.5. "The Aluminium Industry's Sustainability Development Report," International Aluminum Insitute, 2003

7.6. "Aluminum for Future Operations, Sustainablity Update 2006," International Aluminum Institute, 2006.

7.7. P.R. Atkins, Recycling Can Cut Energy Demand Drastically, *Eng. Min. J.,* May 1973

7.8. "Scrap Specifications Circular," Institute of Scrap Recycling Industries, Inc., 1988

7.9. J.H.L. Van Linden, J.R. Herrick, and M.J. Kinosz, Metal Scrap Melting System, U.S. Patent 3,997,366, 1976

7.10. J.H.L. Van Linden, R.J. Claxton, J.R. Herrick, and R.J. Ormesher, Aluminum Scrap Reclamation, U.S. Patent 4,128,415, 1978
7.11. J.H.L. Van Linden and J.B. Gross, Vortex Melting System, U.S. Patent 4,286,985, 1981
7.12. A.G. Szekely, Vortex Reactor and Method for Adding Solids to Molten Metal Therewith, U.S. Patent 4,298,377, 1981
7.13. D.V. Neff, in *Proceedings of TMS Fall Meeting,* AIME, 1985, p 57–72
7.14. R.J. Claxton, Method for Submerging, Entrainment, Melting and Circulating Metal Charge in Molten Media, U.S. Patent 4,322,245, 1982
7.15. J.H.L. Van Linden, Aluminum Recycling—Everybody's Business, *Light Metals 1990,* TMS, 1990
7.16. "2004 Sustainability Report," Alcoa, p 26, www.alcoa.com/global/en/about_alcoa/sustainability_report_2004/images/sr_online_final.pdf (accessed July 2007)
7.17. J.H.L. Van Linden and R.E. Hannula, in *Light Metals 1981,* AIME, p 813–825
7.18. A. Gesing et al., "Separation of Wrought Fraction of Aluminum Recovered from Automobile Shredder Scrap," paper presented at TMS meeting (Pittsburgh, PA), Nov 2000
7.19. A. Gesing, L. Berry, R. Dalton, and R. Wolanski, "Assuring Continued Recyclability of Automotive Aluminum Alloys," presentation at TMS Meeting (Seattle WA), Feb 2002
7.20. J. Butson, A Market Study for the Energy Efficiency Office, *Alum. Recycl.,* 1986
7.21. "Observations at Stamicarbon Pilot Plant at Dalmeyer's Salvage Yard," Nieuwerkerk a/d Yssel, The Netherlands
7.22. "Collection of Aluminum from Buildings in Europe," Delft University of Technology, European Aluminum Association, Brussels; www.eaa.net (accessed July 2007)
7.23. "LEED Fact Sheet—Aluminum Sheet and Plate for the Building and Construction Market," The Aluminum Association, 2004
7.24. G. Kaufman, private communication to J. Green, Dec 2005
7.25. "Aluminum Standards and Data 2003," Section 6–5, Table 6.2, The Aluminum Association
7.26. N. Jeal, High–Performance Magnesium, *Adv. Mater. Process.*, Sept 2005, p 67
7.27. K. McHugh and E. Lavernia, "Spray Rolling Aluminum Strip," presentation to Dept. of Energy, Industries of the Future Review, Oak Ridge National Laboratory, Oct 2005
7.28. G. Long, Explosions of Molten Aluminum and Water—Cause and Prevention, *Met. Prog.*, Vol 71, 1957
7.29. P.D. Hess and K.J. Brondyke, Causes of Molten Aluminum—Water Explosions and Their Prevention, *Met. Prog.*, Vol 95, 1969
7.30. P.D. Hess et al., Molten Aluminum/Water Explosions, *Light Metals 1980,* TMS, 1980
7.31. A.W. Lemmon, Explosions of Molten Aluminum and Water, *Light Metals 1980,* TMS, 1980
7.32. L.S. Nelson, M.J. Eatough, and K.P. Guay, Why Does Molten Aluminum Explode at Underwater or Wet Surfaces?, *Light Metals 1989,* TMS, 1989
7.33. S.G. Epstein, Causes and Prevention of Molten Aluminum-Water Explosions, *Light Metals 1991,* TMS, 1991
7.34. R.T. Richter, D.D. Leon, and T.L. Levendusky, Investigation of Coatings Which Prevent Molten Aluminum/Water Explosions—Progress Report, *Light Metals 1997,* TMS, 1997
7.35. R.P. Taleyarkhan, V. Georgevich, and L. Nelson, Fundamental Experimentation and Theoretical Modeling for Prevention of Molten Metal Explosions in Casting Pits, *Light Met. Age,* June 1997
7.36. D.D. Leon, R.T. Richter, and T.L. Levendusky, Investigation of Coatings Which Prevent Molten Aluminum/Water Explosions, *Light Metals 2001,* TMS, 2001
7.37. S.G. Epstein, A Summary of Findings from Twenty Years of Molten Metal Incident Reporting, *Light Metals 2005,* TMS, 2005
7.38. S.G. Epstein, Recycling Aluminum Safely, *Scrap*, Vol 57, 2000
7.39. *Guidelines for Aluminum Scrap Receiving and Inspection Based on Safety and Health Considerations,* 2nd ed., Aluminum Association Publication GSR, 2002
7.40. D.C. Pierce and C.H. Kenney, Promoting Awareness of Potential Safety, Hazards in Aluminum Scrap, *Light Metals 1997,* TMS, 1997
7.41. M.D. Bertram, F.R. Hubbard, and D.C. Pierce, Scrap Melting Safety—Improving,

but Not Enough, *Light Metals 2005, TMS,* 2005

7.42. J.R. Payne, "*Sow Mold Thermal Analysis,*" final report to The Aluminum Association, 1988

7.43. A. Giron and J.E. Jacoby, "*Experimental Verification of the Aluminum Association's Thermal Analysis Model,*" final report to The Aluminum Association, 1992

7.44. "*Guidelines for Aluminum Sow Casting and Charging,*" Aluminum Association Report GSC, 1998

7.45. *Guidelines for Handling Molten Aluminum,* 3rd ed., Aluminum Association Publication 69, July 2002

7.46. D.D. Leon, Casthouse Safety—A Focus on Dust, *Light Metals 2005,* TMS, 2005

7.47. *Recommendations for Storage and Handling of Aluminum Powders and Paste,* 4th ed., Aluminum Association Publication TR-2, 2006

7.48. *Guidelines for Handling Aluminum Fines Generated during Various Aluminum Fabricating Operations,* Aluminum Association Publication F-1, 2000

SELECTED REFERENCE

- M.E. Schlesinger, *Aluminum Recycling,* Taylor & Francis, 2006

CHAPTER 8

Identification and Sorting of Wrought Aluminum Alloys*

THE IMPORTANCE OF ALUMINUM as a vehicle construction material has been increasing quickly over the last decade. The trend of increasing aluminum content in automobiles is well established and will have a dramatic impact on the recycling industry. On average, aluminum already constitutes ~10% of a vehicle weight and over 50% of the value of the material available for recovery from 0the end-of-life vehicle (Ref 8.1).

Although automotive aluminum is currently completely recycled, recycling technologies need to be improved to assure continued, high-value recyclability of all present and future automotive alloys—particularly wrought alloys, because the use of these alloys increases in the aluminum-intensive vehicle. For this reason, the recycling industry is working diligently to provide technical solutions of new scrap-sorting technologies to deliver high-value recyclability of aluminum scrap from end-of-life vehicles. While dismantling is economically justifiable for part rebuilding and reuse, material recovery is enabled by particle liberation via shredding and bulk separation and sorting techniques. The recycling process for recovery of aluminum from auto shred and its separation into wrought and cast products (Ref 8.2–8.4) has already broadened the market for aluminum recycled from shredder metal concentrate, in particular, options to improve fractions of recovered wrought aluminum alloys, because sheet and extrusion alloys are increasingly used as structural and body components.

This chapter briefly describes a development effort by Huron Valley Steel Corporation for a system to culminate in the sorting of the wrought aluminum alloy concentrate from aluminum-intensive vehicle hulks supplied by auto manufacturers. This development effort is intended to include decoating and precleaning followed by separation of 5182/5754/6111/6061/6082 from an aluminum alloy mix. Three identification/sorting techniques are discussed for further sorting of the mixed wrought aluminum alloys to enable recycling of these scrap alloys into wrought alloy products:

- Color sorting of particles tinted by etching in an NaOH solution
- X-ray absorption imaging
- Chemical-composition-based alloy sorting

These techniques are suitable for use in wrought alloys that are typically batched from prime metal. Solid-particle chemical-composition-based sorting, although technically challenging, is the method of choice for further sorting of the mixed wrought aluminum alloys. This development work has been done as part of Huron Valley Steel Corporation's participation in an Automotive Aluminum Alliance program designed to demonstrate the ability of new scrap-sorting technologies to deliver high-value recyclability of aluminum scrap from end-of-life vehicles.

*Adapted with permission of the Minerals, Metals and Materials Society (TMS) from the article by A. Gesing, L. Berry, R. Dalton, and R. Wolanski, "Assuring Continued Recyclablity of Automotive Aluminum Alloys: Grouping of Wrought Alloys by Color, X-Ray Absorption and Chemical Composition-Based Sorting," *TMS 2002 Annual Meeting: Automotive Alloys and Aluminum Sheet and Plate Rolling and Finishing Technology Symposia*, Feb 18–21, 2002 (Seattle, WA), Minerals, Metals and Materials Society (TMS), 2002, p 3–17.

Sources of Aluminum Raw Material for Alloy Sorting

Wrought aluminum scrap can be separated from several sources of scrap material, such as:

- Shredder product fractions
- Old dealer scrap
- Manufacturing prompt scrap

Shredder fractions include nonmagnetic shredder fraction (NMSF) and aluminum shredder fraction (Al-SF). North American NMSF supply contains approximately 50,000 metric tons/month (~100 million lb/month) of metal at a concentration of 20 to 95%, of which 60 to 70% is aluminum, with 30 to 50% of the aluminum-containing portion comprised of wrought aluminum alloys (9000 to 11,000 metric tons/month, or 20 to 25 million lb/month). North American Al-SF supply (approximately 2000 metric, or 5 million lb, per month) is growing quickly because shredders are installing eddy current separators to separate a portion of aluminum directly from shred. The Al-SF product typically contains >95% metal, >95% Al in this metal, and 40 to 60% wrought alloys in the aluminum (900 to 1400 metric tons, or 2 to 3 million lb, per month).

The categories of the dealer old scrap that contain mixed wrought alloys and thus are suitable for sorting include old sheet and old extrusion. Manufacturing prompt scrap is often commingled for practical reasons and thus becomes a suitable feed for an alloy sorter. Commercially traded categories of wrought aluminum mixed-alloy manufacturing scrap that are suitable as an aluminum alloy sorter feed include mixed stamping scrap, mixed clip, and mixed low-copper clip.

Upgrading automotive scrap to recover and sort aluminum for material recovery involves shredding, followed by magnetic separation of iron and metal-nonmetal separation based on electrical conductivity. The aluminum is then sink-float separated from denser metals. In 2001, a proprietary sorting technology for separation of wrought aluminum shred from the cast aluminum shred was implemented industrially as commercial state-of-the-art automated aluminum sorting (Ref 8.3, 8.4). Typical hand-sort classification of wrought aluminum products is given in Table 8.1. Figure 8.1 shows these products, and the elemental composition of the aluminum melt recovered from these products is given in Table 8.2. These products are now finding application in use as steel deoxidants, 3105 secondary sheet, and in an expanded range of foundry alloys. Because these products are now typically made from segregated manufacturing scrap, the value added to the mixed-alloy aluminum scrap product by wrought-cast sorting is modest. A significantly higher value would be available when the scrap product composition will allow it to replace prime metal in the new alloy batch. In order to achieve that, further upgrading is necessary.

Improving Recovery for Wrought and Cast Fractions

Various alternatives or additional technologies can be considered for further upgrading aluminum recovered from postconsumer scrap. Focusing in particular on the quickly increasing volume of aluminum in auto shred, potential methods to further upgrade wrought and cast aluminum fractions fall into two categories: melt processing and solid-particle sorting. Technically feasible approaches that have been tried are reviewed. This is followed by descriptions of three processes being piloted for commercial development of improvements in recovering from wrought scrap.

Melt Processing

Melt processing routes include fractional crystallization, electrorefining, fluxing, and aluminum chloride chemistry. Fractional crystallization is a commercial process for precipitation of 4N-purity aluminum by controlled cooling of molten prime aluminum. When applied to scrap-based melt, controlled cooling results in precipitation and segregation of complex intermetallic precipitates containing iron, manganese,

Table 8.1 Hand-sort classification of the products of separation of wrought and cast aluminum

Sort of aluminum mix from heavy-media float (HMF)					
			Sort of wrought aluminum concentrate		
	Al mix from HMF, %	Cast Al, %	Wrought Al conc., %	Wrought Al-1, %	Wrought Al-2, %
Al wrought	40.11	10.16	69.74	91.02	95.20
Al cast	54.12	85.67	26.03	3.47	3.31
Al other	2.82	1.50	2.43	0.47	0.82
Other metal	2.34	1.90	1.50	4.72	0.48
Nonmetal	0.61	0.78	0.31	0.33	0.19
Al + Fe	4.86	7.74	2.56	0.54	0.96
Al painted	4.50	3.28	8.44	11.08	14.24

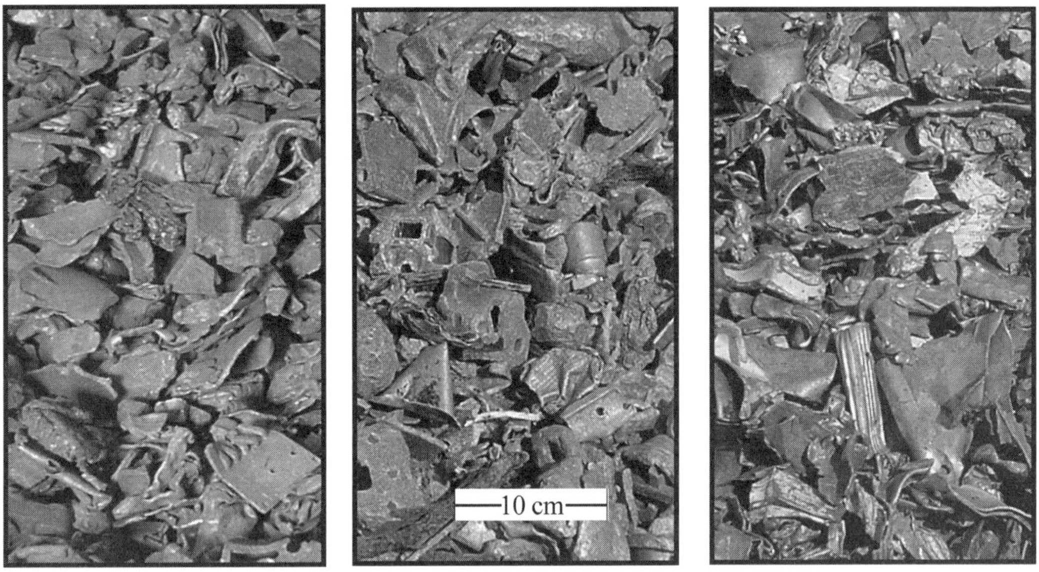

Fig. 8.1 Shredded aluminum products recovered from nonmagnetic shredder fraction. (a) Wrought and cast aluminum mix. (b) Cast aluminum product. (c) Wrought Al-2 product

Table 8.2 Optical emission spectroscopic elemental composition of melts of recovered aluminum products

Element	Al mix from HMF(a),%	Cast Al,%	Wrought Al conc.,%	Wrought Al-2,%
Al	88	83.5	93.1	97.1
Cu	3	4.4	1.2	0.1
Fe	1	1.1	0.7	0.6
Mg	0.6	0.4	0.7	0.8
Mn	0.3	0.3	0.3	0.2
Si	5	8.0	2.6	0.5
Zn	2	1.9	1.2	0.4
Other	...	0.4	0.2	0.2

(a) HMF, heavy-media float

chromium, and titanium in addition to aluminum. The multicomponent phase diagram places limits on achievable upgraded product composition. A considerable fraction of aluminum ends up with the intermetallics, a downgraded residue. Tight control of thermal gradients and convection currents in the melt is critical to obtaining a product composition approaching that predicted by theory, and recovery is invariably significantly lower than expected due to trapping of aluminum melt among the intermetallic crystals.

Electrorefining is a commercial process for production of 5N-purity aluminum from prime aluminum in three-layer electrolytic cells. Both anode and cathode are molten pools of metal, a dense anode of AlCu melt and a low-density cathode of superpurity aluminum product. Because both electrodes are molten, they need to be separated by a layer of electrolyte that is significantly thicker than that in the Hall-Heroult primary aluminum cells. Electrolysis transfers aluminum from the anode to the cathode, and, because the entire product is electrolyzed, energy consumption in electrorefining is higher than that required for prime production. Although it is technically feasible to refine scrap in the three-layer refining cells, scrap electrorefining can never compete economically with prime production.

Melt fluxing refers to the removal of sodium, calcium, lithium, and magnesium by direct chlorination or by exchange reaction with Al^{3+} containing salt flux. Magnesium and lithium alloying elements are converted to chlorides and end up with the dross. Other contaminants and alloying elements are less reactive to chlorine than aluminum and thus cannot be removed from the melt by fluxing. This is the standard commercial way of controlling the levels of alkali and alkaline earth contaminants in the secondary alloy melts. However, the use of the toxic Cl_2 gas flux can be dangerous to the workers and harmful to the environment. Although chlorination is a relatively low-cost commercial procedure, the environmental considerations will favor routes that reduce the need for fluxing in poorly controlled conditions of a melting or holding furnace.

Aluminum chloride chemistry was also explored as a route to making prime aluminum metal by Alcoa, Alcan, and other companies. It is technically feasible to refine impure scrap aluminum melt using the technology developed. One would chlorinate the aluminum out of impure scrap melt to yield either $AlCl_3$ at low temperatures or AlCl at higher temperatures. Pure aluminum metal is recovered by either AlCl decomposition to Al + $AlCl_3$ or by $AlCl_3$ electrolysis in a multipolar inert electrode cell. Although both processes have been piloted and abandoned for primary aluminum production, first by Alcan and second by Alcoa, they should not be summarily dismissed. Toth Aluminum Corporation has recently developed a method of purifying impure $AlCl_3$ produced by low-cost chlorination of clay (Ref 8.5). Toth estimates that, combined with the Alcoa multipolar inert electrode cell, this technology could reduce the cost of production of prime by ~50%. Regardless of whether or not this technology is applied to scrap aluminum melts, the significant production cost reduction for prime metal production would have a large economic impact on the aluminum recycling system.

Solid-Particle Sorting

Solid-particle sorting can be based on a number of particle attributes. Shape and size correlations are widely used in hand-sorting during dismantling and can be used within limits to differentiate between wrought and cast components. These correlations, however, are context-specific and are not generally useful to differentiate between shredded pieces of sheet alloys.

Physical property correlations with density, electrical conductivity, magnetic susceptibility, color, or x-ray translucence form the basis of the mechanized separation of aluminum from other metals and nonmetals. However, because no unique property correlation exists with chemical composition for the wrought aluminum alloys, the usefulness of physical-property-based approaches for further grouping of wrought alloys is quite limited. One of the most promising physical-property-based approaches is the correlation with the chemical reactivity of the alloy. Tinting by selective etching and/or heat treatment allows the commercial color sorter to group the tinted pieces into additional metal-purity categories. Another approach is based on the correlation between the x-ray mass absorption coefficient and the average atomic number of the material. These approaches will be discussed in more detail as follows.

Chemical-composition-based sorting relies on rapid, noncontact quantitative chemical composition analysis. Three different noncontact spectroscopic methods have been applied to the elemental analysis of aluminum alloys: x-ray activation with x-ray fluorescence spectroscopy, neutron activation with gamma-ray fluorescence spectroscopy, and laser activation with optical emission spectroscopy (laser-induced breakdown spectroscopy).

X-ray fluorescence is hampered by the low energy of the light element fluorescence x-rays and their mutual interference. The characteristic Mg-, Al-, and Si-$K\alpha$ x-ray peaks overlap and are all too easily absorbed by any material, including air. This limits x-ray fluorescence to the determination of the concentration of heavy alloying elements (iron, copper, manganese, and zinc) in the aluminum particles. Pilot scrap-sorting units based on x-ray fluorescence analysis were built in the 1980s in Germany and the USSR; however, none were ever commercialized.

Neutron activation analysis depends on the irradiation of the particles with a highly penetrating neutron flux and on detecting the fluorescence of the gamma rays characteristic to the atoms in the particles. Here, the light element characteristic gamma rays are sufficiently distinct and sufficiently energetic to yield quantitative concentration results. The highly penetrating nature of both neutrons and gamma rays permits the analysis of the total volume of the material stream, and commercial units recently have been built that accurately determine the average composition of a metal-particle stream carried on a conveyor through a source-detector unit. However, the neutron flux provided by a radioactive isotope of californium is limited in intensity, leading to long exposure times not suitable for particle-by-particle sorting, which demands high speeds for industrial application.

Laser-induced breakdown spectroscopy (LIBS) is the illumination of a spot on the particle with a focused laser beam to generate a plasma spark, combined with the optical emission spectroscopy of the fluorescence photons. This method can yield quantitative analytical results for all alloying elements of interest in aluminum and magnesium, and at a rate that is suitable for industrial particle-by-particle sorting. There have been a few publications on LIBS analysis of aluminum alloys showing calibration curves with sufficient sensitivity throughout the concentration range of interest for all the major alloying elements (Ref 8.6–8.9). Most of these studies were done by signal averaging a large number of laser shots on a flat, clean, stationary sample surface. The challenge to the application of this technique to industrial particle-by-particle sorting is to:

1. Locate a suitable target spot on a randomly shaped and randomly placed rapidly moving aluminum particle
2. Deliver a laser shot containing both spot cleaning and analysis pulses to this selected single spot
3. Collect the light from the plasma emission from this random location
4. Simultaneously determine the quantitative concentration of all the alloying elements of interest
5. Make the choice of the target output bin
6. Mechanically divert each particle individually into the target bin

This sequence needs to be repeated for each 20 to 50 g particle at a rate that will permit an industrial sorting throughput of 3 to 6 tonnes/h (1 to 2 kg/s). This works out to 40 to 50 particles per second. The following reports both pilot-plant and industrial-prototype results that show how close this challenge is to being overcome.

Pilot Processes for Improved Wrought Recovery

As noted, three processes have been examined for commercial development of improvements in recovery from wrought scrap:

- Wrought aluminum tinting and color grouping
- X-ray absorption imaging
- Aluminum alloy chemical-composition-based sorting

Pilot projects demonstrate that solid-particle chemical-composition-based sorting, although technically challenging, is the method of choice for further sorting of the mixed wrought aluminum alloys to enable recycling of these scrap alloys into wrought alloy products.

Wrought Aluminum Color Grouping

One promising way to obtain further separation of mixed wrought alloys is to tint the particles by etching and then use a commercial color sorter (Fig. 8.2). This Alcoa-patented etching process involves a degreasing wash, hot water rinse, hot caustic solution etch, and a second water rinse followed by hot air drying (Ref 8.10). This process has been further optimized by etching samples of known wrought alloys and choosing the conditions that maximized the color contrast between the various alloys while minimizing the etching time and the weight loss.

The selected etch was a 90s etch at 80 °C (175 °F) in a 10 wt% NaOH-water solution. The dried pieces were sorted into several groups by an industrial color sorter of Huron Valley Steel Corporation. Four color groups were distinguishable by the color sorter. The color-sorting results on the etched known alloys are summarized in Table 8.3, which lists recoveries of a given known alloy in a particular output stream. Recovery is defined here as the weight fraction of the given alloy found in the particular output stream. Figure 8.3 shows the tint-etched particles of the known alloys.

The etching method was then applied to an ~400 kg pilot-scale sample of wrought aluminum concentrate with an unknown mix of alloy feed. See Tables 8.1 and 8.2 for the typical

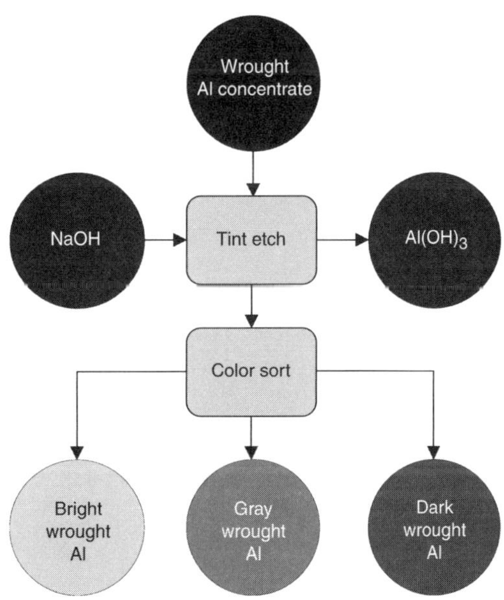

Fig. 8.2 Block diagram of the tint-etching color-sorting process

composition of this feed stream. This time, three color groups were distinguished by the color sorter, as shown in Fig. 8.4. Table 8.4 lists the composition of these three products in terms of hand-sort categories, while Fig. 8.5 shows the elemental composition of ~60 kg crucible melts recovered from the products.

In summary, caustic-etched aluminum alloy particles are tinted gray by silicon, iron, and manganese, and dark by copper and zinc. There is a slight weight loss due to etching and generation of aluminum trihydrate by-product. Color sorting of tinted wrought shred mix resulted in grouping into three or four additional products that have different aluminum purity but not directly correlated to alloy families. Color-sorting results are affected by paint, anodizing, and plating.

Table 8.3 Product stream recoveries for color sorting of known wrought aluminum alloys

	Recovery, number percent			
Component	Product 1, bright	Product 2, light gray	Product 3, gray	Product 4, dark
5182	100	...	...	...
4147	...	100	...	...
6061	...	70	30	...
3003	...	28	72	...
6111	10	20	70	...
7129	10	20	70	...
7016	...	30	70	...
2036	...	...	20	80
7003	...	...	...	100

X-ray Absorption Imaging

X-ray absorption correlates directly to the average atomic number of the material illuminated by the x-ray flux. Figure 8.6 shows a false color image in which the hue is set by the average atomic number, and the luminescence is inversely correlated with the particle thickness. It demonstrates that aluminum particles with a higher content of copper, iron, manganese and zinc (dark gray) can be separated from those containing only light elements—aluminum, magnesium, and silicon (light gray). This type of imaging can give similar alloy groupings as tinting and color sorting without metal loss and sensitivity to surface contamination. Figure 8.7, however, implies that x-ray absorption imaging cannot differentiate between alloys or material mixtures that have the same average atomic number for a different combination of elements. This requires recourse to particle sorting based on quantitative elemental composition analysis.

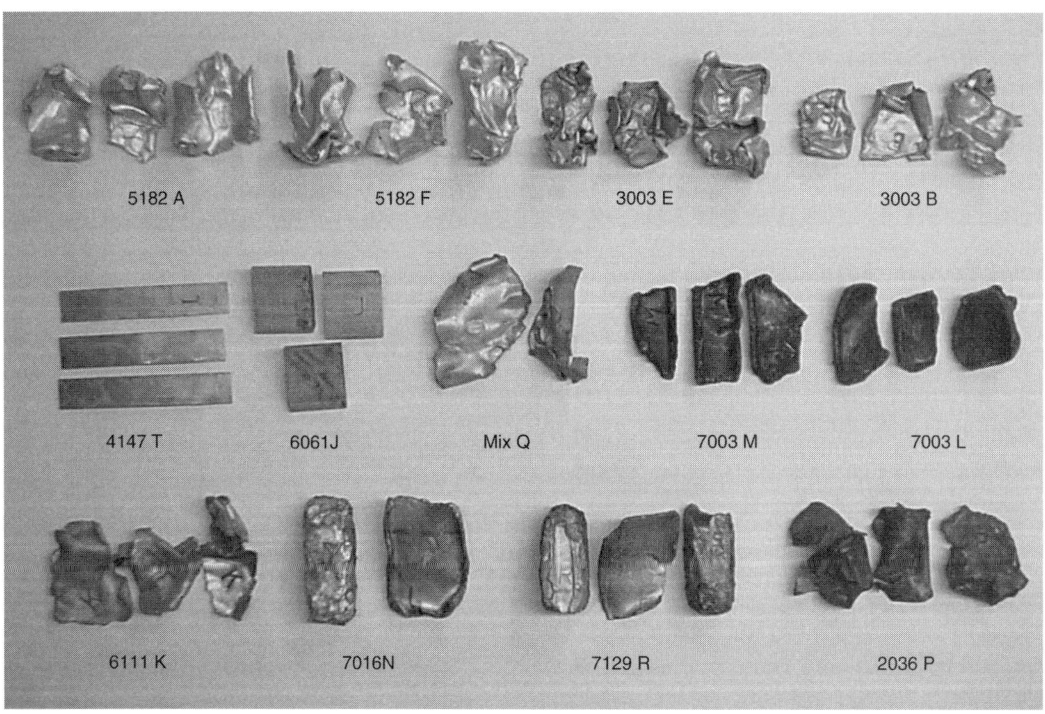

Fig. 8.3 Tint-etched shredded particles of known wrought aluminum alloys

Fig. 8.4 Color-sorted products of tint-etched wrought aluminum concentrate. Labels identify the elemental composition recovered melts

Table 8.4 Hand-sort category classification of the color-sort products from the etch-tinted wrought aluminum concentrate feed

Category	Bright, %	Gray, %	Dark, %
Al wrought	91.69	98.02	87.65
Al cast	2.08	0.25	10.18
Al other	5.64	1.74	1.54
Al irony	3.10	3.27	1.04
Al painted	19.13	21.37	12.94
Cu	0.51	0	0.14
Stainless steel	0.08	0	0
Wire	0	0	0.05
Nonmetal	0	0	0.45

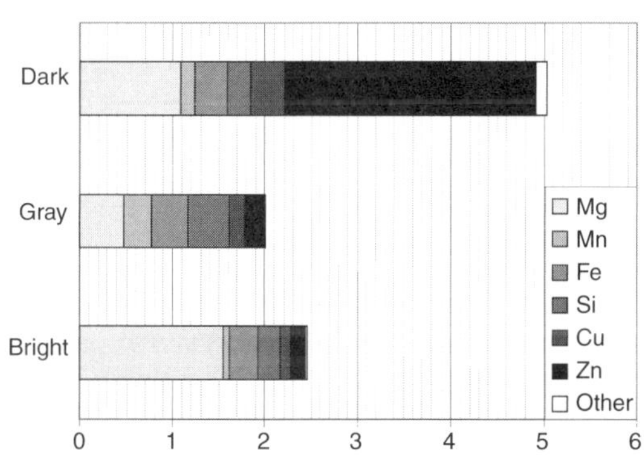

Fig. 8.5 Alloying element concentration in the melt products recovered by color sorting of etch-tinted wrought aluminum concentrate

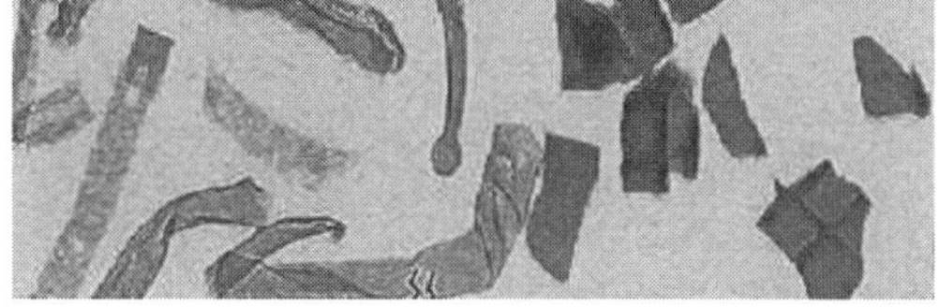

Fig. 8.6 X-ray absorption image showing separation of aluminum alloys based on the average atomic number of each particle. Courtesy of the Technical University of Delft, The Netherlands (see Acknowledgments)

Composition-Based Particle Sorting

Composition-based particle-sorting technology has been in development since 1993, building both on earlier work at Los Alamos National Laboratory (Ref 8.11–8.13), Metallgesselshaft (Ref 8.14–8.16), Alcan (Ref 8.17–8.19), and on industrial color-sorting technology. The development included high-speed, noncontact, quantitative LIBS as an analytical method and the integration of LIBS analysis with robust color-sorter technology developed by Huron

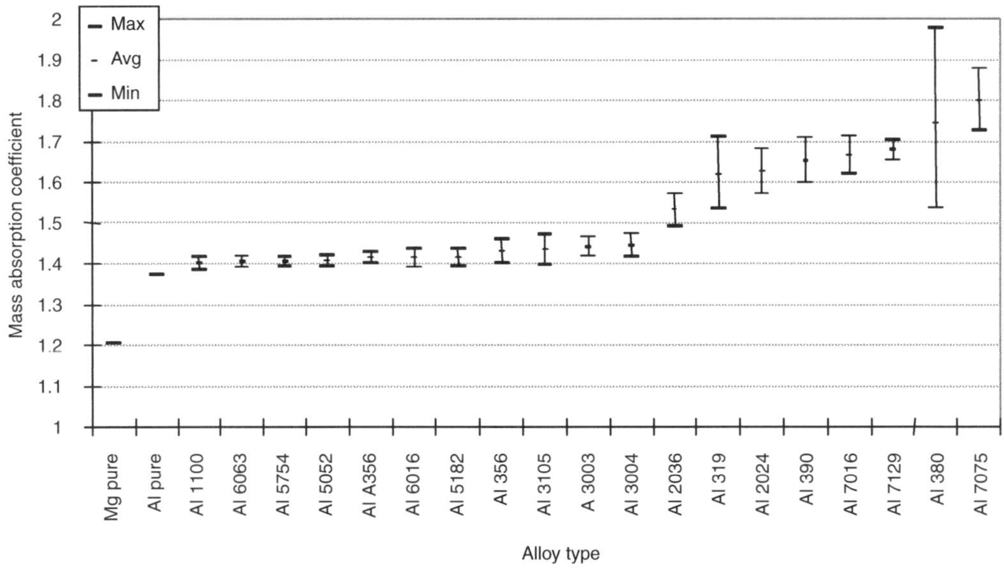

Fig. 8.7 Theoretical alloy grouping based on the correlation of the average atomic number with the x-ray absorption coefficient. Courtesy of the Technical University of Delft, The Netherlands (see Acknowledgments)

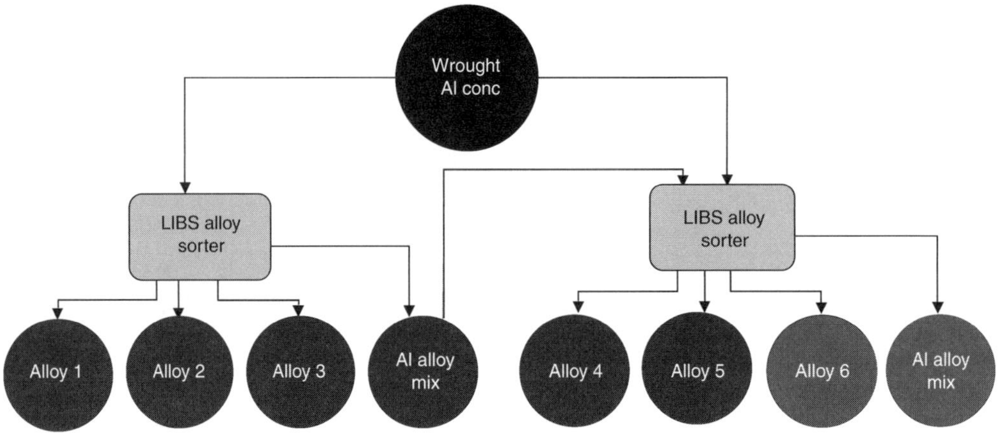

Fig. 8.8 Block diagram of the Huron Valley Steel Corp. full-scale industrial prototype of the laser-induced breakdown spectroscopy (LIBS) alloy sorter

Valley Steel Corporation. In 1998, a full-scale LIBS alloy-sorter pilot plant (see Fig. 8.8 for a block diagram) was commissioned to:

- Calibrate magnesium, copper, and silicon spectral lines on the pilot sorter
- Optimize signal-to-noise ratio
- Simultaneously acquire spectral lines and determine elemental particle composition in < 10 ms
- Achieve a resolution of ~0.1% for ~1% concentration at that rate
- Demonstrate 6111-5182 alloy sort
- Study bulk pre-cleaning and laser spot cleaning

The pilot facility continues to be used for LIBS development, and topics under current investigation include a study of apparent spectral line shifts and a survey of LIBS spectral emission lines for all major aluminum alloying elements. In 2001, an industrial prototype LIBS sorter was commissioned to demonstrate multi-stream diversion on a full industrial scale and speed and for accurate targeting of scrap particles with a laser at production speed. One

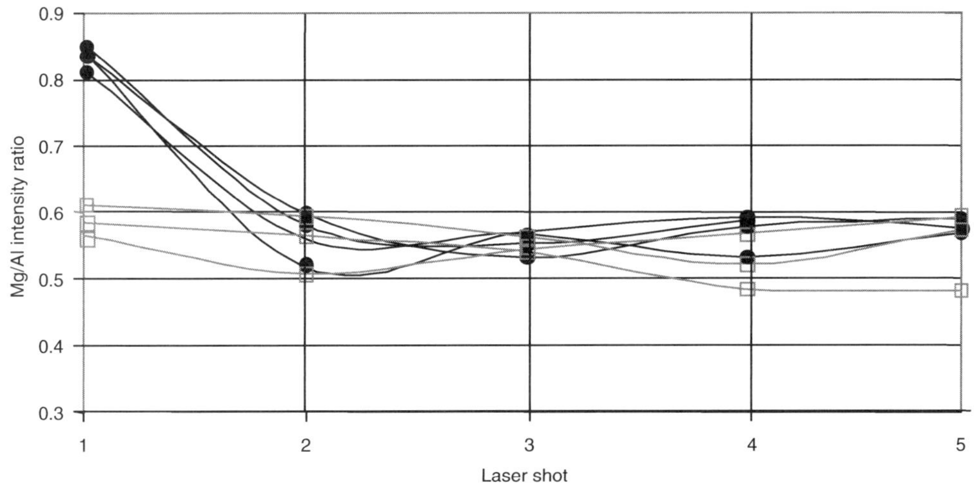

Fig. 8.9 Spectral line intensity ratio as a function of the number of laser shots for • washed and □ ground preparation of a 6111 alloy sample surface

demonstration was for a successful industrial prototype of a 6111-5182 alloy sort.

Bulk-Particle Cleaning. To evaluate bulk-particle-cleaning requirements, LIBS intensity ratios were compared for magnesium-aluminum and silicon-aluminum pieces with bare surfaces cleaned by various methods with the LIBS intensity ratios for a freshly ground spot. As illustrated in Fig. 8.9, a stable intensity ratio is obtained from the first laser shot on a ground surface. With all cleaning methods tried, ranging from water washing to caustic etch, more than one multipulse laser shot is still necessary to achieve an intensity ratio representative of bulk alloy composition that is obtained on a ground surface. A stable value was obtained after a second laser shot on an unpainted 6111 alloy surface precleaned by a proprietary method, leading the development to focus on optimizing laser spot-cleaning conditions rather than searching for more aggressive bulk-cleaning methods. The next challenge for the bulk cleaning is the evaluation of decoated painted surfaces, including the effect of pigment residues and chromate and phosphate pretreatment layers.

6111-5182 Alloy Sort. Alloys 6111 and 5182 are among the most important sheet alloys in the body sheet and body structure of the new aluminum-intensive vehicles. In the sorting of these two alloys, a quantitative analysis that permits the calibration of the optical emission spectral lines is being investigated. In a pilot, a simultaneous analysis based on magnesium, copper, and silicon calibration allowed reproducible sorting of 6111 (< 1% Mg) from 5182 (~5% Mg). This sort was

Table 8.5 Pilot plant laser-induced breakdown spectroscopy sorter spectral line intensity ratio (I_X/I_{Al}) data for the 5182-6111 alloy sort

Element	Intensity ratio data	Alloy	
		5182	6111
Magnesium, I_{Mg}/I_{Al}			
	Concentration, wt%	5	1
	Average	0.934	0.575
	Standard deviation	0.042	0.043
	Coefficient of variance (standard deviation/average), %	5	8
Silicon, I_{Si}/I_{Al}			
	Concentration, wt%	0.1	1
	Average	0.243	0.532
	Standard deviation	0.065	0.143
	Coefficient of variance, %	27	27

accomplished under the following conditions. Sheet samples of known 6111 ~1% Mg and 5182 ~5% Mg samples were placed randomly on a belt moving at 1.5 m/s (5 ft/s) on the pilot plant sorter or 2 m/s (6.5 ft/s) on the industrial prototype sorter. During the sort, aluminum, magnesium, copper, and silicon spectral line intensities were monitored simultaneously, and the aluminum line was used as the internal standard to normalize the remaining spectral lines. Because the composition of these two alloys differed primarily in the magnesium concentration, the sort strategy was based mainly on this element.

Table 8.5 lists the spectral line intensity ratio data and their precision for the measurements done on the pilot plant sorter. Figure 8.10 illustrates that, even with this limited precision, it is quite easy to separate these two alloys based on

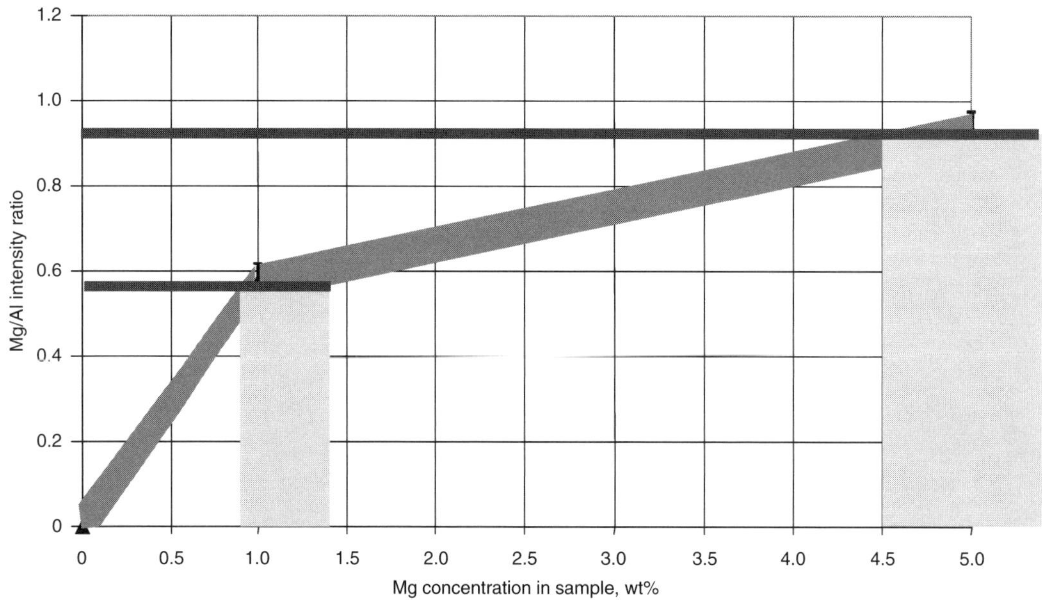

Fig. 8.10 Magnesium calibration of pilot plant laser-induced breakdown spectroscopy sorter based on known 6111-1% Mg and 5182-5% Mg alloys. The width of the calibration line represents confidence limits, and the wide vertical bars show the precision of concentration determination for a given value of the intensity ratio

LIBS measurement of magnesium concentration. Table 8.3 and Fig. 8.7 show that such clean separation is not possible by tint-etching or x-ray absorption methods. Both pilot plant and industrial prototype sorters were system integrated to combine image acquisition, laser targeting, LIBS analysis, sort decision, and particle diversion. During the sort on both machines, the 5182 alloy particles were consistently diverted to bin 1 and 6111 alloy particles to bin 2.

Full-Scale Industrial Prototype LIBS Sorter. A full-scale and operational prototype sorter (built in 2001 by Huron Valley Steel Corporation) has been used to demonstrate accurate, reproducible particle image acquisition and laser targeting at a belt speed of 2 m/s (6.5 ft/s), with 50 laser shots being delivered to up to 70 particles passing the target area in a second. Further, high-accuracy particle diversion into four output streams was also demonstrated at these production rates. Three diverters were used, with the first one giving 98% diversion of the targeted particles into the first product stream, the second one 95%, and the third one ~90%. The particles not diverted end up in the fourth output stream, which is an alloy mix or a recycle stream, and thus have only a minor effect on the composition of the three sorted products. Detailed analytical calibration of the prototype for all alloying elements is in progress. It is already possible to reproduce the 6111 and 5182 sort on the prototype.

Continued Alloy Sorting Process Development. To satisfy the objectives of the Auto Aluminum Alliance project on Automotive Aluminum Scrap Sorting (Ref 8.19–8.23), alloy sorting process developments are aimed at separation of known bare alloys. Separation objectives are between the following alloy families and also separation within the families. Alloy families for separation are:

- 1100
- 2036
- 3003
- 4147
- 6111
- 5182
- 7129

Separation objectives within these families are:

- Manganese: 3003/3004/3105
- Magnesium: 5052/5754/5182
- MgSi: 6063/6061/6082

When these sorting objectives are met, the next step is sorting of bare mixed-alloy shred of unknown composition. Grouping by major alloying elements (e.g., low zinc, copper, silicon) would be the first step, then moving to sorting out particles compatible with a particular alloy, and finally attempting real-time particle-by-particle batching of particular alloy composition. Software for real-time optimized batching is currently under development. The accuracy of the sorting will be verified by elemental analysis of 60 kg crucible test melts of the sorted products. This development effort is intended to culminate in the sorting of the wrought aluminum alloy concentrate from aluminum-intensive vehicle hulks supplied by auto manufacturers. It will include decoating and precleaning followed by separation of 5182/5754/6111/6061/6082 from the remaining aluminum alloy mix.

ACKNOWLEDGMENTS

The editors thank the Huron Valley Steel (HVS) Corp. and the Minerals, Metals and Materials Society (TMS) for permission to publish this paper. Information presented here is a compilation of the work of several teams generated over an extended development timeframe. These teams included:

- ***Wrought-cast aluminum sorting:*** George Hopson, Tim Good, Richard Grieve, Scott Costello, Sam Lambert (HVS, Belleville, MI)
- ***Wrought aluminum color grouping:*** George Hopson, Sam Lambert, Scott Costello (HVS, Belleville, MI) and Paul Schultz (Alcoa Technical Center, Pittsburgh, PA)
- ***X-ray absorption imaging:*** Tako de Jong, Wijnand Kuilman, and Wijnand Dalmijn (Technical University of Delft, The Netherlands), and Bahram Farahbakhsh (Columbus, OH)
- ***Composition-based particle sorting:*** Chris Stewart, Ben AuBuchon, Sam Lambert, Arnold Petz, and Kerry Lang (HVS, Belleville, MI); Philip MacNutt (Austin, Tx); David Hoadley (Farmington Hills, MI); and Paul Torek (Ann Arbor, MI)

REFERENCES

8.1. "Passenger and Light Truck Aluminum Content Report," Ducker Research Company, 1999

8.2. A.J. Gesing et al., Scrap Preparation for Aluminum Alloy Sorting, *2000 TMS Fall Extractive Meeting (Recycling of Metals and Engineered Materials)*, Oct 22–25, 2000 (Pittsburgh, PA), p 1233–1249

8.3. A.J. Gesing and R. Wolanski, New Developments in Recycling of Light Metals from End-of-Life Vehicles, *J. Met.*, Nov 2001, p 21–23

8.4. A.J. Gesing et al., Separation of Wrought Fraction of Aluminum Recovered from Automobile Shredder Scrap, *Aluminum 2001 (Proceedings of the TMS 2001 Annual Meeting, Aluminum Automotive and Joining Symposia)*, p 31–42

8.5. C. Toth et al., "TAC's Bipolar Clay-to-Aluminum Technology and Project: The New Way to Make Aluminum," seminar edition, Toth Aluminum Corporation, Feb 2001

8.6. M. Sabsabi and P. Ceilo, Quantitative Analysis of Aluminum Alloys by Laser-Induced Breakdown Spectroscopy and Plasma Characterization, *Appl. Spectrosc.*, Vol 49 (No. 4), 1995

8.7. M. Sabsabi and P. Ceilo, Quantitative Analysis of Copper Alloys by Laser-Produced Plasma Spectrometry, *J. Anal.* At. Spectrom., Vol 10, Sept 1995

8.8. A.K. Rai et al., Parametric Study of a Fiber-Optic Laser-Induced Breakdown Spectroscopy Probe for Analysis of Aluminum Alloys, *Spectrochim. Acta B, At. Spectrosc.*, Vol 56, 2001, p 2371–2383

8.9. B. Német and L. Kozma, Time-Resolved Optical Emission Spectrometry of Q-Switched Nd:YAG Laser-Induced Plasmas from Copper Targets in Air at Atmospheric Pressure, *Spectrochim. Acta B, At. Spectrosc.*, Vol 50, 1995, p 1869–1888

8.10. R.K. Wyss and P.B. Schultz, Color Sorting Aluminum Alloy Scrap for Recycling, *Light Metals 1999, The CD ROM Collection,* The Minerals, Metals and Materials Society, 1999

8.11. "Laser-Induced Breakdown Spectroscopy: A New Spectrochemical Technique," Technical Report LA-UR-82-465, Los Alamos National Laboratory, 1982

8.12. L.J. Radziemski and D.A. Cremers, *Laser-Induced Plasmas and Applications,* M. Dekker, New York, 1989

8.13. D.A. Cremers, The Analysis of Metals at a Distance Using Laser-Induced Breakdown Spectroscopy, *Appl. Spectrosc.*, Vol 41, 1987, p 572–579

8.14. M. Potzschke et al., Scrap Detector, U.S. Patent 5,042,947, 1991

8.15. H.-P. Sattler, Automatic Sorting of Non-Ferrous Metals from Automobile Shredders, *Second International Symposium on Recycling of Metals and Engineered Materials*, The Minerals, Metals and Materials Society, 1990, p 333–341

8.16. H.-P. Sattler and T. Yoshida, New Sorting System for Recycling of Magnesium and Its Alloy After Use, *First International Conference on Processing Materials for Properties*, The Minerals, Metals and Materials Society, 1993, p 861–864

8.17. A.J. Gesing and T. Shaw, Method of Sorting Pieces of Material, U.S. Patent 5,813,543, 1998

8.18. A.J. Gesing and A. Rosenfeld, Composition Based Sorting of Aluminum Scrap from Aluminum Intensive Vehicles, *Developments in Aluminum Use for Vehicle Design*, SP-1164, Society of Automotive Engineers, Inc., Feb 1996, p 29–38

8.19. "Automotive Aluminum Scrap Sorting," Project Fact Sheet, Office of Industrial Technologies, Energy Efficiency and Renewable Energy, U.S. Department of Energy, July 2001

8.20. J.L. Broge, Managing and Remaking Metals, *Automot. Eng. Int.*, Vol 109 (No. 8), 2001, p20–31

8.21. L.E. D'Astolfo, Jr. and P.R. Bruggink, Preserving the Value Chain in Automotive Aluminum Recycling, *Light Metals 1994,* The Minerals, Metals and Materials Society, 1994, p 1121–1127

8.22. L.E. D'Astolfo, Jr. et al., Recycling Automotive Aluminum Scrap into New Products, *First International Conference on Processing Materials for Properties,* The Minerals, Metals and Materials Society, 1993, p 839–842

8.32. *Aluminum Industry Roadmap for the Automotive Market: Enabling Technologies and Challenges for Body Structures and Closures,* The Aluminum Association, Inc., Energetics, Incorporated, May 1999, p 51

CHAPTER 9

Emerging Trends in Aluminum Recycling*

Subodh K. Das, Secat, Inc.

THE ALUMINUM that goes into the production of aluminum alloy products for many applications comes increasingly from recycled products. In a recent modeling study of the industry, Choate and Green (Ref 9.1) illustrated that most of the increase comes from recycled automotive components, which, in 2005, was expected to exceed for the first time the recycled metal coming from used beverage cans. The Aluminum Industry Roadmap (Ref 9.2) also illustrates the importance of these trends and of the efforts to address the technology from primary production to finished products. Fielding (Ref 9.3) illustrates how one segment of the industry, the extrusion business, is approaching the challenge.

As Choate and Green (Ref 9.1) demonstrated, the increase in available recycled metal is a positive trend, because secondary metal produced from recycled metal requires only approximately 2.8 kWh/kg of metal produced, while primary aluminum production requires approximately 45 kWh/kg of metal produced. It is to the industry and national advantage to maximize the amount of recycled metal, both from the standpoint of energy savings and reduction of dependence on overseas sources (now approximately 40% of U.S. consumption) and also from the ecological standpoint, because recycling emits only approximately 4% as much CO_2 as primary production.

*Adapted with permission of the Minerals, Metals and Materials Society (TMS) from the article by Subodh K. Das, "Emerging Trends in Aluminum Recycling: Reasons and Responses," *Light Metals 2006*, TMS, 2006.

A joint study of a representative American community by Secat, Inc., the Center for Aluminum Technology, and the Sloan Industry Center for a Sustainable Aluminum Industry found that for each 1% increase in the amount of aluminum cans recycled, the economic savings to the U.S. economy is $12 million/year. This will approach $600 million/yr if all the available aluminum is recycled. The additional recycling also contributes to energy savings of 1 trillion Btu/year. For the United States as a whole, recycling also has the potential to significantly decrease reliance on overseas sources of primary aluminum metal.

Aluminum recycling in North America is one of the well-developed and mature metals recycling economies in the world. While today's (2007) recycling metals markets also include ferrous metals such as iron and steel and nonferrous metals such as copper and brass, aluminum recycling is the engine of recycling economics. However, it is clear that more should be done to maximize the advantages of a recycling-friendly world. The growth in aluminum usage in transportation applications, the decline in aluminum beverage can recycling, and the increasing reliance of the domestic fabrication industry on secondary aluminum have combined to create new needs in both the materials design and processing realm.

Today (2007), in order to meet the performance requirements of many alloy and product specifications, much of the recycled metal must be "sweetened" with the more costly and energy-intense primary metal before it can be re-employed in many applications. The specialty alloys required for a number of applications require such strict controls on impurities that recycled metal cannot be used without modification. The result is that, in many cases (except beverage cans), recycled metal tends to be used primarily for lower-grade casting alloys and products. While a certain amount of this is acceptable, the recycling-friendly world will only be truly optimized when the recycle loop is closer to a closed loop within a number of product lines.

This chapter details the history and future projections for aluminum recycling, emphasizing the increasing importance of mixed-scrap streams in the makeup of secondary aluminum. For the most economical use of these scrap streams, new approaches are needed to develop acceptable materials processed to control properties suitable for an expanded range of applications. This chapter discusses how the aluminum enterprise, including industry, academia, and government, can work together to meet these important but aggressive targets and transform recycling from strictly an environmental imperative to an economic development opportunity.

Objectives and Challenges

A robust recycling regime for aluminum would include:

- Recycled products to approach the total required for new consumption, with less dependence on primary production
- Use of automatic sorting, shredding, and separation technology to facilitate its reuse in new products
- Having a variety of existing and new aluminum alloys with compositions suitable for direct reuse of most recycled metal, so the opportunities for direct use of the recycled, shredded, and sorted metal would be optimized
- Developing a number of high-value applications, such as beverage cans, into which the recycled metal would flow. In such situations, product made directly from the recycled metal would readily meet both specification composition and mechanical property limits of the intended applications.

Among the key challenges to be met in creating this ideal recycling-friendly world are:

- Maximize recovery of used aluminum products and components for recycling
- Identify more useful by-products to handle residual elements; for example, aluminum alloys containing relatively high iron content identified for deoxidizing steel (de-ox)
- Automate and optimize the presorting, shredding, and separation technologies
- Broaden the number of available aluminum alloys whose specifications will readily accept recycled metal and will perform well in high-quality, value-added products

Progress has been and is still being made in addressing the first three challenges. One significant new development in terms of automated sorting is the application of laser-induced breakdown spectroscopy (LIBS) for sorting of aluminum and aluminum alloys (Ref 9.4). An overview of this technology, developed and applied by Huron Valley Steel Corp., is described in Chapter 8, "Identification and Sorting of Wrought Aluminum Alloys" in this book.

The fourth challenge, the identification of new alloys that more readily accept recycled aluminum, has received little attention. However, the potential economic and environmental benefits are sufficiently great for more attention to this challenge. There are a number of detailed challenges facing any effort to increase the number of aluminum alloys and applications suitable for direct production from recycled metal.

The Nature of Recycled Metal

As a starting point in considering the challenges faced in directly using recycled aluminum scrap, it is appropriate to look at representative compositions observed in such metal. Representative compositions of recycled aluminum are shown in Table 9.1 from presorting wrought and cast alloy scrap (Ref 9.4), so two samples were of wrought separations, two of cast separations, and one mixed. These representative compositions illustrate several of the fundamental complications in directly reusing scrap aluminum:

- Even segregated wrought scrap can have relatively widely varying compositions. Wrought lots 3 and 4 in Table 9.1, for example, have higher copper (from more 2*xxx* alloys) and higher zinc (from more 7*xxx* alloys) in the mix than did wrought lots 1 and 2. It appears that auto bumper alloys such as

7029 and auto body sheet alloys such as 2036 were more highly represented in wrought lots 3 and 4.

- Some lots of wrought recycled metal (lots 1 and 2) match existing wrought alloys reasonably well, for example, 3005, 3104, 3105, and 6061, and can be readily reused; others, such as lots 3 and 4, will be more difficult to use directly.
- Cast alloy scrap can vary greatly in composition. Cast alloy scrap also differs significantly from wrought alloy scrap, notably with higher total alloy content, higher silicon content, and, depending on which cast alloys are involved, higher copper (from 380.0 and 390.0) and zinc (from 7*xx*.0 cast alloys).
- Compositions resulting from mixed wrought and cast scrap will be more difficult to use directly because of their combinations of higher silicon, copper, and zinc.

With the exception of recycled beverage cans, most recycled aluminum involves a mixture of alloys from a fairly wide variety of applications, including a selection of castings containing rather high percentages of silicon. While there is generally no problem recycling most of this metal as castings, there is a significant challenge in shredding, sorting, and, in some cases, further refinement of the metal to achieve acceptable impurity levels for products other than castings, including sheet, plate, forgings, and extrusions.

This is particularly true for any of the specialized alloys produced today (2007), for example, those used in the aerospace industry where requirements for exceptionally high ductility and toughness are common. Such performance requirements call for very tight composition controls on both iron and silicon. Impurity levels above 0.15% Fe or 0.25% Si are unacceptable in premium aerospace alloys such as 7050, 7055, and 7475. Similarly, some high-performance automotive alloys (e.g., 5457 and 6111) restrict both silicon and iron to 0.40% maximum. Both of these elements (iron and silicon) are difficult to control in recycled metal and tend to increase modestly the more often the metal has been recycled.

Iron, in particular, can be a significant challenge because of its tendency to increase gradually in metal recycled over and over again, primarily from pickup from scrap-handling system equipment. As a result, iron is an ideal candidate for application to alternative products, an excellent example being the use of high-iron-bearing aluminum as a deoxidizing agent for steel production. It should be noted that elements other than iron may also be expected to increase with repeated recycling and to require special attention, for example, magnesium, nickel, and vanadium.

Cast Alloy Scrap. As noted earlier, recycled metal from castings can often be used directly in new cast products, usually of the 3*xx*.0 and 4*xx*.0 series, because the impurity limits tend to be relatively high, and almost all contain relatively high silicon, which improves their flow characteristics in the dies. Examples of scrap-tolerant casting alloys are shown in Table 9.2. Even these relatively tolerant limits pose a challenge for direct recycling reuse. For all except 336.0, for which no "Others" limit exists, the "Others" contents noted in scrap samples are higher than desired. In the 4*xx*.*x* series, the tight magnesium contents will be a challenge. Nevertheless, casting alloys as a whole have higher impurity limits than wrought alloys and will be more tolerant for direct recycling.

Table 9.1 Representative compositions of presorted wrought and cast scrap

Lot	Al	Cu	Fe	Mg	Mn	Si	Zn	Others
Wrought 1	97.1	0.11	0.59	0.82	0.21	0.51	0.45	0.19
Wrought 2	96.7	0.30	0.60	0.60	0.20	0.90	0.50	0.10
Wrought 3	93.1	0.95	1.01	0.89	0.12	2.41	1.25	0.27
Wrought 4	93.1	1.20	0.70	0.70	0.30	2.60	1.20	0.20
Cast 1	83.5	4.40	1.10	0.40	0.30	8.0	1.90	0.40
Cast 2	86.0	3.90	1.00	0.10	0.20	6.30	2.30	0.30
Cast 3	88.4	2.50	0.75	0.58	0.26	5.18	1.27	1.09
Mixed wrought and cast	90.1	2.30	0.80	0.50	0.20	4.50	1.20	0.30

Table 9.2 Sample of cast alloy scrap compositions

Alloy	Cu	Fe	Mg	Mn	Si	Zn	Others
B319.0	3.0–4.0	1.2 max	0.10–0.50	0.8 max	5.5–6.5	1.0 max	0.50 max
336.0	0.50–1.5	1.2 max	0.70–1.3	0.35 max	11.0–13.0	0.35 max	...
C443.0	0.6 max	2.0 max	0.10 max	0.35 max	4.5–6.0	0.50 max	0.25 max

Wrought Alloy Scrap. The extent of the current opportunity as well as the challenge in directly reusing recycled wrought alloy scrap without "sweetening" with primary metal is much greater. This can be illustrated by comparing the compositions in Table 9.2 with those of several wrought commercial alloys already recognized as good consumers of scrap (Table 9.3).

As noted earlier, material from wrought lots 1 and 2 could reasonably be used in alloys such as 3105 and 6061. Even for these alloys, the maximum limit on "Others" may be a challenge, because that value for wrought 1 is slightly in excess of the limit. Scrap from wrought lots 3 and 4 could not be directly reused without flexible impurity limits (see the section "Developing Recycling-Friendly Compositions" in this chapter). Material from the cast lots could not be used directly for any of these wrought alloys or any others, for that matter.

An additional elemental impurity problem needs to be recognized, and that is the general trend for iron content in scrap reused over and over again to increase gradually, primarily through pickup from scrap-handling system equipment. With only a few exceptions, iron is an impurity in wrought alloys today (2007) and is an ideal candidate for application to alternate products (e.g., a deoxidizing agent in steel production). Maximization of this capability will benefit both the aluminum and steel industries and add to the life-cycle benefits of aluminum operations. Another possible approach to the increased iron content is to make use of the affinity of zirconium for iron, resulting in a heavy particle that sinks to the bottom of crucibles during processing. Combining this iron-zirconium product with magnesium and perhaps other undesirable impurities such as nickel and vanadium may also improve their impact on the resulting recycling content.

Recycling Aluminum Aerospace Alloys (Ref 9.5)

As an example, consider the principal challenges for effective recycling of aircraft. For decades, thousands of obsolete aircraft have been sitting in "graveyards," while the demand for recycled aluminum continues to increase. The discarded aircraft provide a large source of valuable metal. However, cost-effective recycling of aircraft alloys is complex, because aircraft alloys are typically relatively high in alloying elements and contain very low levels of impurities to optimize toughness and other performance characteristics.

Thus, recycling of aluminum aerospace alloys represents a major challenge to both the aluminum and aerospace industries. While the recycling of high percentages of aluminum from packaging and automotive applications has been commercialized and has become economically attractive (Ref 9.1–9.3), the unique compositions and performance requirements of aerospace alloys have resulted in the delay of directly addressing techniques for cost-effectively recycling ignore those alloys. Aluminum remains the most economically attractive material from which to make aircraft and space vehicles, and new construction proceeds at a prodigious rate. However, the development of newer aircraft structures has proceeded at such a pace that thousands of obsolete civil and military aircraft stand idle in "graveyards" around the United States. Yet, it has been impractical to reuse the metal in these planes because of the combination of the differences in compositions of older, obsolete aircraft and those of new aircraft, which often have special performance requirements for specialized alloy compositions.

Cost-effective means is the subject of a study (Ref 9.5) for recycling aluminum alloys used in the production of private, civil, and military aircraft. The principal challenges that must be dealt with in creating this ideal aircraft-recycling scenario include:

- Identifying decision options for dismantling aircraft to simplify recycling
- Identifying and optimizing technologies for automated shredding, sorting, and remelting of those 2*xxx* and 7*xxx* alloys with relatively high levels of alloying elements (sometimes in excess of 10%)

Table 9.3 Sample of wrought alloy scrap compositions

Alloy	Cu	Fe	Mg	Mn	Si	Zn	Others
3005	0.30 max	0.7 max	0.20–0.8	1.0–1.5	0.6 max	0.25 max	0.15 max
3104	0.8 max	0.6 max	0.8–1.3	0.8–1.4	0.6 max	0.25 max	0.15 max
3105	0.30 max	0.7 max	0.20–0.8	0.30–0.8	0.6 max	0.40 max	0.15 max
6061	0.15–0.40	0.7 max	0.8–1.2	0.15 max	0.40–0.8	0.25 max	0.15 max

- Identifying the range of representative compositions likely to be obtained from recycling aircraft components, dependent on the amount of presorting that proves practical
- Identifying the combination of performance requirements and compositions that would make useful aircraft components from recycled metal, even though they may not achieve the highest attainable levels of toughness
- Identifying useful by-products to handle elemental residue unable to be used in recycled metal, for example, iron

These challenges include many of the general issues of aluminum recycling. As noted, progress is being made in the application of LIBS for shredding and sorting of some aluminum alloys. In general, however, little or nothing has yet been done to apply recycling technology to shredding, sorting, and reuse of recycled metal from obsolete aircraft and space vehicle components. Implementation issues addressed in Ref 9.5 include:

- Dismantling and presorting strategies
- Automated shredding, sorting, and remelting
- Identifying the resulting compositions of recycled aircraft components
- Options for reuse of the metal from recycled aircraft components in new aircraft
- Options for reuse of the metal from recycled aircraft components in nonaircraft applications
- Options for reuse of the metal from recycled aircraft components in aluminum castings

Of these issues, the latter three options are described as follows.

Options for Reuse of the Recycled Aircraft Components in Aircraft. Table 9.4 illustrates composition examples from presorting of 2*xxx* and 7*xxx* alloys. Assuming that these estimated compositions are reasonably correct, it would appear that the resultant alloys could be used for a number of noncritical aircraft components, such as stiffeners, flaps, and other relatively low-to-moderately stressed components made of sheet, plate, or extrusions. These may be used in private, civil, and many military aircraft. Typically, these would be components that are not designed based on fracture mechanics concepts employing fatigue crack growth rates and fracture toughness parameters. Alloys for fracture-critical areas may still have to be fabricated using primary metal. Further study is needed to estimate the percentage of aircraft components that do not require fracture-critical design and whether it is broad enough to justify the reuse of compositions likely to result from recycling.

Options for Reuse of the Recycled Aircraft Components in Nonaircraft Applications. If the use of these compositions for non-fracture-critical components in new aircraft is too tightly limited, that is, if the number of non-fracture-critical components in civil and military aircraft is not large enough to justify reuse of the recycle compositions, it is useful to look at the other opportunities that may exist for use of the compositions (Tables 9.5, 9.6).

To aid in addressing that question, the compositions of several wrought 2*xxx* and 7*xxx* alloys used in other applications (including 2014, also an aircraft alloy) are presented in Table 9.7. Comparing compositions in Tables 9.5 and 9.7

Table 9.4 Nominal compositions of some 2*xxx* and 7*xxx* alloys

Alloy	Al	Cu	Fe	Mg	Mn	Si	Zn
2014	~93	4.4	0.7 max	0.50	0.8	0.8	0.15 max
2214	~93	4.4	0.3 max	0.50	0.8	0.8	0.15 max
2024	~93	4.4	0.5 max	1.5	0.6	0.5 max	0.25 max
2324	~94	4.1	0.12 max	1.5	0.6	0.1 max	0.15 max
7050	~89	2.3	0.15 max	2.2	0.1 max	0.12 max	6.2
7075	~90	1.6	0.5 max	2.5	0.3 max	0.4 max	5.6
7475	~90	1.6	0.12 max	2.2	0.06 max	0.1 max	5.7
7178	~89	2.0	0.5 max	2.8	0.3 max	0.4 max	6.8

Table 9.5 Potential compositions of some recycled aircraft alloys, assuming presorting

Alloy	Al	Cu	Fe	Mg	Mn	Si	Zn	Others
R2*xxx*	~93	4.4	0.5	1.0	0.7	0.5	0.1	0.2
R7*xxx*	~90	2.0	0.4	2.5	0.2	0.2	6.0	0.2

Table 9.6 Potential composition of some recycled aircraft alloys, assuming no presorting

Alloy	Al	Cu	Fe	Mg	Mn	Si	Zn
R2 + 7*xxx*	~92	3.0	0.4	1.8	0.4	0.4	3.0

illustrates that it may be possible to reuse recycled aircraft metal in certain other products. Further study of this option is justified.

Options for Reuse of the Recycled Aircraft Components in Castings. Other opportunities for the use of recycled metal from aircraft may include aluminum alloy castings, especially those of the 2*xx*.0 and 7*xx*.0 series, aluminum-copper and aluminum-zinc, respectively. Examples of such alloys are included in Table 9.8. Even these relatively tolerant limits pose challenges for direct reuse of recycled metal. Nevertheless, opportunity for study of new alloy options remains, and the properties of the alloys in Table 9.5, when produced as casting, should be studied.

Alloys Designed for Recycling

As noted, an ideal component of resource maximization in recycling would be the availability of a larger number of aluminum alloys suitable for a wide variety of applications that do not require any extra purification step before they are recycled into high-performance products. That is largely the case with beverage cans, if the recycle scrap is not mixed with any other material.

Such an approach calls for "tailored" alloys. The goal of identifying new recycling-friendly aluminum alloy compositions is to increase the opportunities to directly, or with only minor modification, reuse recycled scrap aluminum products. Such an approach requires compositions with relatively broad specification limits on major alloying elements, such as copper and magnesium, plus more tolerant (i.e., higher) limits on iron, silicon, and other impurities, without significant restriction on performance characteristics for many applications.

Full development of this approach requires several important steps. Application of LIBS, developed by Huron Valley Steel Corp., to screen scrap with certain combinations of the desired elemental additions may permit relaxation of the broadest interpretation of the aforementioned guidelines, but it is beneficial to look at the most useful long-term trends when such technology may not be available. Adopting the approach of alloy optimization for recycling requires several steps, which are potentially phases in a development program. The phases may include:

- Identify with increasing precision the range of expected current and future recycled metal content, using feedback from organizations that are already capitalizing on the economics of recycling. Perform a mass balance to the extent practical, indicating the relative volumes of various scrap compositions to be expected
- Identify approximately five to seven basic candidate compositions of alloys that would accept recycled metal directly and have acceptable/desirable performance characteristics for a wide variety of applications, including structural components for bridges and buildings, high-temperature applications, and architectural usage
- Evaluate the performance of these candidate alloys in representative production lots, including, in particular, the following, along with the usual tensile and design properties, in order to assess their abilities to meet the requirements of representative applications as compared to existing alloys (Ref 9.6–9.11):

  a. Atmospheric corrosion resistance
  b. Stress-corrosion crack growth

Table 9.7 Nominal compositions of some 2*xxx* and 7*xxx* alloys used in nonaerospace applications

Alloy	Application	Cu	Fe	Mg	Mn	Si	Zn
2014	Railroad; truck bodies	4.4	0.7 max	0.50	0.8	0.8	0.15 max
2017	Rivets	4.0	0.7 max	0.60	0.7	0.5	0.25 max
7129	Auto bumpers	0.7	0.3 max	1.6	0.1 max	0.15 max	4.7

Table 9.8 Some aluminum casting alloy compositions

Alloy	Al	Cu	Fe	Mg	Mn	Si	Zn	Others
201.0	~95	4.6	0.15 max	0.35	0.35	0.10	0.25	0.05 max
242.0	~94	4.1	0.6 max	1.4	0.10 max	0.6 max	0.10	0.05 max
295.0	~94	4.5	1.0 max	0.03	0.35 max	1.1	0.25	0.05 max
710.0	~93	0.5	0.5 max	0.7	0.05 max	0.15 max	6.5	0.05 max
713.0	~91	0.7	1.1 max	0.35	0.6 max	0.25	7.5	0.05 max

c. Toughness, with tear tests and/or fracture toughness tests (for thick sections)
d. Formability tests, with bulge, minimum bend, and hemming tests

It is recognized that there will be some negative effects; the question is the degree to which such alloys are still useful for high-volume applications.

Developing Recycling-Friendly Compositions

Based on what is known about recycled aluminum metal, preliminary candidates for recycling-friendly alloys can be considered. Attention is given here primarily to wrought alloy compositions, because, as noted earlier, casting alloys already can be produced in rather large quantities from recycled cast products. The greatest challenge is direct use of recycled wrought products. The rationale is to define alloy specifications that can be readily met using recycled aluminum with no addition of primary metal. The target applications for the new recycling-friendly aluminum alloys include many of the same as for their existing counterparts, with tighter limits. Examples may include:

- 3*xxx*: heat-exchanger tubing, chemical piping
- 4*xxx*: forged or cast engine parts
- 5*xxx*: tankage plate; housing components
- 6*xxx*: extruded structural components

While suitable performance requirements with the higher level of impurities may not be entirely successful, steps in that direction will enable the aluminum industry to maximize its recycling opportunities. Therefore, addressing the challenge of a new approach to alloy design is warranted.

Candidate Compositions for Recycling-Friendly Aluminum Alloys. In developing candidate compositions, the following rather basic guidelines have been used:

- For major alloying elements in a particular series (e.g., copper in the 2*xx*x series, silicon and magnesium in the 6*xx*x series, etc.), propose relatively broad specification limits. Select alloying elements that are commonly and successfully used in alloys of the various series, for example, 2024 or 2219 in the 2*xx*x series, or 7005 in the 7*xx*x series.
- For limits on elements not usually added intentionally or on undesirable impurities (e.g., iron, nickel, and vanadium), propose more tolerant (i.e., higher) limits to the degree potentially practical, to the levels of those elements typically found in recycled metal.

Table 9.9 shows six preliminary candidate wrought alloy compositions that may reasonably be made from recycled and shred-sorted wrought products with, at minimum, the addition of some alloying elements. In this initial list, one composition has been selected from each major alloy series. Other candidates may well be devised by adjustments in the major alloying elements and/or the addition of other minor alloying elements.

These are only preliminary candidates, and it is recognized that the completion of phase 1 (described in the preceding section as identifying to a higher precision the compositions of incoming scrap, current and future) may result in significant changes in these candidate alloys. It may also lead to a focus on several different candidates from specific series that show maximum fit with the incoming metal (e.g., the 3*xxx*, 5*xxx*, and 6*xxx* series representing the highest volumes of recycled metal).

The application targets for these candidate compositions are much the same as their existing counterparts with tighter limits, recognizing that these are not likely to be suitable for the more fracture-critical items. However, the possibility exists that they may perform quite satisfactorily in such applications as chemical plant piping (A-2*xxx*), heat-exchanger tubing (B-3*xxx*), forged or cast engine parts (C-4*xxx*), rolled and extruded structural components

Table 9.9 Candidate alloys for recycling

Alloy	Si	Fe	Cu	Mn	Mg	Zn	Others
A (2*xxx*)	0.7	0.6	5.5–7.0	0.2–0.4	0.7	0.5	0.3
B (3*xxx*)	0.7	0.6	0.4	1.0–1.5	0.8–1.5	0.5	0.3
C (4*xxx*)	10.0–14.0	1.0	0.5–1.5	0.3	0.8–1.5	0.5	0.3
D (5*xxx*)	0.7	0.6	0.3	0.05–0.35	2.0–3.0	0.5	0.3
E (6*xxx*)	0.3–1.0	0.6	0.3	0.3	0.4–1.0	0.5	0.3
F (7*xxx*)	0.5	0.6	0.5–1.2	0.3	2.0–2.8	4.0–6.0	0.3

(D-5*xxx* and E-6*xxx*), and even for noncritical aircraft components (F-7*xxx*).

The application of LIBS technology to screen scrap with certain combinations of the desired elemental additions may permit relaxation of the broadest interpretation of the aforementioned guidelines, but it is beneficial to look at the most useful long-term trends when such technology may not be widely available.

Unialloy. Another approach that may be considered, and one that has been studied to considerable degree in the past, is the development of one or two "unialloys," that is, alloys that meet all the requirements for a large application, such as aluminum beverage packaging, automotive components, or architectural components. The aforementioned approach of selecting "recycle-optimized" alloys from each series may lead to a master list of several recycling-friendly alloys.

The unialloy concept was first developed by Golden Aluminum (Ref 9.12) in the late 1980s and focused on alloy AA5017 with 2% Mg and 0.7% Mn. The concept was derived from the idea that unialloy had an average weighted composition of can body alloy AA3004 and can end alloy AA5182. This recycling idea achieved limited application due to economic and commercial factors. In the current environment of rising prices of primary aluminum and its alloying elements, such as magnesium, manganese, and copper, coupled with the societal desire for enhancing recycling rates of products, it is time to rethink applicability and commercialization of the unialloy concept.

This has proven difficult to achieve because of the diverging performance requirements of different applications. Even within autobody panels, for example, the differing requirements for dent resistance in outer panels and optimized formability for inner panels continue to lead to two different types of alloys being used (for example, 6111 heat treated for high outer panel dent resistance, and 5754 annealed for maximum formability for inner panels).

Some Caveats. The large bank of alloy design experience in the aluminum industry may well result in skepticism about the probabilities of success and the seeming backward movement in the aforementioned approach of permitting higher levels of impurities in new alloy candidates. There is indeed some reason for the skepticism, because it is well known that optimizing fracture toughness, for example, requires tight impurity controls, especially on iron and silicon. So, it is probably appropriate to acknowledge at the outset that it may be impossible to reach the stage where recycled metal will be used untreated for all aluminum alloy products, such as fracture-critical-component aerospace wings and wing spars; the fracture toughness requirements on these are simply too stringent to be met without tight impurity controls.

However, aerospace applications account for only approximately 8% of the total approximated 14 to 18 $\times$ 10^6 metric tons (30 to 40 billion lb) of aluminum used annually. Therefore, the adoption of a recycling-friendly alloy system applicable to most other applications will still have a very great economic and ecological benefit in world consumption.

The performance requirements of many high-volume aluminum products, such as building and highway structures and chemical industry components, may well be satisfied by alloys with higher levels of impurities than presently mandated. In fact, most composition limits were set when the greater volume of production was primary metal and there was no need for higher impurity levels. The limits were not set by performance requirements but by anticipated incoming metal compositions.

Finally, it must be acknowledged that attempts at obtaining suitable performance requirements with higher levels of impurities may not be successful. However, any step in that direction will better enable the aluminum industry to maximize its recycling opportunities; therefore, the challenge of a new approach to alloy design is warranted.

Conclusions and Looking Ahead

There are significant economic and ecological advantages in maximizing the recycling rates of aluminum and the ready reuse of recycled metal. Several important conclusions for the aluminum industry, both in the United States and throughout the world, include:

- Avenues for recovery of aluminum scrap from as many products as possible should continue to be exploited; there are massive economic, energy, and ecological advantages to the communities and to the aluminum industry.
- Development and application of enhanced shredding and sorting technologies should continue.

- A focus on the most cost-effective remelting processes is justified, including the possibility of combining iron with zirconium and other impurities such as nickel and vanadium in high-density particles that could be easily separated from the melt.
- The production of alternative products such as aluminum-iron deoxidizing agents should be pursued to use that part of recycled aluminum products that cannot cost-effectively be used in the production of new aluminum alloys, to the benefit of both the steel and the aluminum industries.
- The most overlooked aspect of maximizing recycled metal appears to be the development of new alloys tailored to meet composition and performance criteria when produced directly from recycled metal. A study of the type suggested in the section "Alloys Designed for Recycling" in this chapter should be carried out to potentially add to the number of alloys available for direct recycling. Some candidate compositions have been suggested for such alloys, with the intent to broaden this part of the discussion and perhaps lead to interesting and economically attractive components to maximize recycling efficiency and effectiveness.

Looking ahead, the challenge is to identify the most successful means of implementing the additional studies and development programs needed to maximize the benefits of recycling. Accordingly, one step may be the formation of an aluminum recycling consortium to consider options and opportunities to increase the overall effectiveness and efficiency of aluminum recycling and its benefits. The consortium could include representatives of:

- The municipalities and their representatives at the first line of collecting recycled metal
- The recyclers themselves (e.g., members of the Aluminum Association Recycling Division)
- Fabricators and distributors of recycled products (e.g., auto and beverage can producers)
- End-users of such recycled products

In this regard, Secat, Inc. is in the process of forming an aluminum recycling consortium to further the goals outlined in this chapter. The first two charges of the consortium should be to:

- Identify all means of maximizing the amount of aluminum products entering the recycling chain. Carry out the study of alloy design optimization for recycling as described in the section "Alloys Designed for Recycling," identifying the most likely high-volume recycling compositions, carrying out the resultant mass balance, and fine-tuning several candidate compositions that would take advantage of the anticipated recycling content
- Explore opportunities to improve the quality of the melt by removal of high-density particles formed by the combination of iron with zirconium and possibly other high-density elements such as nickel and vanadium.

The next logical step would be a complete evaluation of the physical, mechanical, corrosion, and fabricating characteristics of whatever optimized recycling candidate alloys are generated. Successful application of these various approaches to maximizing the cost-effectiveness and efficiency of recycling processes should lead to increased opportunity to extend the life-cycle advantages of aluminum alloys, increase the usefulness of directly recycled alloys, and therefore increase the amount of metal that is directly reused without the addition of primary metal.

ACKNOWLEDGMENTS

The author gratefully acknowledges valuable input and advice from Gil Kaufman, Vice President of Technology, The Aluminum Association (Retired), Dr. John A.S. Green, Vice President of Technology, The Aluminum Association (Retired); Dr. Warren Hunt, President, TMS; and Dr. Wayne Hayden, Oak Ridge National Laboratory (Retired).

REFERENCES

9.1. W.T. Choate and J.A.S. Green, *Modeling the Impact of Secondary Recovery (Recycling) on the U.S. Aluminum Supply and Nominal Energy Requirements,* TMS, 2004

9.2. *Aluminum Industry Technology Roadmap,* The Aluminum Association, Washington, D.C., 2003

9.3. R.A.P. Fielding, Recycling Aluminum, Especially Processing Extrusion Scrap, *Light Met. Age,* Aug 2005, p. 20–35

9.4. A. Gesing, L. Berry, R. Dalton, and R. Wolanski, Assuring Continued Recyclability of Automotive Aluminum Alloys: Grouping of Wrought Alloys by Color, X-Ray Absorption and Chemical Composition-Based Sorting, *TMS 2002 Annual Meeting: Automotive Alloys and Aluminum Sheet and Plate Rolling and Finishing Technology Symposia,* Feb 18–21, 2002 (Seattle, WA), Minerals, Metals and Materials Society (TMS), 2002, p. 3–17

9.5, S.K. Das and J.G. Kaufman, *Reycling Aluminum Aerospace Alloys,* TMS, 2007

9.6. *Properties and Selection: Nonferrous Alloys and Special–Purpose Materials, Vol 2, ASM Handbook*, ASM International, 1990

9.7. J.E. Hatch, Ed., *Properties and Physical Metallurgy,* American Society for Metals, 1984

9.8. J.R. Davis, Ed., *Aluminum and Aluminum Alloys,* ASM International, 1993

9.9. J.G. Kaufman, Ed., *Properties of Aluminum Alloys—Tensile, Creep, and Fatigue Data at High and Low Temperatures,* ASM International, 1999

9.10. J.G. Kaufsman, *Fracture Resistance of Aluminum Alloys—Notch Toughness, Tear Resistance, and Fracture Toughness,* ASM International, 2001

9.11. J.G. Kaufman and E.L. Rooy, *Aluminum Alloy Castings—Properties, Processes, and Applications,* ASM International, 2004

9.12. D. McAuliffe and I.M. Marsh, *Light Metals 1989,* p739–741

CHAPTER 10

U.S. Energy Requirements for Aluminum Production: Historical Perspective, Theoretical Limits, and New Opportunities*

William Choate, BCS Inc.

THIS CHAPTER, based on a 2005 revised version of a 2003 report (Ref 10.1) for the U.S. Department of Energy, provides reliable and comprehensive statistical data over the period 1960 to 2005 for the evaluation of energy trends and issues in the U.S. aluminum industry. It should be noted, however, that these trends need careful interpretation because they incorporate unusual circumstances of a single year, that is, 2001. During the summer of 2001, the extensive heat wave in the western United States produced an increased demand for electricity. Simultaneously, the ability to generate hydroelectric power was reduced due to historically low snow packs in the Columbia River basin and new regulations mandating the spill of water to aid migrating salmon. The combination of high electricity demand and limited water supply contributed to a significant increase in the market price of electricity during this time. This price increase in the Pacific Northwest made it more economical for aluminum smelters to stop metal production and sell back power from their low-cost, fixed-price electric contracts to aid in minimizing the shortfall in energy supply. As a result, the majority of aluminum smelting capacity in the Pacific Northwest, representing approximately 43% of all U.S. primary aluminum capacity, shut down.

The dramatic and relatively quick shutdown of a substantial amount of the U.S. primary aluminum capacity can make certain trend numbers appear misleading. Throughout this chapter, percentages are given for 10-year trends in various sectors throughout the aluminum industry. For example, U.S. primary aluminum production has a 10-year annual growth rate of −4.2%, decreasing from 3375 thousand metric tons in 1995 to 2480 thousand metric tons in 2005. These numbers would seem to imply a steady decline throughout the specified time period, but this is not the case. In fact, the total drop in primary production over this time period (895 thousand metric tons) is less than the 1031 thousand metric ton drop from 2000 (3668 thousand metric tons) to 2001 (2637 metric tons) (Ref 10.2). This one-year drop coincides with the majority of the Pacific Northwest shutdown. Other primary production numbers in this chapter have similar trends, and this drastic shutdown should be taken into account when considering these numbers.

It is currently too early to accurately assess the long-term impact of these sudden shutdowns and changing conditions on the aluminum industry. It remains to be seen whether the shutdowns will

*Adapted with permission from Ref 10.1. The original reference 10.1 was prepared for Industrial Technologies Program, Energy Efficiency and Renewal Energy, and funded by the U.S. Department of Energy.

lead to a permanent decline of primary metal production in the Pacific Northwest, or whether the industry will emerge based on higher aluminum prices, new power contracts, or additional self-generated power capacity and energy efficiency improvements. Between October 8, 2004, and January 2006, +110,000 metric tons/yr of the Pacific Northwest capacity has come back on line. It is difficult to judge whether or not this will be the common trend. Whatever the future of the industry, it is clear that the local and global pressures to increase overall energy efficiency will determine its vitality. The energy-efficiency opportunities discussed in this chapter are pertinent to the future of the aluminum industry.

Another recent trend to make note of is China's rapid growth in both primary and secondary aluminum production. While the United States has slipped from being the largest producer of primary metal in 2000 to being the fourth largest in 2005, China has taken the worldwide lead in primary aluminum production, producing 7,800,000 metric tons of primary aluminum in 2005. Additionally, China has been aggressively buying up much of the aluminum scrap supply. In 2003, the United States exported 568,721 metric tons of scrap and dross, with 43% (244,374 metric tons) going to China (Ref 10.2). These trends are manifested in the slight decline of secondary aluminum production within the United States.

Summary

The United States aluminum industry, processed 10.7 million metric tons of metal and produced over \$40 billion in products and exports in 2005. It operates more than 400 plants in 41 states and employs more than 145,000 people. Aluminum impacts every community and person in the country, through either its use and recycling or the economic benefits of manufacturing facilities.

Energy reduction in the U.S. aluminum industry is the result of technical progress and the growth of recycling. These two factors have contributed 22 and 42%, respectively, to the total 64% energy reduction over the past 45 years. By many measures, aluminum remains one of the most energy-intensive materials to produce. Only paper, gasoline, steel, and ethylene manufacturing consume more total energy in the United States than aluminum. Aluminum production is the largest consumer of energy on a per-weight basis and is the largest electric energy consumer of all manufactured products. The U.S. aluminum industry directly consumes 42.3×10^9 kWh (0.16 quad, or 0.16×10^{15} Btu) of electricity annually, or 1.1% of all the electricity consumed by the residential, commercial, and industrial sectors of the U.S. economy. This is equivalent to the electricity consumed by 4,826,000 U.S. households annually, or the energy equivalent of nearly 50 million barrels of oil.

The aluminum industry has large opportunities to further reduce its energy intensity. The annual sum of all the energy required in the production of aluminum metal and products in the United States is equivalent to 173×10^9 kWh (0.59 quad). The difference between the gross annual energy required and the theoretical minimum requirement amounts to over 141×10^9 kWh (0.48 quad). This difference is a measure of the theoretical potential opportunity for reducing energy consumption in the industry, although achievable cost-effective savings are smaller.

This chapter provides energy performance benchmarks for evaluating new process developments, tracking progress toward performance targets, and facilitating comparisons of energy use. It provides a basic description of the processes and equipment involved, their interrelationship, and their effects on the energy consumed and environmental impact of manufacturing aluminum and aluminum products. This knowledge can help identify and understand process areas where significant energy reductions and environmental impact improvements can be made.

This chapter examines and carefully distinguishes between the actual on-site energy consumption values and gross or tacit energy values. The tacit or gross energy value accounts for the generation and transmission energy losses associated with electricity production, the feedstock energy of fuels used as materials, and the process energy used to produce fuels. On-site energy improvements provide concomitant gross energy savings.

Primary aluminum is produced globally by mining bauxite ore, refining the ore to alumina, and combining the alumina and carbon in an electrolytic cell to produce aluminum metal. Secondary aluminum is produced globally from recycled aluminum scrap. Primary and secondary aluminum metal are cast into large ingots, billets, T-bar, slab, or strip and then rolled, extruded, shape cast, or otherwise formed into the components and products used daily.

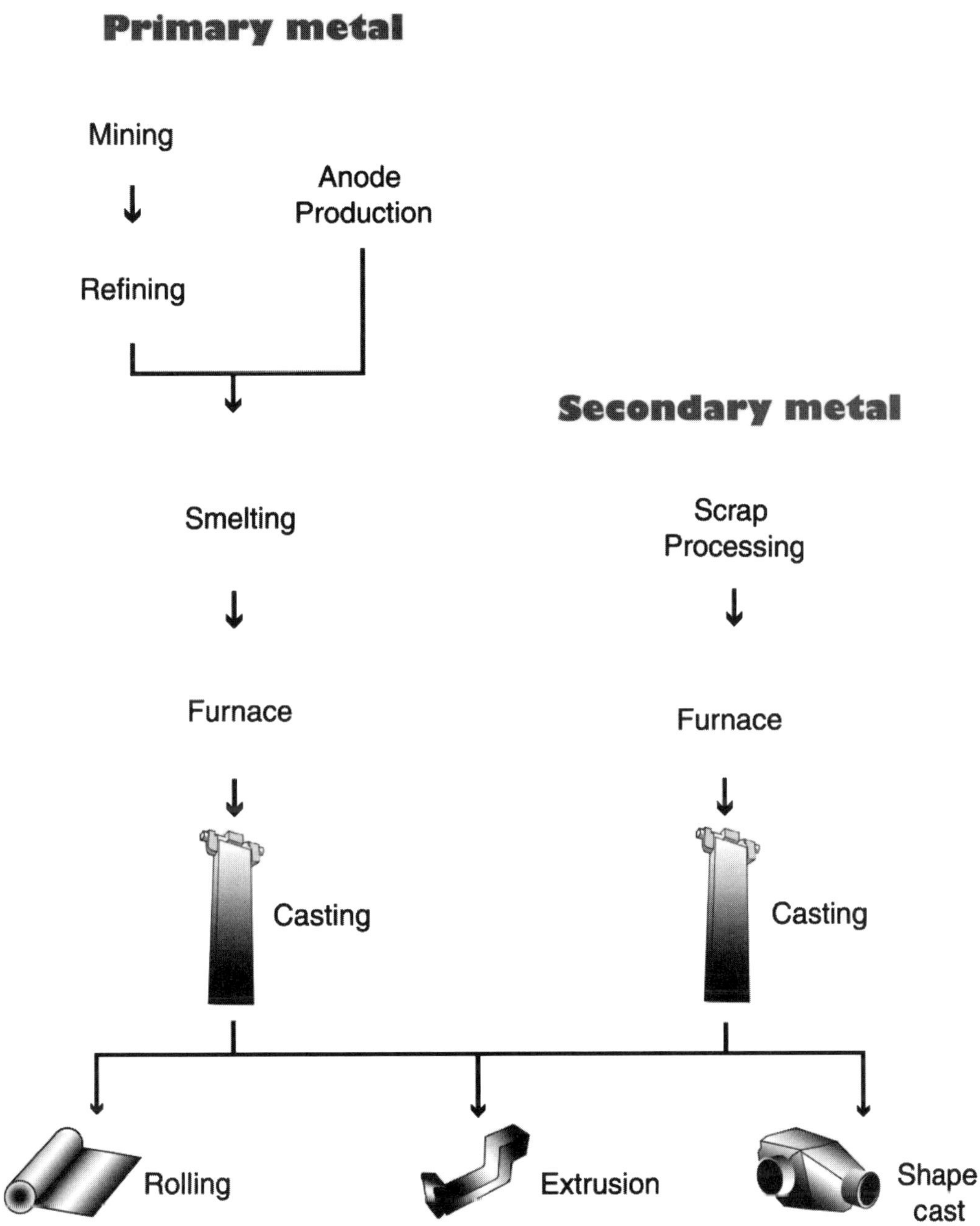

Fig. 10.1 Aluminum industry flow diagram

Figure 10.1 shows the major processing operations required to produce aluminum and aluminum products. This chapter examines these processes and the energy they require.

Note on Tacit Energy Consumption. A complete accounting of the energy consumed in the production of any product should include the energy required to produce the fuels and electricity that are used plus the feedstock energy associated with any fuels that are used as materials (e.g., carbon used in anode production). This chapter, in many places, reports energy as a set of two numbers, for example, 1.0 (2.0^{tf}) kWh. The first value represents the energy consumed within a facility (on-site energy consumption, or primary energy), while the value with "tf" superscript (tacit-feedstock energy) is a measure of energy that includes the energy used to produce and transmit the energy consumed within a facility and the raw material feedstock energy. This adjustment is very significant in processes that consume electricity or use fuels as materials.

The U.S. average grid connection requires approximately 3 kWh (9890 Btu) of fuel energy to deliver 1 kWh (3412 Btu) of electrical energy (Table 10.1). The average U.S. grid is supplied with approximately 7% hydroelectric generation. The aluminum industry is located near low-cost power sources, and many of these are hydroelectric. The average U.S. primary aluminum plant connection is 39% hydroelectric generation. If it is assumed that the heat rate for hydroelectric power is 3412 Btu/kWh, then the average primary aluminum plant grid connection requires 2.2 kWh (7570 Btu) of fuel energy to deliver 1 kWh (3412 Btu) of electrical energy. It should be noted that values reported in this chapter use the U.S. average grid connection values. The use of U.S. average grid values results in a higher reported tacit energy consumption value than the actual tacit value required for primary aluminum production. The advantage of reporting primary production energy consumption based on the U.S. average grid is that it allows easier and same-basis comparisons to other U.S. manufacturing industries.

Aluminum Production and Energy Consumption

Aluminum is an essential material for modern manufacturing. It is a lightweight, high-strength, corrosion-resistant metal with high electrical and thermal conductivity, and it is easy to recycle. The U.S. aluminum industry is the largest in the world in terms of consumption. The U.S. aluminum industry used 9,266,000 metric tons of metal in 2005 to produce an enormous variety of products. The U.S. per capita consumption was 66 lb. The industry operated more than 400 plants in 41 states and employed over 145,000 people to make aluminum products (Ref 10.2). These products are shipped to thousands of businesses in the United States, from which they are distributed or are incorporated into other products. Aluminum, per unit mass, is the most energy-intensive material produced in large quantities in the United States. Only paper, gasoline, steel, and ethylene manufacturing consume more total energy for manufacturing in the United States than aluminum (Appendix A).

Research and development (R&D) efforts to reduce energy consumption are important, because energy consumption correlates to manufacturing economics, environmental impact, and U.S. dependence on imported energy sources. Identifying process areas where opportunities for energy-use reduction exist and applying resources to capture these opportunities will benefit the industry and the nation.

The science and technologies associated with the production of aluminum and aluminum products are complex. This report attempts to provide the reader with a basic understanding of the science, technology, and energy usage of the aluminum industry. More detailed books (Ref 10.3, 10.4) are available for the reader who requires further in-depth study of the subject.

Energy and Environmental Overview. The energy consumption and environmental effects associated with product manufacturing and use are important measures of the impact of the product on society. Energy consumption and environmental-impact measures are becoming key decision tools for consumers and corporations when choosing a product. In the near future, manufactured products will compete not only on price and performance but also on their impact on society.

Aluminum manufacturing is energy-intensive, and approximately one-third of the cost to

Table 10.1 Electric tacit energy and emission data for fuels used for aluminum production

Electrical energy source	U. S. primary aluminum capacity: Metric tons	U. S. primary aluminum capacity: %	Heat rate (2005), Btu/kWh	Carbon emission coefficient, Mt/Qbtu
Hydro	1,633,696	39.4	3412	0
Coal	2,415,826	58.2	10,303	21.25
Oil	8,245	0.2	10,090	19.08
Natural gas	31,120	0.8	7937	12.50
Nuclear	60,112	1.4	10.420	0
Total	4,149,000	100.0		
	Weighted averages based on aluminum capacity		7570	12.51
	Average U.S. grid		9890	13.56

See Appendix C for additional details

produce aluminum from ore is associated with the use of energy and environmental compliance. The aluminum industry, in the past 40 years, reduced its overall energy intensity by nearly 64% (Table 10.2). However, even with the large reduction in energy intensity, the industry consumes nearly three times the theoretical minimum energy required. Significant opportunities for further energy improvements still remain.

Technologies, practices, and product use determine the energy consumption and environmental impact of aluminum. Many of the current technologies and practices used to produce aluminum metal and aluminum products are mature. New technologies and practices are being proposed and studied to improve aluminum manufacturing from an energy and environmental standpoint. The history and explanation of current state-of-the-art technologies and practices are presented so the reader can appreciate the values and benefits that new technologies or practices may bring to the aluminum industry. Current U.S.

Table 10.2 Impact of secondary metal production on energy requirements for U.S. aluminum production

Year	U.S. primary aluminum production, thousand metric tonnes	Smelting energy, kWh/kg	U.S. secondary aluminum production, thousand metric tonnes	Market percent of secondary metal, %	Total energy ore-to-metal to produce combined metals, kWh/kg	Effective smelting energy to produce combined metals, kWh/kg
1960	1828	23.1	401	18	67.6	19.2
1961	1727	22.9	445	...	65.1	18.5
1962	1921	22.7	533	...	63.6	18.1
1963	2098	22.6	601	...	62.7	17.9
1964	2316	22.4	648	...	62.6	17.8
1965	2499	22.3	774	...	60.7	17.3
1966	2693	22.1	833	...	60.3	17.2
1967	2966	21.9	821	...	61.4	17.5
1968	2953	21.8	935	...	59.1	16.8
1969	3441	21.6	1067	...	58.9	16.8
1970	3607	21.4	937	21	60.7	17.3
1971	3561	21.0	1004	...	58.6	16.7
1972	3740	20.6	1022	...	57.9	16.5
1973	4109	20.2	1127	...	56.8	16.1
1974	4448	19.9	1163	...	56.2	16.0
1975	3519	19.5	1121	...	52.8	15.0
1976	3857	19.1	1334	...	50.7	14.5
1977	4117	18.7	1456	...	49.4	14.1
1978	4358	18.3	1518	...	48.5	13.8
1979	4557	17.9	1612	...	47.3	13.5
1980	4653	17.5	1577	25	46.8	13.3
1981	4489	17.4	1790	...	44.4	12.7
1982	3274	17.2	1666	...	40.9	11.8
1983	3353	17.1	1773	...	40.1	11.5
1984	4099	16.9	1760	...	42.4	12.1
1985	3500	16.8	1762	...	40.0	11.5
1986	3039	16.6	1773	...	37.8	10.9
1987	3347	16.5	1986	...	37.2	10.7
1988	3945	16.3	2122	...	38.2	11.0
1989	4030	16.2	2054	...	38.5	11.1
1990	4048	16.1	2393	37	36.3	10.5
1991	4121	16.0	2286	...	36.9	10.6
1992	4042	15.9	2756		34.0	9.8
1993	3695	15.8	2944	...	31.7	9.2
1994	3299	15.7	3086	...	29.3	8.6
1995	3375	15.6	3188	...	29.0	8.5
1996	3577	15.5	3307	...	29.1	8.5
1997	3603	15.4	3547	...	28.1	8.2
1998	3713	15.3	3442	...	28.7	8.4
1999	3779	15.2	3695	...	27.8	8.1
2000	3668	15.1	3450	48	28.1	8.2
2001	2637	15.0	2970	...	25.6	7.5
2002	2705	14.9	2927	...	25.9	7.6
2003	2704	14.8	2820	...	26.2	7.7
2004	2517	14.6	3025	...	24.1	7.1
2005	2480	14.4	2990	55	23.7	7.0

The nominal U.S. energy required to produce aluminum metal has been rapidly declining as secondary aluminum production has grown Secondary aluminum requires only 6% of the energy necessary to manufacture primary aluminum (Appendix E, Table E.6). The total U.S. ore-to-metal primary energy values include: refining of half the alumina supply, anode manufacture, electrolysis, and ingot casting. Combining the energy requirements for U.S. production of primary and secondary metals lowers the average energy associated with U.S. aluminum metal from over 40 kWh/kg for primary alone to approximately 21 kWh/kg for the combined metals.

production levels, historical production levels, and projected growth rates of aluminum are presented. These production values are needed to measure the magnitude of the impact of a change in technology or practice. The energy and environmental impacts from the use of aluminum products are generally low and, in some applications such as transpontation, may be significantly better than the impacts of alternative materials. As significant as these impacts are on the use of aluminum products, they are beyond the scope of this chapter.

The greatest impact on the future energy intensity of aluminum has been the structural change in the industry itself. More than 55% of the aluminum produced by U.S. industry in 2005 came from recycled material. In 1960, recycled material was used to generate less than 18% of U.S.-produced aluminum (Table 10.3). Recovering aluminum from wastes and scraps requires less than 6% of the energy of aluminum production from bauxite mining (Table 10.4). This report examines how "urban mining" (recycling) will continue to change the structure of the aluminum industry and continue to lower the overall energy associated with aluminum production. Recycling is the largest contributor to the reduction of the energy intensity of U.S-produced aluminum.

Aluminum is an "energy bank" in that nearly all of the original energy stored in the metal can be recovered again and again every time the product is recycled. Small fractions of the recycled metal are lost to oxidation (melt loss) and entrapment in purifying fluxes (dross) during the recycling process. Aluminum can be recycled indefinitely, allowing this saved energy to be collected again and again.

Greenhouse gas (GHG) emission reduction is a key environmental and sustainability issue for the 21st century. Energy-intensive manufactured materials (such as aluminum) could be significantly affected, both in terms of price and use, by GHG emission-reduction policies. However, contrary to common belief, aluminum production could be positively affected by GHG emission-reduction policies. A combination of emission mitigation in production and significant GHG emission reduction further down the product chain enhance the attractiveness of aluminum for end-use applications, particularly in the fastest growing sector of transportation (5.4% annual growth from 2001 to 2005).

Additional energy and environmental savings can be achieved in the aluminum product chain through the introduction of new alloys and

Table 10.3 U.S. supply of aluminum from 1960 to 2005 (in thousand metric dry tons)

Year	U.S. primary aluminum	U.S. secondary aluminum	U.S. aluminum imports	U.S. total aluminum
1960	1828	401	178	2407
1961	1727	445	230	2402
1962	1921	533	341	2795
1963	2098	601	419	3118
1964	2316	648	409	3373
1965	2499	774	543	3816
1966	2693	833	591	4117
1967	2966	821	466	4253
1968	2953	935	684	4572
1969	3441	1067	484	4992
1970	3607	937	406	4950
1971	3561	1004	582	5147
1972	3740	1022	684	5446
1973	4109	1127	523	5759
1974	4448	1163	511	6122
1975	3519	1121	453	5093
1976	3857	1334	608	5799
1977	4117	1456	683	6256
1978	4358	1518	907	6783
1979	4557	1612	711	6880
1980	4653	1577	603	6833
1981	4489	1790	782	7061
1982	3274	1666	823	5763
1983	3353	1773	1023	6149
1984	4099	1760	1376	7235
1985	3500	1762	1332	6594
1986	3039	1773	1843	6655
1987	3347	1986	1702	7035
1988	3945	2122	1467	7534
1989	4030	2054	1353	7437
1990	4048	2393	1421	7862
1991	4121	2286	1398	7805
1992	4042	2756	1573	8371
1993	3695	2944	2327	8966
1994	3299	3086	3136	9521
1995	3375	3188	2701	9264
1996	3577	3307	2572	9456
1997	3603	3547	2804	9954
1998	3713	3442	3264	10,419
1999	3779	3695	3680	11,154
2000	3668	3450	3580	10,698
2001	2637	2970	3487	9094
2002	2705	2927	3947	9579
2003	2704	2820	4068	9592
2004	2517	3025	4565	10,107
2005	2480	2990	5221	10,691
Average growth rate 1995 through 2005, %				
	−4.2	−1.7	6.9	0.4
Percent of total supply (2005), %				
	28	29	42	100

Source: *Aluminum Statistical Review for 2003*, The Aluminum Association, 2004, p 7

Table 10.4 Energy saved with recycling (tacit energy values)

60.5	kWh/kg to produce primary metal ingot
2.8	kWh/kg to produce secondary metal ingot
5.0%	secondary-to-primary energy
2,990,000	kg of secondary metal produced in 2003
1.72×10^{11}	kWh/yr energy saved 2003
0.59	quads saved per year 2003
19,700	MW saved 2003

improved (lightweight) product design. These options are not considered in this chapter, but their potential is at least of the same order of magnitude as changes to production practices and processes.

Identifying Energy-Reduction Opportunities. Energy performance benchmarks, current practice, and theoretical minimums provide the basis for evaluating energy-reduction opportunities. These benchmarks and gross energy consumed during aluminum production in the United States are summarized in Table 10.5.

Industrial processes that consume energy at significantly higher rates than their theoretical requirements are, on the surface, obvious targets for potential improvement. However, energy performance is only one factor in identifying the best opportunities for improving energy efficiency. Other factors, particularly market dynamics, process economics, and forecasting of future demand are very significant in identifying real opportunities. This report examines the energy performance of the operations involved in manufacturing aluminum products.

The amounts of energy used onsite in the major processing operations of the U.S. aluminum industry are shown in Fig. 10.2. The bottom band on each bar shows the theoretical energy requirement, while the top band of each bar shows the energy used above the theoretical minimum. The size of the top band is an indication of how large the opportunity is for energy reduction in that process step.

Smelting requires 43% of the total primary energy consumed in U.S. manufacturing of aluminum (66% when electric generation and

Table 10.5 U.S. energy requirements and potential savings

	U.S. annual production 2005, metric tons	Theoretical minimum energy requirement, kWh (10^9)/yr (quad)	U.S. process energy required, kWh (10^9)/yr (quad)	Potential process U.S. energy savings, kWh (10^9)/yr (quad)	Total U.S. gross energy[tf] required(a), kWh (10^9)/yr (quad)	Potential gross U.S. energy[tf] savings(a), kWh (10^9)/yr (quad)
Bauxite mining alumina refining	4,419,000	0.62 (0.002)	16.64 (0.057)	16.02 (0.055)	18.03 (0.062)	17.41 (0.059)
Anode production	1,128,000	11.12 (0.038)	14.44 (0.049)	3.33 (0.011)	15.20 (0.052)	4.08 (0.014)
Al smelting	2,530,000	15.15 (0.052)	39.42 (0.135)	24.27 (0.083)	113.41 (0.387)	98.26 (0.335)
Primary casting	2,480,400	0.82 (0.003)	2.50 (0.009)	1.68 (0.006)	3.55 (0.012)	2.73 (0.009)
Secondary casting	2,990,000	0.99 (0.003)	7.47 (0.026)	6.48 (0.022)	8.37 (0.029)	7.37 (0.025)
Rolling	4,842,600	1.55 (0.005)	3.04 (0.010)	1.49 (0.005)	5.91 (0.020)	4.36 (0.015)
Extrusion	1,826,000	0.80 (0.003)	2.37 (0.008)	1.57 (0.005)	2.75 (0.009)	1.96 (0.007)
Shape casting	2,289,000	0.76 (0.003)	5.85 (0.020)	5.09 (0.017)	6.04 (0.021)	5.28 (0.018)
Total		**31.81 (0.109)**	**91.74 (0.313)**	**59.93 (0.204)**	**173.3 (0.591)**	**141.4 (0.483)**

(a) tf, tacit feedstock

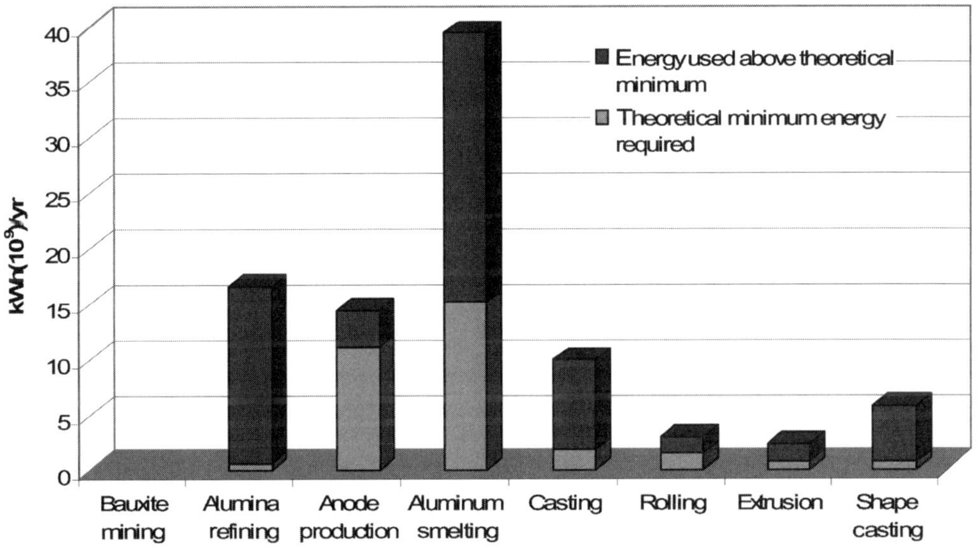

Fig. 10.2 Process energy used in U.S manufacturing of aluminum products

transmission losses are considered). This process is the largest consumer of energy and the most technically complex operation. Smelting requires more than twice its theoretical minimum energy and has the potential for the greatest energy reduction of all operations. Electricity is required for smelting and accounts for over 98% of the energy used in the process. Current R&D efforts to advance existing technology and to develop alternatives to the existing smelting process have the potential to lower smelting energy consumption by more than 30%.

Process heating accounts for 27% of the total primary energy consumed in U.S. manufacturing of aluminum. Process heating is required for holding, melting, purifying, alloying, and heat treating. It is used in nearly all aluminum production operations. Heating is the second largest energy-consuming operation.

Recycled aluminum now accounts for over half of all U.S-produced aluminum. It requires less than 5% of the energy to produce aluminum from mined bauxite and provides significant environmental benefits. The R&D efforts that improve the ability to recycle aluminum offer some of the greatest opportunities for energy reduction in the industry, because recycling displaces aluminum produced by smelting.

The magnitude of the top bands of the energy bars in Fig. 10.2 shows that large opportunities exist for lowering energy consumption in the industry. The *Aluminum Industry Vision: Sustainable Solutions for a Dynamic World* (Ref 10.5), published by the Aluminum Association in 2001, recognizes these opportunities and sets industry goals for achieving further energy reduction. In Hall-Heroult smelting technology, the most energy-intensive process, the industry has set a target for reducing electrical energy usage to 11 kWh/kg of aluminum produced by the year 2020, a 27% reduction from 2000 practices.

Evaluation of the many opportunities that exist for reducing energy consumption in the industry can only be made by comparing processes using consistent system boundaries and measures. This chapter provides data and information necessary for the reader to understand opportunities for energy savings in the aluminum industry.

Methodology, Metrics, and Benchmarks

This chapter focuses on the most energy-intensive manufacturing operations for aluminum: electrolysis (smelting) and process heating operations. These two operations account for over 70% of the primary energy used by the industry (Table 10.6). There is a large difference between the theoretical minimum energy requirements and current practice energy values in electrolysis and melting. The magnitude of the energy consumed and the difference between current practice and theoretical energy levels means improvement in electrolysis and process heating will have the largest impact on the performance of the industry. This chapter documents existing operations and explores potential new technology opportunities.

The purpose of this chapter is to:

- Provide an understanding of the processes involved, the energy consumed, and the environmental impact of manufacturing aluminum and aluminum products
- Provide a common set of terms, benchmarks, and values for comparing processes and issues related to the aluminum industry
- Identify process areas in which significant energy reductions and environmental impact improvements could be made
- Strengthen public and workforce awareness, education, and training (as an identified industry goal in Ref 10.5)

Table 10.6 Smelting and heating fractions of total U.S. aluminum industry energy consumed

Smelting fraction of total U.S. aluminum industry energy consumed

		Electrolysis	Total industry	Percent of industry
Onsite	kWh/yr	3.94×10^{10}	9.17×10^{10}	43
Tacit	kWh/yr	1.13×10^{11}	1.73×10^{11}	66

Aluminum metal process heating/melting fraction of total U.S. aluminum industry energy consumed

		Casting	Rolling: Hot	Rolling: Cold	Extrusion	Shape casting	Total process heating	Total industry	Percent of industry
	Btu/kg Al	9244	1212	905	4096	8704	2.42×10^4		
	kWh/kg Al	2.71	0.36	0.27	1.20	2.55	7.08		
Onsite	kWh/yr	1.48×10^{10}	8.60×10^8	6.42×10^8	2.19×10^9	5.84×10^9	2.44×10^{10}	9.17×10^{10}	27
Tacit								1.73×10^{11}	14

Note: The tacit increase for process heating is negligible.

There are a variety of metrics, measurements, benchmarks, boundaries, systems, and units that are used differently by various analytical groups. These variations can cause confusion when comparing values stated in one report to those in another. Two commonly confused values are the relationship between on-site and tacit energy values, and between U.S. energy requirements and worldwide energy requirements. On-site energy values are based on physical measurements. Tacit energy values involve assumptions, which can create large differences in the reported values. The on-site and tacit values used here are explained in the section "Tacit, Process, Feedstock, and Secondary Energies" in this chapter. The United States does not mine ore for aluminum production but does refine ore domestically. Energy consumption within the United States is the focus here. The total energy associated with production of metal from ore is an important value and is reported as the "worldwide" energy requirement in this chapter.

Theoretical, Practical Minimum, and Current Practice Benchmarks. When examining industrial processes, two metric values for energy requirement are obtainable with little debate: the process theoretical minimum value and the current practice value. The theoretical minimum energy requirement for chemically transforming a material is based on the net chemical reaction used to manufacture the product. In the case of aluminum made from alumina ($2Al_2O_3 \rightleftharpoons 4Al + 3O_2$), the theoretical minimum energy is 9.03 kWh/kg of aluminum produced (see Appendix F). This minimum value is simplistic and represents the thermodynamically ideal energy consumption. It requires any reaction to proceed infinitely slowly. The theoretical minimum energy to transform a material from one shape to another shape is based on the mechanical properties of the material. It is also an idealized value. Neither chemical nor mechanical theoretical minimums can be realized in practice; however, these provide the benchmarks that no process can surpass. (Analogy: The theoretical minimum score for a full round of golf is 18.)

The current practice value is the average of the actual measurements of existing processes and practices. (Analogy: The current practice value for golf is the average score of every player, which is well above par.) The boundaries drawn around the process or practice, the number of samples, sampling techniques, and so on determine the precision and accuracy of this value. The difference between the theoretical minimum and current practice metric is a valuable measure of the opportunities for energy efficiency improvement in that process or practice.

Practical minimum energy is a term in common use. However, its definition varies widely. In some instances, it is used to describe the process energy value that represents the combination of integrated unit operations using best available technology and best energy management practices. In other instances, practical minimum energy is defined as the optimal design value projected with the adoption of new, advanced technology. Practical minimum energy values are, in reality, a moving target because it is not possible to predict the new technologies, practices, and materials that will impact an industrial process. What is known about the practical minimum energy value is that it lies somewhere between the current best available value and the theoretical minimum value. (Analogy: The practical minimum score for golf is some value below par and over 18.)

The *Aluminum Industry Vision* (Ref 10.5) has selected a goal of 11 kWh/kg of aluminum as its smelting current practice value for the year 2020. This represents a 27% reduction over the 1995 value of 15.4 kWh/kg of aluminum. The industry envisions this as an obtainable and practical minimum smelting energy goal for 2020.

Tacit, Process, Feedstock, and 'Secondary' Energies. Current practice process measurements are actual measurements taken within a facility on existing operations. These on-site process measurements are valuable because they are the benchmarks that industry uses to compare performance between facilities and companies. More importantly, these on-site process measurements are used to assess the value of new processes and practices. These are the critical values used in the decision-making process to adopt new technologies and practices. On-site process measurements, however, do not account for the complete energy and environmental impact of manufacturing a product. A full accounting of the impact of manufacturing must also include the energy used to produce the electricity, as well as the fuels and the raw materials used within a manufacturing facility. These secondary energy requirements for electric power generation and transmission, for the energy needed to produce fuels, and for the energy values of feedstock materials are very important from a regional, national, and global energy perspective, but they are seldom analyzed or accounted for within an individual plant site.

The process energy or secondary energy associated with the fuels used in aluminum processing is presented in Table B.1 of Appendix B. The process energy adds approximately 3% to the energy values of the fuels used (Appendix B, Table B.2). Feedstock energy represents the energy inherent in fuels that are taken into a manufacturing process but used as materials rather than fuels. Aluminum production uses coke as a raw material in the production of carbon anodes. The feedstock energy of coke is significant and is equivalent to a 30% increase in the on-site energy consumption of the Hall-Heroult process (Appendix E, Table E.1). The energy contribution of feedstocks is expressed in terms of calorific or fuel value plus the secondary energy used to produce the feedstock. (Note: Fuel and feedstock tacit energy values used here are the calorific fuel value plus the fuel processing energy, [Appendix B, Table B.1]).

Tacit energy is a term frequently used to describe the combined total of on-site energy and the secondary energy requirements. Tacit electrical energy and environmental impact measurements account for the fact that substantial electrical generation inefficiencies and transmission losses occur outside the facility. It can take as much as four units of hydrocarbon or coal calorific energy to produce one unit of electric energy. Saving 1 kWh of on-site electricity is equivalent to saving nearly 2.9 kWh of the energy contained in the petroleum or coal-based fuels used to generate electrical power.

Tacit electric conversion factors are variable because they are dependent on the sources of the energy used to produce electricity. Each manufacturing facility has a different tacit conversion factor depending on its location. Typical U.S. grid electricity requires approximately 9890 Btu of energy to deliver 1 kWh of on-site electricity (3412 Btu) for use. Electricity production from coal requires 10,303 Btu to deliver 1 kWh of on-site electricity (3412 Btu). Water has no fuel value, and hydroelectric facilities can be assumed to have a tacit energy requirement of 3412 Btu to deliver 1 kWh of on-site electricity (3412 Btu) and near zero GHG emissions (Appendix C). The on-site and tacit electric energy requirements for a facility operating on hydroelectric power are equal.

Comparing energy values for the various steps used in the production of aluminum products is simpler when a common unit is used for all processing steps. Because electricity is the single largest source of energy consumed in the manufacture of aluminum, the common units of a kilowatt-hour (kWh) are used in this chapter. Process energy values for production steps that consume fuels are converted to kWh using the conversion factor of 3412 Btu/kWh.

The large variations in tacit electric energy conversion values, 10,303 Btu per on-site kWh for coal compared to 3412 Btu per on-site kWh for hydroelectric, have a dramatic influence on the reported tacit energy profile of an industry. Aluminum smelting energy is 98% electric energy. A modern smelter operating from a hydroelectric utility requires on-site energy of 14.4 kWh/kg of aluminum produced and tacit energy of 14.4 kWh/kg of aluminum, whereas an identical smelter operating from a coal-fired utility requires on-site energy of 14.4 kWh/kg of aluminum and tacit energy of 43.5 kWh/kg of aluminum (Appendix B, Table B.3). The U.S. primary aluminum industry has approximately 39% of its capacity connected to hydroelectric facilities.

All values reported in this document use the U.S. average grid connection values (i.e., 9890 Btu/kWh). The use of U.S. average grid values results in a higher energy consumption value than the actual tacit value required for primary aluminum production. The advantage of reporting primary production energy consumption based on the U.S. average grid is that it allows easier and same-basis comparisons to other U.S. manufacturing industries. For clarity, on-site operating energy values are distinguished from the secondary energy values (which include tacit/feedstock contributions with the use of a superscript "tf").

Life-cycle assessment (LCA) is recognized as the most complete analysis model of the product impact on energy, environmental, economic, and social values. The LCA of an industrial product extends from "cradle-to-grave," that is, from material acquisition and production; through manufacturing, product use, and maintenance; and finally, through the end of the product life in disposal or recycling. The LCA recognizes the importance of considering energy, economic, and environmental factors not only during the production of a product but also over the complete life cycle of the product, including use and disposal. The LCA is particularly useful in ensuring that the benefits derived in one area do not shift the impact burden to other places within a product life cycle.

The LCA "use and maintenance" factor for aluminum varies by end-product and, in many applications, is more significant in terms of energy and environmental impact than production. Aluminum, in some cases, provides LCA "use and maintenance" energy savings that are significantly greater than the energy used in its production. For example, production of an equal-strength but lighter aluminum product in the transportation sector saves significant amounts of transportation fuel and provides substantial reductions in GHGs during the "use" phase of the product when compared to traditional materials. In 2001, an estimated 2.2 billion gallons (18.3 billion liters) of gasoline use and 20 million metric tons of CO_2 emissions were reduced due to the use of lightweight aluminum castings in automobiles (Ref 10.6).

Complete LCA for aluminum products must account for the significant portion of aluminum that, in the acquisition phase, comes from "urban mining" (recycling). The ability of aluminum to be easily recycled is reflected in the fact that over half of the U.S.-produced aluminum now originates from recycled material. Recycling is the best option for disposal of nearly every product made from aluminum. This makes aluminum a "cradle-to-cradle" LCA product.

Energy Value Chain Analysis. The energy values studied and presented are based on an energy "value chain" analysis. The value chain analysis or "cradle-to-shipping dock" analysis provided is an integral part of an LCA. It provides valuable information and data values for organizations performing LCA on aluminum products. Value chain analyses are similar to LCA; however, they cover only a portion of a total LCA. Figure 10.3 shows the global boundaries of an LCA study and the boundaries for this study's value chain.

Value chain analysis allows for the capture of the direct energy and feedstock inputs of each processing step (link) and builds a cumulative value of each product along the chain. This chapter looks at a portion of the LCA: the energy value chain from cradle-to-shipping dock. It does not account for the LCA use and maintenance phase energy or for tertiary energy inputs

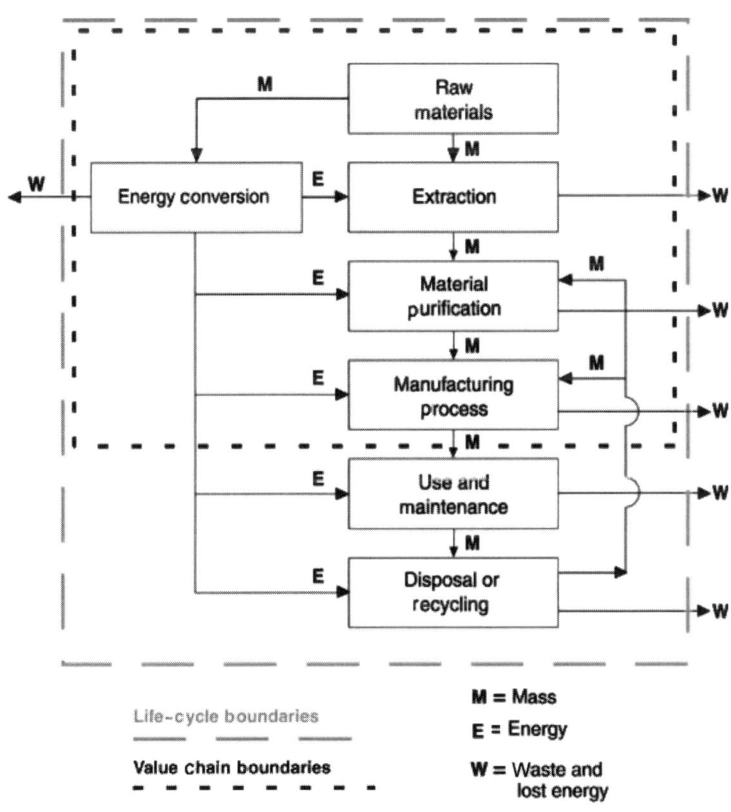

Fig. 10.3 Boundaries for life cycle and value chain

(i.e., the energy used to make the equipment or buildings that house the process steps). The cradle-to-shipping dock approach is valuable for providing decision-making analyses within the manufacturing sphere.

Transportation Energy. The transportation energy associated with acquiring raw materials and distribution of intermediate products is important for a full LCA. Transportation energy can account for a significant portion of the total energy associated with manufacturing a final product. The energy required to transport mined bauxite to refining operations, alumina to smelting operations, ingots to metal processors, and scrap from collection to melting is not accounted for in the process energy requirements that are developed here. This chapter focuses on the energy associated with the processing of raw materials and the processes employed in aluminum production. The transportation energy associated with these raw materials and processes is small in relation to the total energy consumed in production.

Transportation energy calculations for raw materials that are mined globally are highly variable. They are a function of the location and multiple modes of transportation, for example, conveyors, trucks, trains, and ocean freight. Transportation energy requirements are evaluated in Chapter 3, "Life-Cycle Inventory Analysis of the North American Aluminum Industry" in this book. Transportation of raw materials accounted for 2% of the total energy associated with primary aluminum production in the United States.

Evaluation of the transportation energy requirements associated with secondary aluminum production is complicated. Consumer scrap can require considerable transportation energy resulting from individual consumer drop-off, curbside collection, transfer station collection, and the actual transportation to a secondary processor. Transportation energy, associated with industrial manufacturing scrap and scrap originating at large automotive and white-good scrap processing centers, is more easily estimated because its boundaries are easier to define. Chapter 3 estimates transportation energy from these sources to account for 6 to 8% of the total energy associated with the production of secondary aluminum products.

Emissions. Energy use and GHG emissions are closely related. This chapter provides overall carbon emission data associated with fuels used for aluminum operations. Other fuel-related emissions (e.g., nitrous oxides, sulfur dioxide, volatile organic compounds) are not considered because their quantities are typically small as compared to the carbon-base emissions. Emissions that are aluminum process-related (e.g., perfluorocarbons, from cryolite) are reported. *The Energy and Environmental Profile of the U.S. Aluminum Industry* (Ref 10.8) provides detailed emission data for aluminum operations.

Emission calculations for this chapter are shown in Appendix D. Greenhouse gases contribute to climate change by increasing the ability of the atmosphere to trap heat. Gases differ in their ability to trap heat. To express the greenhouse effect of different gases in a comparable way, atmospheric scientists use a weighting factor, global-warming potential. The heat-trapping ability of 1 metric ton of CO_2 is the standard, and emissions are expressed in terms of a million metric tons of CO_2 equivalent, or 10^6 tonnes CO_{2e}. This report uses carbon dioxide equivalents (CO_{2e}). Emissions are also commonly expressed in terms of a million metric tons of carbon equivalent (10^6 tonne C_e). Carbon comprises $^{12}/_{44}$ of the mass of CO_2; to convert from CO_2 equivalent to carbon equivalent, multiply the CO_2 equivalent by 0.273.

Aluminum Production

Aluminum metal is classified as primary aluminum if it is produced from ore and as secondary aluminum if it is produced predominantly from recycled scrap material. Primary aluminum metal production consists of bauxite mining, refining bauxite to produce alumina, and finally, smelting alumina to produce aluminum. Secondary aluminum is produced by sorting, melting, and treating scrap aluminum. Primary and secondary aluminum metal are further processed using traditional metalworking technologies—rolling, extrusion, forging, shaping, and casting into thousands of products.

Aluminum is the most abundant metallic element in the Earth's crust. However, it is never found in natural deposits as a free metal, such as copper and gold. Aluminum is typically found as one of several aluminum oxides or silicates mixed with other minerals and must be processed to be recovered in its pure form. All commercial primary aluminum is produced from one raw material, bauxite, and by one process, electrolytic reduction. For economic and strategic reasons, the aluminum industry continues to perform research and development on alternative raw materials (e.g., kaolin clay) and processes (e.g., chemical reduction). Although these alternatives hold promise for reducing costs, energy consumption, and environmental impacts, none are near commercialization.

The markets for the aluminum industry's raw materials and products are global. Global primary aluminum production has been growing at a rate of 4.5% annually from 1995 to 2005 (Ref 10.2). The U.S. aluminum total supply grew at an annual rate of 0.5% over the period of 1995 to 2005 (Table 10.3). Aluminum is still in the growth phase of the product cycle. Demand for aluminum is increasing, mainly due to aluminum substitution for other materials in the transportation sector and other lightweight applications. Its light weight, corrosion resistance, and processing possibilities, coupled with its ease and value for recycling, strengthen its position as the material of choice in many applications. Measured in either mass produced or economic value, the use of aluminum exceeds that of any other metal except iron. It is important in virtually all segments of worldwide manufacturing.

The global estimate for economically recoverable bauxite reserves is 22,000,000,000 metric tons. This quantity can address the demands for the next century. Two countries have nearly half of the world's identified bauxite resources (Guinea has 25%, and Australia has 20%). Bauxite is no longer mined in the United States as a commercial feedstock for aluminum production. Domestic ore, which accounts for less than 1% of the U.S. requirement for bauxite, is used in the production of nonmetallurgical products, such as abrasives, chemicals, flame retardants, and refractories.

Alumina is produced by refining bauxite in a wet caustic chemical leaching process (Bayer process). Imported bauxite is refined in the United States, the largest importer of bauxite and the second largest bauxite refiner after Australia. Alumina production is continuing to rise in Australia, Brazil, Jamaica, Surinam, Venezuela, and India, all countries with large indigenous bauxite reserves. The trend in alumina production is toward placing refining capacity near the mineral resources, thereby reducing transportation energy and costs and adding more value to exports.

Primary aluminum (aluminum from ore) is globally produced by the Hall-Hèroult process, a method that involves electrolysis or smelting of alumina. Companies choose their smelting locations where production conditions are favorable, such as the availability of skilled labor, proximity to a consumer market, and provision for a highly-developed infrastructure, and, especially, for low-cost and reliable energy. Hydroelectric power accounts for 50% of the energy used worldwide for electrolysis of aluminum (Table 10.7). The bulk of energy use in aluminum production is related to the electricity required for primary electrolysis. Because energy costs are approximately one-third of the total cost of smelting primary aluminum, smelter production has been moving from sites close to consumer markets to sites with low electricity costs. Most of the primary aluminum industry restructuring began in the late 1970s and continues to this day. China and Russia have emerged as major metal producers; other countries entering the world market include Canada, Australia, Brazil, Norway, and countries in the Persian Gulf area, all areas with low energy costs.

Secondary aluminum is produced from scrap or recycled aluminum. The world's average share of secondary aluminum production is roughly one-quarter of total aluminum production. The United States produces over half of its aluminum from recycled aluminum scrap (Table 10.3). Aluminum recycling is concentrated in the countries where the scrap is generated, with the exception of Asia, which imports significant amounts of aluminum scrap (driven by the demand for cast aluminum in the Asian car industry).

Primary and secondary aluminum are used to manufacture numerous products, ranging from aircraft components to household and packaging foils. Each product requires processing, such as heating, melting, alloying, and mechanical

Table 10.7 Energy sources of electrical power in 2002 (electrical power used in gigawatt hours)

Electric energy source	Africa, GWh	North America		Latin America, GWh	Asia, GWh	Europe, GWh	Oceania, GWh	Total, GWh	Grand total, %
		GWh	%						
Hydro	6444	50,312	64	32,207	3030	29,969	7380	129,342	50
Coal	13,443	27,248	35	0	10,080	15,773	24,120	90,664	35
Oil	0	93	0	0	109	1279	0	1481	1
Natural gas	38	351	0	1343	16,336	6199	0	24,267	9
Nuclear	189	678	1	0	0	12,672	0	13,539	5
Total	**20,114**	**78,682**	**100**	**33,550**	**29,555**	**65,892**	**31,500**	**259,293**	**100**

working. Figure 10.4 shows the tacit energy consumption of the major processes in the aluminum production chain. Production of primary aluminum accounts for 86.7[tf]% of the energy consumed by the U.S. industry; production of secondary aluminum for 4.8[tf]%; rolling for 3.4[tf]%; extrusion for 1.6[tf]%; and shape casting for the remaining 3.5[tf]% (Appendix E, Table E.4).

Two operations, electrolysis and the heating/melting of aluminum, account for over 70 (80[tf])% of energy consumed in aluminum processing (Appendix E, Table E.8). Heating and melting technologies are used for holding, alloying, and treating metal as well as for recycling. Programs that improve thermal efficiency of heating and melting while minimizing the formation of aluminum oxide and/or dross provide a much larger impact on decreasing industry energy usage than their energy consumption indicates.

The U.S. aluminum supply of 10,692,000 metric tons in 2005 originated from three basic sources: primary aluminum (domestically produced), secondary aluminum (recycled domestic material), and aluminum imports. This consisted of 2,480,000 metric tons of primary aluminum, 2,990,000 metric tons of secondary aluminum, and 5,221,000 metric tons of imported aluminum (Ref 10.9). From 1995 to 2005, the annual U.S. growths of these supplies were –4.2, –1.7, and 6.9%, respectively. Since 1995, the total U.S. supply has risen at an annual rate of approximately 0.5%. Figure 10.5 shows the distribution of these supplies over the past 45 years.

The United States is the fourth leading producer of primary aluminum metal in the world. However, its dominance in the global industry has declined. The U.S. share of world production in 1960 accounted for slightly more than 40% of the primary aluminum produced. By 2005, the U.S. share of world production had decreased to 7.8%. The U.S. primary production peaked in 1980 and over the past 25 years has been gradually declining. Significant year-to-year variations occur as a result of U.S. electrical costs and global market changes.

Secondary (recycled) aluminum is of growing importance to the U.S. supply. In 1960, only 401,000 metric tons of aluminum were recovered. In 2005, 2,990,000 metric tons of aluminum were recovered. For the years 1995 through 2005, the secondary production of aluminum has dropped at an annual rate of –1.7% (Table 10.3). Recently, the U.S. secondary aluminum growth rate has been slowing due to a combination of factors. Scrap collection programs are beginning to reach their maturation stage, and the market growth of scrap sources has slowed. Additionally, the U.S. has now become a net exporter of scrap and dross. In 1993, the U.S. had net imports of scrap and dross of 98,540 metric tons, while in 2003, the U.S. had net scrap and dross exports of 152,000 metric tons. The main cause for this change is

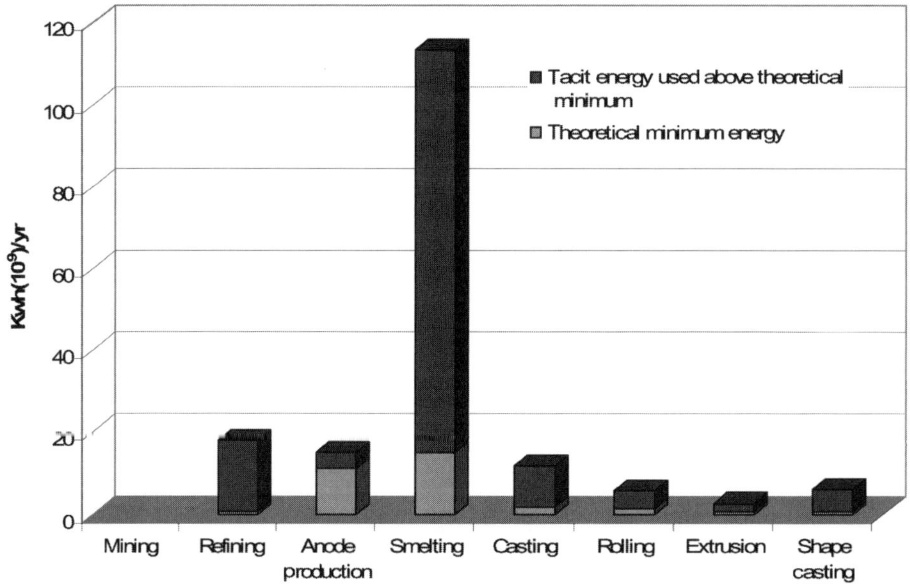

Fig. 10.4 Energy[tf] consumption of U.S. aluminum operations

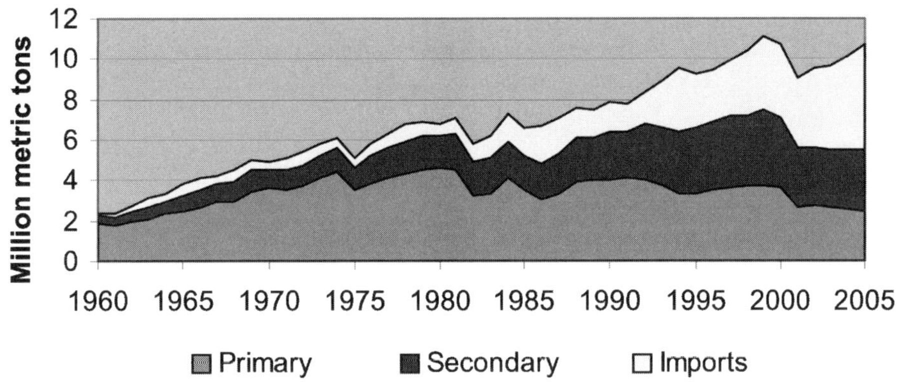

Fig. 10.5 U.S. aluminum supply from 1960 to 2005

again China's radical increase in demand for scrap. In 2003, 244,000 metric tons of the 567,700 metric tons of scrap and dross exported by the U.S. (43%) went to China (Ref 10.9). These trends of decreasing U.S. secondary aluminum production are, however, expected to change. Use of aluminum in the automotive industry grew at over 5.4% annually between 1995 and 2005. This large and growing supply is now beginning to enter the scrap markets and will spur new growth in secondary aluminum.

Imported aluminum is the fastest growing source of U.S. supply, with an annual rate of 6.9% over the 1995 to 2005 time frame (Table 10.3). New primary aluminum facilities are being located outside the United States, near new sources of low-cost electricity.

Primary Aluminum Raw Materials

The total energy associated with producing the raw materials required for aluminum production from bauxite ore was approximately 8.20 (14.22^{tf}) kWh/kg of aluminum. This accounts for 28% of the total energy required to produce primary aluminum metal and consists of:

- 0.32 (0.34^{tf}) kWh/kg aluminum for bauxite mining
- 7.27 (7.87^{tf}) kWh/kg aluminum for bauxite refining to alumina
- 0.61 (6.01^{tf}) kWh/kg aluminum for carbon anode production

A complete account of the energy requirements and environmental impacts to produce any product must include the energy requirements and environmental impact associated with the production of the raw materials used. The raw material energy requirements and environmental impacts associated with primary aluminum production can be divided into the major operations required to produce it. These are bauxite mining, bauxite refining, and carbon anode manufacturing. Approximately 5900 kg of earth are mined to produce 5100 kg of bauxite, which is refined into 1930 kg of alumina. The 1930 kg of alumina are electrolytically processed with 446 kg of carbon to produce 1 metric ton (1000 kg) of aluminum (Appendix E, Table E.1).

Cryolite and other fluoride salts are used as the electrolytic bath for aluminum production. These materials are theoretically not consumed in the process or combined as part of the final product. However, approximately 19 kg of bath material is lost for every metric ton of aluminum produced (Appendix E, Table E.1). These losses are a result of process upsets and bath drag-out when molten aluminum is removed from the smelting operation. These salts represent only a small portion of the energy requirement for producing the raw materials required for aluminum production and are not addressed here.

Bauxite

Aluminum, never found as a free metal, occurs naturally in the form of hydrated aluminum oxides or silicates. Because the silicates are mixed with other metals such as sodium, potassium, iron, calcium, and magnesium, and it is chemically difficult and expensive to extract aluminum from them, the silicates are not a practical source of aluminum. The oxides are, therefore, used for producing aluminum. The aluminum oxides commonly found as naturally occurring minerals include:

- Corundum (alumina, Al_2O_3)
- Böehmite (α-$Al_2O_3 \times H_2O$, a monohydrate containing 85 wt% alumina)
- Diaspore (β-$Al_2O_3 \times H_2O$, chemically the same as böehmite but with a different crystal structure)
- Gibbsite (γ-$Al_2O_3 \times 3H_2O$, a trihydrate containing 65.4 wt% alumina)
- Alumina, used for the production of aluminum, is obtained from bauxite deposits.

Bauxite is not a true mineral but a rock that contains mostly böehmite and gibbsite along with diaspore, corundum, and numerous impurities (mostly compounds of iron, silicon, and titanium). Bauxite commonly appears as a collection of small, reddish-brown nodules in a light-brown, earthy matrix. The alumina available in commercial bauxite ranges from 30 to 60 wt%. Bauxite is typically classified according to its intended commercial application: abrasive, cement, chemical, metallurgical, refractory, and other end uses. The bulk of world bauxite production (approximately 85%) is metallurgical and used as feedstock for the manufacture of aluminum.

The United States mines less than 1% of the bauxite it uses annually, virtually all of which is used in the production of nonmetallurgical products, such as abrasives, chemicals, and refractories (Ref 10.10). Nearly all bauxite consumed in the United States is imported. Figure 10.6 tracks the domestic and imported components of U.S. bauxite supply from 1960 to 2005 (Table 10.8).

In 2005, the United States imported a total of 10,400,000 metric tons of bauxite. Approximately 95% of the imported bauxite is refined to produce alumina, and approximately 91% of the refined alumina is used to produce primary aluminum.

Bauxite Energy Requirements (Onsite and Theoretical). Approximately 0.32 (0.34[tf]) kWh of process energy were required to produce the 5.1 kg of bauxite needed to produce 1.0 kg of aluminum. Approximately 16.7 kg of carbon dioxide equivalent (CO_{2e}) were released for each metric ton of bauxite mined.

The energy demand associated with the extraction of bauxite is typical of most mining operations. Bauxite ore is generally strip-mined by removing the overburden (the soil on top of the deposit) and excavating it with mechanical equipment. The overburden is saved for reclamation operations, which are extensively practiced to ecologically restore mined areas. The soft, earthy nature of many bauxite deposits generally does not require drilling or blasting operations. After mining, the bauxite is crushed, sometimes washed and dried, and transported to refining plants via ship, barge, rail, truck, or conveyor belt.

Approximately 5.1 kg of bauxite are required to produce a kilogram of aluminum. The energy requirement per kilogram of mined bauxite is 0.06 kWh for typical extraction (Ref 10.11). Because electricity accounts for less than 1% of the energy used in bauxite production, the tacit addition is negligible (Appendix E, Table E.1).

Calculation of a theoretical minimum energy requirement for mining bauxite is dependent on the system boundaries applied and the processes used. The laws of thermodynamics state that separating the constituents of a mixture, such as bauxite from bauxite-rich soil, requires a certain minimum expenditure of energy. Bauxite is the major constituent of bauxite-rich soils, and there is no change in the chemical nature of bauxite in the mining process. So, the theoretical minimum

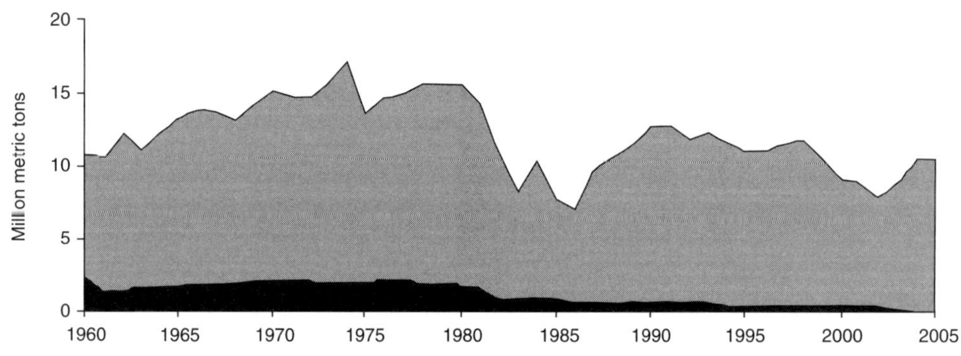

Fig. 10.6 U.S. bauxite supply from 1960 to 2005

Table 10.8 U.S. supply of bauxite and alumina from 1960 to 2005 (in thousand metric dry tons)

A Year	B Imports of bauxite	C U.S. mined bauxite	D Exports of bauxite	E Net bauxite supply	F Estimated alumina production from bauxite(a)	G Imports of alumina	H Exports of alumina	I Estimated net alumina for electrolysis(b)
1960	8879	2030	29	10,880	4651	80	...	4266
1961	9354	1248	153	10,449	4467	171	...	4191
1962	10,745	1391	263	11,873	5076	158	...	4726
1963	9408	1549	206	10,751	4596	173	...	4309
1964	10,518	1627	283	11,862	5071	191	...	4755
1965	11,601	1681	149	13,133	5614	206	290	4969
1966	11,928	1825	63	13,690	5852	443	290	5420
1967	12,010	1681	2	13,689	5852	865	499	5633
1968	11,359	1692	7	13,044	5576	1190	780	5429
1969	12,355	1873	5	14,223	6080	1730	885	6317
1970	13,039	2115	3	15,151	6477	2840	998	7171
1971	12,837	2020	35	14,822	6336	2190	980	6913
1972	12,803	1841	29	14,615	6248	2590	797	7416
1973	13,618	1909	12	15,515	6633	3090	694	8365
1974	15,216	1980	16	17,180	7344	3290	927	8973
1975	11,714	1801	20	13,495	5769	3180	933	7439
1976	12,749	1989	15	14,723	6294	3290	1050	7905
1977	12,989	2013	26	14,976	6402	3760	857	8665
1978	13,847	1669	13	15,503	6628	3970	878	9057
1979	13,780	1821	15	15,586	6663	3840	849	8988
1980	14,087	1559	21	15,625	6680	4360	1140	9232
1981	12,802	1510	20	14,292	6110	3980	730	8749
1982	10,122	732	49	10,805	4619	3180	590	6747
1983	7601	679	74	8206	3508	4030	602	6585
1984	9435	856	82	10,209	4364	4290	648	7570
1985	7158	674	56	7776	3324	3830	316	6506
1986	6456	510	69	6897	2948	3600	487	5767
1987	9156	576	201	9531	4075	4070	1130	6607
1988	9944	588	63	10,469	4475	4630	1040	7618
1989	10,893	525	44	10,849	4638	4310	1330	7154
1990	12,142	490	53	12,089	5168	4070	1260	7461
1991	12,300	460	58	12,242	5233	4590	1350	7950
1992	11,400	400	68	11,332	4844	4700	1140	7920
1993	11,900	378	92	11,808	5048	3940	1240	7243
1994	11,200	350	137	11,063	4729	3120	1040	6336
1995	10,800	300	120	10,680	4566	4000	1040	7069
1996	10,700	270	154	10,546	4508	4320	918	7460
1997	11,300	230	97	11,203	4789	3830	1270	6870
1998	11,600	200	108	11,492	4913	4050	1280	7192
1999	10,400	180	168	10,232	4374	3810	1230	6517
2000	9030	150	147	8883	3797	3820	1090	6186
2001	8670	125	88	8582	3669	3100	1250	5189
2002	7710	125	52	7658	3274	3010	1270	4719
2003	8860	W(c)	89	8771	3750	2310	1090	4632
2004	10,500	W(c)	75	10,425	4457	1650	1230	4476
2005	10,400	W(c)	62	10,338	4419	1860	1200	4682

Note: 4,419,000 metric tons or 94% of the 4,682,000 metric tons of alumina needed to produce aluminum in the United States was refined in the United States in 2005. (a) Calculated based on United States Geological Survey (USGS) data. The average alumina content for bauxite is 45%, and 95% of bauxite is converted to alumina. The calculation is column F = 0.95 × 0.45 × column E. The numbers may be slightly overestimating because slightly less than 100% of the alumina content of bauxite is actually recovered. (b) Net is calculated based on USGS data. 90% of alumina produced is metallurgical alumina used in aluminum production. (c) W, data withheld. Source: United States Geological Survey, Minerals Information, Statistical Compendium

energy for preparing bauxite is negligible. In addition, because it is theoretically possible to find bauxite on the surface, the theoretical minimum energy requirement to produce bauxite is very close to zero. In the interest of simplicity, the theoretical minimum energy requirement for mining bauxite is assumed to be zero.

Emissions from fuels used in the extraction of bauxite are listed in Appendix D, Table D.2. These emissions are typically from surface-mining operations and result from a variety of fuels used in the production of bauxite. Nearly 0.0167 kg CO_{2e} are emitted for each kilogram of bauxite mined.

Alumina (Al_2O_3)

Theoretically, from the stoichiometric equation ($2Al_2O_3 + 3C \rightleftharpoons 4Al + 3CO_2$), 1.89 kg of alumina is required to produce 1 kg of aluminum. In practice, a very small portion of the alumina supply is lost, and the industry requires approximately 1.93 kg of alumina for production of each kilogram of aluminum. In 2005, the United States produced 4,419,000 metric tons of alumina from bauxite and imported an additional 1,860,000 metric tons of alumina to make aluminum.

Figure 10.7 shows the alumina supply sources from 1960 to 2005. The United States had four Bayer refineries in operation in 2005. These refiners processed approximately 10,400,000 metric tons of bauxite into 4,419,000 metric tons of alumina. Approximately 9% of the alumina produced is used to manufacture abrasive, refractory, and other products. Approximately 4,883,000 metric tons of U.S.-refined alumina were transferred to the primary aluminum industry. This quantity of alumina was not sufficient to supply the U.S. demand for alumina; therefore, an additional 1,200,000 (net) metric tons of alumina were imported.

All commercial alumina is refined from bauxite using the Bayer refining process. The process, developed by Karl Bayer in 1888, consists of four major steps: digestion, clarification, precipitation, and calcination Bauxite composition varies, and refining plant designs are slightly different to account for the site-specific quality of the bauxite.

In digestion, crushed, ground, and sized bauxite is dissolved under pressure with a hot (180 to 250 °C, or 360 to 480 °F) sodium hydroxide and sodium carbonate solution in a series of steam-heated digesters. The concentrations, temperatures, and pressures employed vary depending on the properties of the bauxite. Gibbsite is soluble in caustic soda above 100 °C (212 °F), while böehmite and diaspore are soluble in caustic soda above 200 °C (390 °F). Because the treatments of böehmite and diaspore require higher temperatures and longer digestion times, they are more expensive than the treatment of gibbsite. The aluminum oxides in the bauxite react to form soluble sodium aluminate or "green liquor":

$$Al_2O_3 \times H_2O + 2NaOH \rightleftharpoons 2NaAlO_2 + 2H_2O \text{ and}$$

$$Al_2O_3 \times 3H_2O + 2NaOH \rightleftharpoons 2NaAlO_2 + 4H_2O$$

Silicas in the bauxite are detrimental to the digestion efficiency. They react to form sodium aluminum silicate, which precipitates. This precipitate chemically binds the aluminum from the bauxite and the sodium from the sodium hydroxide into a solid from which the alumina cannot be economically recovered. This decreases the yield of alumina and increases the costs associated with sodium hydroxide. Chemical additions and the adjustment of refining practices can effectively provide desilication and decalcification of specific alumina streams.

In clarification, the green liquor produced by digestion is clarified to remove sand, undissolved iron oxides, titanium oxides, silica, and other impurities. The insoluble materials, called bauxite residue or red mud, are thickened, washed, and dewatered to recover sodium hydroxide. Bauxite residue is a large-quantity waste product that is generally stored adjacent to refinery sites in landfills or lagoons. After weathering, the landfills can sustain vegetation.

In precipitation, the clarified liquid that results from clarification is cooled and "seeded" with crystals of gibbsite to aid precipitation of alumina trihydrate ($Al_2O_3 \times 3H_2O$). This is the reverse of the digesting step ($2NaAlO_2 + 4H_2O \rightleftharpoons (Al_2O_3 \times 3H_2O + 2NaOH)$). However, by carefully controlling the seeding, temperature, and cooling rate, specific physical properties can be given to the precipitating alumina trihydrate.

In calcination, alumina trihydrate is typically calcined in a fluid bed or rotary kiln at approximately 980 to 1300 °C (1800 to 2370 °F) to

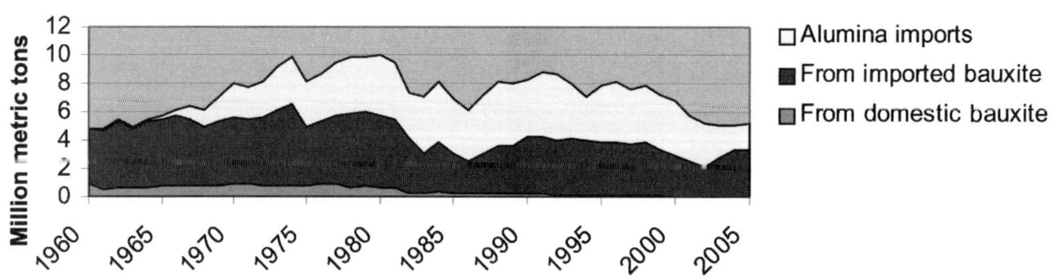

Fig. 10.7 U.S. alumina supply from 1960 to 2005

remove the water of crystallization and produce the dry white powder, alumina ($Al_2O_3 \times 3H_2O \rightleftharpoons Al_2O_3 + 3H_2O$). Calcining rates and temperatures are carefully controlled and vary depending on the final physical properties specified for the alumina.

Alumina used for electrolysis not only has a chemical purity specification but also a physical specification on particle size, surface area, bulk density, and attrition behavior. These properties affect the free-flowing properties of alumina (how it flows in feeders), the rate at which it dissolves in cryolite, dust levels, the strength of the alumina crust, its insulating properties, and other properties important in the aluminum electrolysis cell operation. The bulk density of alumina is 880 to 1100 kg/m^3, while its specific gravity is 3.9 g/cm^3.

Bauxite residue (red mud) is a by-product of the Bayer process and contains the insoluble impurities of bauxite. The amount of residue generated per kilogram of alumina produced varies greatly depending on the type of bauxite used, from 0.3 kg for high-grade bauxite to 2.5 kg for low-grade bauxite. Its chemical and physical properties depend primarily on the bauxite used and, to a lesser extent, the manner in which it is processed.

Although a great deal of effort has been expended over several decades to find and develop uses for bauxite residue, a cost-effective, large-scale bulk application has yet to be found (Ref 10.12). Numerous attempts have been made to recover additional metals from the residue, such as iron, titanium, and gallium. Other possible uses for the residue have included production of ceramic bricks or tiles, use as roadbed material or as filler material for plastics, or production of cement. Accordingly, the current industry efforts focus on minimizing the amount of residue generated and improving its storage conditions.

Probably the most promising and recent application for the residue has occurred in Western Australia, where it is being evaluated as a soil amendment or conditioner. The soils in Western Australia are sandy and drain freely, allowing fertilizers to leach into waterways where they boost nutrient levels and can lead to problems such as algal blooms. The application of bauxite residue to these sandy soils aids the retention of phosphates and moisture and reduces the need to apply lime for soil pH adjustments. As part of the Bauxite Roadmap project on bauxite residue, a model farm has been planned to run for 5 to 7 years with the application of red mud, and all its aspects and results will be studied in great detail.

Alumina Energy Requirements. Approximately 7.27 (7.87^{tf}) kWh of energy were required and 1.62 kg CO_{2e} were released to refine the 1.93 kg of alumina from bauxite needed to produce 1 kg of aluminum in 2003.

The energy required to produce alumina from bauxite in 1985 was estimated to range from 2 to 9 kWh/kg of alumina (Ref 10.3). This broad range of energy intensity reflects both bauxite quality (alumina content) and refinery design. It was estimated that in 1991, U.S. refiners averaged 3.66 kWh/kg of alumina produced (Ref 10.13). The most recent available refining data list 3.76 kWh/kg of alumina for 1995 (Ref 10.11). Calcination is the most energy-intensive operation of the Bayer process. On average, 1.93 kg of alumina are consumed to make 1 kg of aluminum (Ref 10.11). The alumina energy requirement for 1 kg of aluminum can be estimated as 3.76 kWh/kg of alumina times 1.93 kg alumina, or 7.27 kWh/kg of aluminum, a tacit value of 7.87^{tf} kWh/kg of aluminum.

The *Alumina Technology Roadmap* (Ref 10.14) provides insight into the high-priority research and development needs of the global alumina industry. The Roadmap recognizes the need and the opportunity for a 25% energy reduction by 2020 and improved, more sustainable handling of bauxite residues. Better chemical process knowledge, waste heat utilization, and cogeneration are opportunities for energy reduction in the refining process.

Emissions from fuels used in the refining process are listed in Appendix D, Table D.2. These emissions are predominantly related to natural gas and coal consumption for digestion and calcination. Nearly 1.62 kg CO_{2e} are emitted for each kilogram of refined alumina.

Alumina Theoretical Minimum Energy Requirements. The theoretical minimum energy required to produce alumina is 0.13 kWh/kg of aluminum produced.

The theoretical minimum energy requirements to produce metallurgical-grade alumina from bauxite can be calculated from the reactions required in the process. The minimum energy requirements for the digestion, clarification, and precipitation steps are related to the two chemical reactions that take place during these processing steps:

$$Al_2O_3 \times 3H_2O + 2NaOH \rightleftharpoons 2NaAlO_2 + 4H_2O \rightleftharpoons Al_2O_3 \times 3H_2O + 2NaOH$$

However, because there is no net chemical or temperature change for the combined reactions, the theoretical minimum energy requirement for these steps is near zero. The final reaction, calcination, requires energy to remove the hydrated water from alumina ($Al_2O_3 \cdot 3H_2O \rightleftharpoons Al_2O_3 + 3H_2O$). The theoretical minimum energy requirement to calcine (dehydrate) the alumina is 0.13 kWh/kg of aluminum produced (Appendix F, Table F.10). This value assumes that the precipitated alumina trihydrate is completely bone dry before entering the calcining process. At 8% moisture, this would add approximately 0.01 kWh/kg of aluminum to the minimum requirement.

Carbon Anode

The United States consumed 1,128,000 metric tons of carbon anode in 2005. Approximately 0.45 kg of carbon anode were needed to produce 1 kg of aluminum (Appendix E, Table E.1). All commercial production of aluminum uses carbon as the anode material for the electrical reduction of alumina to aluminum. The carbon anode net reduction reaction ($2Al_2O_3 + 3C \rightleftharpoons 4Al + 3CO_2$) requires three carbon atoms for the reaction to free four aluminum atoms. The theoretical minimum anode consumption is 0.33 kg of carbon per kilogram aluminum [(3 × 12.01 carbon molecular weight) (4 × 26.98 aluminum molecular weight)]. Anode material quality is important because all the impurities dissolve into the bath and ultimately contaminate the molten aluminum. The physical quality of the anode also affects both the energy efficiency and productivity of smelting cells.

Anode consumption rates in practice, typically approximately 0.45 kg carbon per kilogram of aluminum, are 35% higher than the theoretical requirement. Excess carbon usage results from the need to protect the iron electrical connection within the carbon anode, and from air burning and dusting. The surface of the carbon is hot enough at cell operating temperatures to oxidize at a slow rate. This is minimized by coating the anode surface and covering the anode with alumina, which insulates it from air exposure.

Dusting, which is breaking off small particles of carbon into the air or bath, can account for half of the excess carbon used in the smelting process. Dusting is a direct function of the anode material uniformity. It is caused by selective electrolytic oxidation and air burning of the binder pitch, which releases aggregate carbon particles into the bath. Anode carbon dust is unavailable for aluminum production.

Two different anode technologies are used by the U.S. industry. Prebaked carbon anodes account for more than 86% of the U.S. capacity. Older, in situ baked Soderberg anodes account for the remainder of the capacity. New prebaked anode reduction cells have surpassed Soderberg anodes in terms of current efficiency and emissions control. Only three operational smelters in the United States are currently using Soderberg anodes. No new Soderberg cells are being built, and those that exist are progressively being replaced, converted, or shut down. This chapter focuses mainly on prebaked anodes for smelting technology.

Carbon prebaked anodes are made by mixing ground used carbon anodes, calcined petroleum coke, and coal tar or petroleum pitch. Pitch acts as a binder to hold the anode mass in a "green" formed shape. Compacting the anode by using vacuum and vibrating the mixture when forming produces a denser, more conductive and lower-dusting anode. Baking carbonizes the pitch and creates a solid bond between the particles of calcined coke and used anode material. Cast iron is poured into preformed sockets in the baked anode to form an electrical connection. Prebaked anodes can weigh as much as 1250 kg and have a working face size of approximately 0.70 by 1.25 m (2.3 by 4.1 ft) and a 0.5 m (1.6 ft) height. Prebaked anodes are removed before they are completely consumed. Used anodes are recycled into the anode production system to recover the carbon and the iron rods used for electrical connections. Used anodes can account for 15 to 30% of the mass used in green anode makeup.

Calcined petroleum coke is a by-product of the crude oil refining industry. Green or raw coke contains 8 to 10% moisture and 5 to 15% volatile organic materials. Raw coke must be calcined at approximately 1200 to 1350 °C (2190 to 2460 °F) in gas-fired kilns or rotary hearths to remove the moisture, drive off volatile matter, and to increase the density, strength, and conductivity of the product (Ref 10.15). Calcined coke has a bulk density of approximately 800 kg/m^3. Worldwide,

Table 10.9 Energy associated with aluminum industry carbon anode manufacturing

	Mass of material input(a), kg/1000 kg anode	Material input energy(b), Btu/kg	kWh/kg anode	kWh/kg aluminum
Pitch				
Mass	231	...	...	...
Feedstock energy	...	8813	2.58	1.15
Process energy	...	9	0.003	0.001
Calcined coke				
Mass	820	...	...	...
Feedstock energy	...	27,569	8.08	3.60
Process energy	...	395	0.12	0.05
Green coke				
Mass	85	...	...	...
Feedstock energy	...	2664	0.78	0.35
Process energy	...	43	0.01	0.01
Carbon anode baking				
Process energy(a)	...	...	1.90	0.85
Total raw materials and energy				
Mass (kg/1000 kg anode)	1136	...	...	...
Feedstock energy (kWh)	...	39,046	11.44	5.10
Process energy (kWh)	...	446	2.03	0.90
Total tacit energy input			**13.47**	**6.01**

(a) Values from Appendix E, Table E.1. (b) Values from Appendix B, Table B.1

approximately 25% of all raw coke is calcined, and approximately 70% of all calcined coke goes to aluminum production. Modern calcining hearth and kiln designs capture and use the volatile organic matter in raw coke as their major fuel source.

Carbon Anode Energy Requirements. Approximately 0.61(6.01^{tf}) kWh of energy were required and 0.12 kg CO_{2e} were released in the manufacturing of the 0.45 kg of carbon anode needed to produce 1 kg of aluminum in 2003.

Anode blocks are typically baked in a natural gas-fired furnace for several weeks. Quality anodes depend on careful baking controls to gradually raise the temperature to approximately 1250 °C (2280 °F). Volatile hydrocarbons from the pitch are gradually released during the baking process. Theoretically, these volatile compounds could provide sufficient heat for anode baking and no additional energy would be required, However, in practice, volatile organic compounds account for only 46% of the energy input to the prebake ovens. The remaining 54% of the energy needed comes from fuel. Only approximately 30% of the input energy goes into making the anode; 24% is lost from oven surfaces, 29% goes up the stack, and 17% is lost in other ways. New prebake furnace ovens with computer controls are more efficient, with both regenerative and recuperative elements (Ref 10.3).

Emissions associated with prebaked anodes result mainly from the combustion of natural gas and the volatile organic compounds contained in the pitch. These amount to 0.27 kg CO_{2e} per kilogram of anode or 0.12 kg CO_{2e} per kilogram of aluminum (Appendix D, Table D.2).

Table 10.10 Energy associated with carbon anode manufacturing

	kWh/kg of anode A	kWh/kg of aluminum 0.45 × A
Pitch	0.003 (2.58^{tf})	0.001 (1.15^{tf})
Coke	0.13 (8.99^{tf})	0.06 (4.01^{tf})
Anode	1.36 (1.90^{tf})	0.61 (0.85^{tf})
Total	**1.49 (13.47^{tf})**	**0.66 (6.01^{tf})**

The process energy used to produce a carbon anode is 1.36 (1.66^{tf}) kWh/kg of anode, from the most recent available U.S. data for 1995 (Ref 10.15). The total energy (Table 10.9) to produce a carbon anode is shown in Table 10.10. The energy per kilogram of aluminum produced is obtained by multiplying 0.45 times the total energy required to produce a kilogram of carbon anode.

Soderberg anodes use green coke and are baked in situ. These anodes can be baked only to the maximum cell-operating temperature, which results in an anode with 30% higher electrical resistivity and a greater dusting propensity than a prebaked anode (Ref 10.3). The cell emission control system is also more complex than for a prebake cell, because emission systems must be designed to handle the 5 to 15% volatile organic material content of the green coke (Ref 10.15). Some plants employ both wet and dry gas scrubbing systems to meet the environmental regulations.

Carbon Anode Theoretical Energy Values. The minimum theoretical energy requirement to manufacture a carbon anode is the energy necessary to convert the coal tar pitch by destructive distillation to a coke-based binder. Approximately one-third of the pitch binder mass is lost in the baking process (Ref 10.16). A portion of this loss consists of volatile components, while the remainder is the carbonization of the pitch. Pitch contains approximately 85% C. Approximately 79% of the pitch is actually carbonized, while 21% is volatilized. The fuel energy value of the pitch that is carbonized, 0.75 kWh/kg of anode, is a measure of the theoretical energy required for anode manufacturing. In addition, the tacit or inherent energy of 11.55^{tf} kWh/kg of anode must be accounted for as part of the total theoretical requirement. Therefore, the total theoretical minimum energy associated with the production of prebaked carbon anodes is 12.30^{tf} kWh/kg of carbon anode. Further, because 0.33 kg of carbon anode are required to produce a kilogram of aluminum in theory, the minimum energy requirement to produce anodes is 4.1 kWh/kg of aluminum.

Primary Aluminum Production

The total energy associated with primary aluminum production from bauxite ore was approximately 23.78 (45.21^{tf}) kWh/kg of aluminum. This consisted of:

- 8.20 (14.22^{tf}) kWh/kg aluminum for raw materials
- 15.58 (44.82^{tf}) kWh/kg aluminum for electrolytic reduction

Alumina is insoluble in all ordinary chemical reagents at room temperature and has a high melting point (above 2000 °C or 3630 °F). These properties make conventional chemical processes used for reducing oxides difficult and impractical for conversion of alumina into aluminum.

Commercial primary aluminum is produced by the electrochemical reduction of alumina. Charles Martin Hall in the United States and Paul Lewis Toussaint Heroult in France independently developed and patented a commercially successful process for alumina reduction in 1886. This process, commonly referred to as the Hall-Heroult process, is still the primary method in use for aluminum production. Although the engineering has improved vastly, the process fundamentals are basically unchanged today. The Hall-Heroult process takes place in an electrolytic cell or pot. The cell consists of two electrodes (an anode and a cathode) and contains a molten bath of fluoride compounds (cryolite), which serves as an electrolyte and solvent for alumina. An electric current is passed through the bath, which reduces the alumina to form liquid aluminum and oxygen gas. The oxygen gas reacts with the carbon anode to form carbon dioxide. Molten aluminum collects at the cathode in the bottom of the cell and is removed by siphon.

Production, Capacity, and Growth

In 2005, 9 companies operated 19 primary aluminum production facilities in the United States. These facilities, having a production capacity of approximately 3,708,000 metric tons, operated at approximately 67% capacity to produce 2,480,000 metric tons of aluminum in 2005 (Ref 10.9).

Figure 10.8 shows variations in the production of primary aluminum in the United States

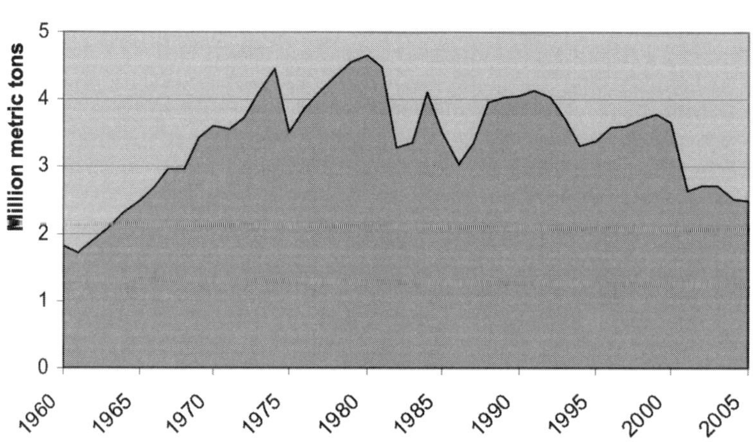

Fig. 10.8 U.S. production of primary aluminum from 1960 to 2005

from 1960 to 2003. These production variations are more representative of the costs to produce aluminum than of the domestic demand. Primary aluminum is traded on a global market, and global supply has been growing steadily at a rate of 4.5% annually for the past ten years. The United States accounted for 7.8% of the world's primary aluminum production in 2005.

Historical Hall-Heroult Energy Utilization

"The first commercial aluminum cells at Neuhausen, Switzerland (Heroult), and Pittsburgh, Pennsylvania (Hall), required more than 40 kWh/kg of aluminum produced and had current efficiencies ranging from 75 to 78%" (Ref 10.17). The Hall-Heroult process is still electric energy-intensive. Because electricity costs are an important component (approximately one-third) of the total production costs, energy efficiency continues to be a major area of focus for the aluminum industry.

The electrical energy consumed in a primary aluminum cell is measured by the number of watts consumed over a period of time. Wattage is determined by multiplying the cell voltage by cell amperage. Figure 10.9 shows the significant electrical energy improvements made between 1900 and 2000. Total electricity usage (excluding tacit generation and transmission losses) varies from less than 13 kWh/kg of aluminum for the state-of-the-art plants up to more than 20 kWh/kg for older Soderberg facilities (U.S. plants in 1995 averaged approximately 15.4 kWh/kg of aluminum). The theoretical minimum energy requirement for carbon anode aluminum electrolysis is approximately 5.99 kWh/kg of aluminum. Compared to theoretical values, U.S. facilities are operating at approximately 38% energy efficiency.

Significant engineering changes in cell design and operation have occurred over the past 50 years. Table 10.11 shows the changes in operating parameters of a typical cell from 1948 to 1999 (Ref 10.18). Each new or updated primary facility tries to increase productivity and incorporate energy-reducing technologies to lower production costs. This results in gradual changes in the industry. The most significant change is the new equipment and techniques allowing for smaller and more frequent alumina additions (point feeders). This, combined with higher amperage, lower current density, and larger cells, has dramatically improved current efficiencies and productivity.

There is a minimum cell amperage (electrical current) required to produce aluminum. Production in the United States now operates at approximately 95% current efficiency, a significant improvement over the past several years, as shown in Fig. 10.10. The high current efficiency of existing technologies leaves little opportunity for process or technology improvements to further reduce amperage and save additional energy. Because current efficiency is high, lowering the voltage requirements of cells presents the largest challenge and the best opportunity for improving Hall-Heroult efficiencies. The voltage requirements of a cell are described in the section "Voltage Requirements" of the "Hall-Heroult Reduction Process" in this chapter.

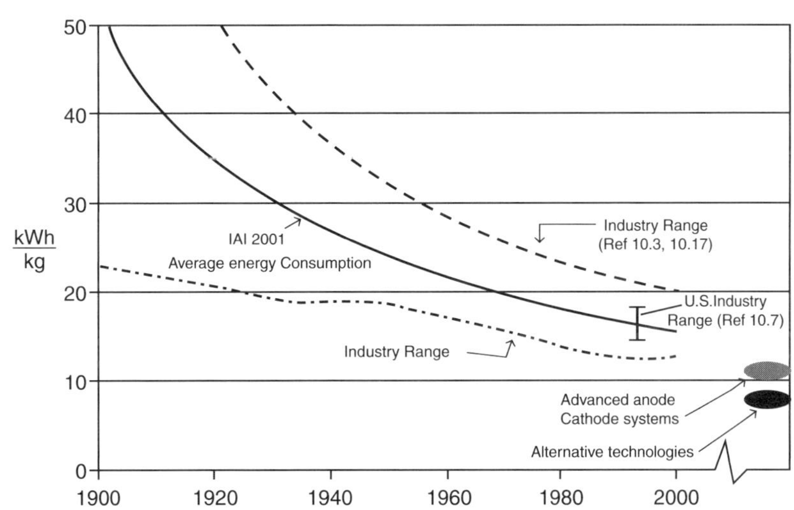

Fig. 10.9 Primary aluminum electric energy consumption from 1900 to 2000

Table 10.11 Ore-to-metal comparison of near-and midterm technology improvements

Energy input kWh/kg Al	Modern prebaked Hall-Heroult	Wetted cathode ACD=2.0	Inert anode wetted cathode ACD=2.0	Carbothermic reduction	Kaolinite ($AlCl_3$) reduction
On-site energy demands					
Raw materials					
Bauxite-alumina	7.59	7.59	7.59	7.59	...
Kaolinite	...	...	...	...	8.14
Anode materials	0.61	0.61	0.76	...	0.76
Reaction carbon	...	...	...	0	0
Total	8.20	8.20	8.35	7.59	8.91
Reaction energy					
Reaction thermal	...	...	...	7.71	–1.90
Furnace losses	...	...	...	1.36	0.40
Reaction	3.76	3.76	6.90	...	6.48
Cell ohmic	10.67	7.62	6.20	...	2.93
Total reaction	14.43	11.38	13.11	9.07	7.91
Total onsite	22.63	19.58	21.46	16.66	16.82
Percent energy savings	Reactions	21%	9%	37%	45%
	Reactions and anode	20%	8%	40%	42%
	Reactions, anodes, and ore	13%	5%	26%	26%
Tacit energy demands					
Raw materials					
Bauxite-alumina	8.21	8.21	8.21	8.21	...
Kaolinite	...	...	...	...	8.81
Anode materials	6.01	6.01	0.76	...	0.76
Reaction carbon	...	...	...	8.41	11.34
Total	14.22	14.22	8.98	16.63	20.92
Reaction energy					
Reaction thermal	...	...	...	17.10	–1.90
Furnace losses	...	...	...	3.02	0.60
Reaction	10.91	10.91	20.00	...	18.79
Cell ohmic	30.91	22.07	17.98	...	8.48
Total	41.83	32.99	37.99	20.12	25.98
Total tacit	56.05	47.21	46.96	36.74	46.90
Percent energy savings	Reactions	21%	9%	52%	38%
	Reactions and anode	18%	19%	58%	44%
	Reactions, anodes, and ore	16%	16%	34%	16%

ACD, anode-cathode distance

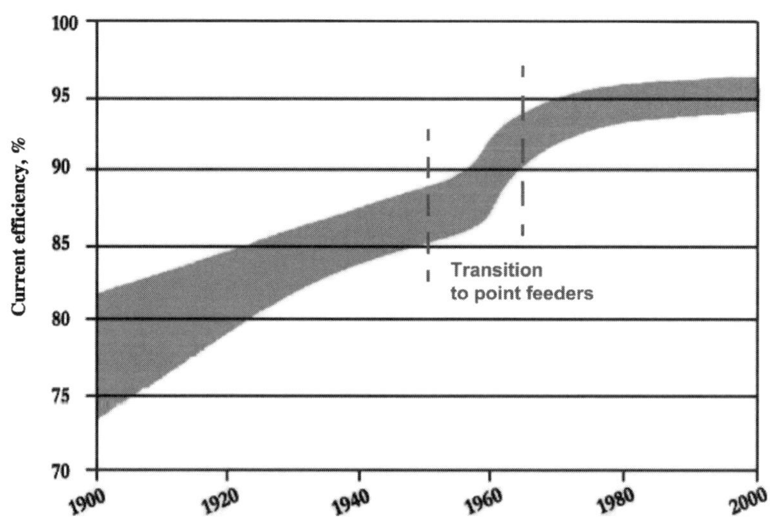

Fig. 10.10 Primary aluminum current efficiency from 1900 to 2000

Theoretical Minimum Energy Requirement for Reduction

All current primary production facilities and most alternative processes for aluminum production use alumina as their raw material. Any process that starts with alumina to produce aluminum has the same theoretical energy-requirement. Different processes do not offer any theoretical energy advantage. However, they do offer significant tradeoffs between efficiencies, emissions, footprints, and sources of input energy (electricity, carbon, and fuels). The theoretical limits required to manufacture aluminum provide a valuable insight into Hall-Heroult cell operation and potential future reduction processes.

The product of primary reduction process is molten aluminum. This report calculates the theoretical minimum energy by assuming the reactants enter and the by-products leave the system at room temperature and that molten aluminum leaves the system at 960 °C (1760 °F). The molten metal temperature, 960 °C (1233 °K), is an approximation of an average commercial operating cell. Figure 10.11 illustrates the theoretical boundaries for a system that reduces alumina to form aluminum and oxygen. Changes in the operating temperature of a cell have a minor effect on the theoretical energy requirements. Operating changes of 100 °C (180 °F) in a Hall-Heroult cell, operating in the range of 700 to 1100 °C (1290 to 2010 °F), result in less than a 1% change in the theoretical minimum energy requirements (Appendix F, Table F.7).

Some studies assume that the gases evolved during reduction leave the system at the molten metal temperature. In these studies, the theoretical minimum is 2.5 to 3% higher than the values presented in this chapter (Appendix F). Theoretically, it is possible to capture all the energy associated with these gaseous emissions. In practice, however, the gas stream is collected from hundreds of hooded pots and treated before release to the atmosphere. Only a very small portion of the heat is actually absorbed and returned to the system.

Three energy factors must be examined in the production of aluminum: the energy required to drive the reduction reaction forward, the energy required to maintain the system at constant pressure and temperature, and the energy required to change the temperature of the reactant and/or product. The thermodynamics and chemical equilibrium of the reactions are described by the following equation:

$$\Delta G = \Delta H - T\Delta S$$

The energy required to drive the reaction forward is the energy for the electrolytic reduction of alumina ($2Al_2O_3 \rightleftharpoons 4Al + 3O_2$) and is given by the change in the Gibbs free energy value (ΔG). The energy required to maintain system equilibrium is the difference between the heat of reaction (ΔH) and the Gibbs free energy value (ΔG), which equals the entropy term ($T\Delta S$). Because the Gibbs free energy requirement is less than the heat of reaction for

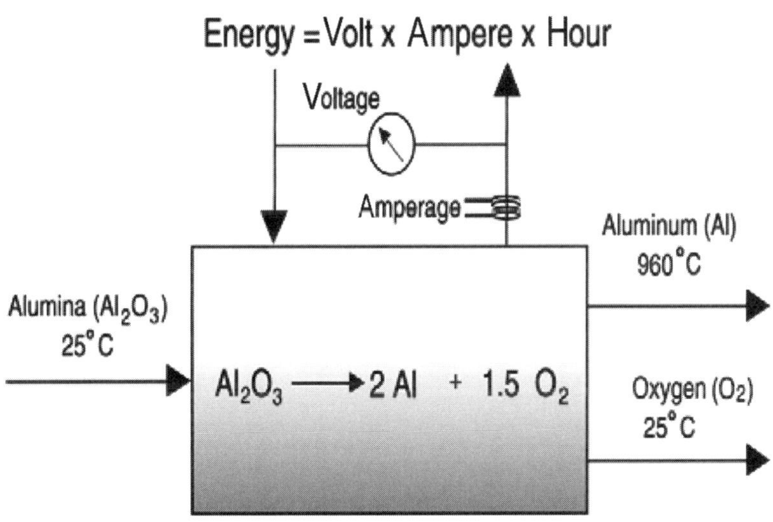

Fig. 10.11 Alumina-to-aluminum theoretical minimum energy

alumina reduction, additional energy must be added to the system to maintain the system temperature. Otherwise, the system will cool as the reaction proceeds. Hence, for the alumina reduction reaction, the ΔH term provides the minimum theoretical energy requirement (reaction and equilibrium). Reduction cells operate at atmospheric conditions, and no pressure change results from the reduction. Numeric values for these thermodynamic measures for the elements and compounds common to aluminum processing and the calculations used to determine the theoretical minimum energy requirement are given in Appendix F. The energy required to change the temperature of reactants and products is calculated from their heat capacities (C_p), which are provided in Appendix F.

Faraday's law provides the minimum amperage requirement for electrolytic reduction. This law states that 96,485 coulombs of electricity are passed through a cell to produce a 1 g equivalent of an element or compound. Aluminum has an atomic weight of 26.98, a charge of 3+, and therefore has an 8.99 g equivalent weight. Faraday's law is converted to more common measurements:

$$\left[\frac{96{,}485 \text{ coulombs}}{\text{g equivalent}}\right]\left[\frac{\text{Ampere/s}}{\text{Coulomb}}\right]\left[\frac{\text{Hour}}{3600 \text{ s}}\right] \times \left[\frac{\text{g equivalent}}{8.99 \text{ g}}\right]\left[\frac{1000 \text{ g}}{\text{kg}}\right] = 2980 \text{ Ah/kg}$$

The value 2980 Ah/kg of aluminum is the theoretical minimum amperage (current) required for production. This value assumes perfect conditions, where there are no reverse or parasitic reactions that consume amperage and no limitation to the ionic species availability to react at the electrodes (no concentration gradients or gas bubbles). The Gibbs free energy (ΔG) divided by the Faraday amperage provides the minimum voltage required to drive the reaction forward. Cell voltage and current efficiency are variables that are controllable by design, and they determine the electrical power required for reducing alumina. In practice, electrolytic cells have significant inefficiencies and operate above the minimum voltage requirement. This excess voltage provides the thermal energy required to maintain system equilibrium ($\Delta H - \Delta G$) and to produce molten material (C_p).

In the case of aluminum made directly from alumina ($2Al_2O_3 \rightleftharpoons 4Al + 3O_2$), shown in Fig. 10.11, the energy required to drive the reaction forward (ΔG) is 8.16 kWh/kg, the thermal energy ($\Delta H - \Delta G$) required to maintain thermal equilibrium is 0.48 kWh/kg, and the thermal energy (C_p) associated with producing the molten aluminum is 0.39 kWh/kg of aluminum. The theoretical minimum energy requirement is 9.03 kWh/kg of aluminum. (Note: If the gas emission at 960 °C, or 1760 °F, is included, the total theoretical minimum energy requirement is 9.30 kWh/kg of aluminum, (Appendix F, Table F.2).

Theoretical Energy for Hall-Heroult Carbon Anode Reduction. The theoretical minimum energy requirement for producing molten aluminum at 960 °C (1760 °F) in a Hall-Heroult cell with a carbon anode is 5.99 kWh/kg.

All commercial aluminum production uses a carbon anode in a Hall-Heroult cell. The carbon is consumed during the electrolytic process and supplies part of the energy necessary for the reduction of alumina. This gives the Hall-Heroult carbon anode process a lower energy requirement than the direct reduction of alumina to aluminum. The theoretical energy required for reduction is the same for prebaked or Soderberg carbon anodes.

The net reaction for the carbon anode Hall-Heroult process is $2Al_2O_3 + 3C \rightleftharpoons 4Al + 3CO_2$. Figure 10.12 shows an idealized Hall-Heroult cell for the production of aluminum. In this cell, it is assumed that the reactants (alumina and carbon) enter the cell at 25 °C (75 °F), the carbon dioxide by-product leaves the cell at 25 °C (75 °F), and the aluminum product leaves as molten metal at the cell operating temperature of 960 °C (1760 °F). The reaction is assumed to occur under perfect conditions, where there are no reverse reactions, no parasitic reactions consuming additional anode carbon, no limitations to the ionic species reacting at the electrodes, and no heat or energy losses external to the system.

Appendix F, (Table F.1) details the calculation of theoretical minimum energy for this reaction. The results show that the energy required to drive the reaction forward (ΔG) is 5.11 kWh/kg, the thermal energy required to maintain equilibrium is 0.49 kWh/kg, and the thermal energy associated with the molten aluminum is 0.39 kWh/kg of aluminum. Therefore, the theoretical minimum energy requirement for the reduction of alumina in a carbon anode cell adds up to 5.99 kWh/kg of aluminum. (Note: If the CO_2 gas emission at 960 °C, or 1760 °F, is included, the total theoretical minimum energy requirement is 6.16 kWh/kg of aluminum).

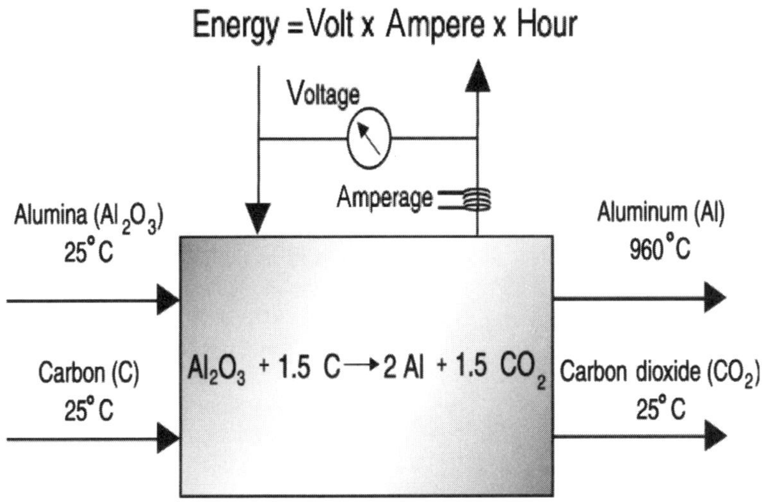

Fig. 10.12 Alumina-and-carbon-to-aluminum theoretical minimum energy

In actual carbon anode cell operations, current efficiencies of less than 100% result from reverse oxidation reactions between part of the aluminum metal that is dissolved in the cryolite and carbon dioxide gas produced ($2Al + 3CO_2 \rightleftharpoons Al_2O_3 + 3CO$) and, to a lesser extent, between the carbon dioxide gas and the carbon anode ($CO_2 + C_2 \rightleftharpoons CO$). Current efficiency losses can also result from direct shorting of the anode to the aluminum pad.

Today's (2007) state-of-the-art reduction cells are achieving current efficiency levels of more than 96% and energy consumption levels of less than 13.0 kWh/kg of aluminum (Ref 10.19). The theoretical minimum energy requirement at 100% current efficiency is 5.99 kWh/kg of aluminum. The energy efficiency levels of present state-of-the-art carbon anode reduction cells are approximately 46%.

Hall-Heroult Reduction Process

The engineering, materials, and process knowledge of existing components and processes form the foundation for developing new components, processes, and techniques for producing aluminum. The Hall-Heroult cell works as a system where all the components perform together. Therefore, improving one component may not necessarily result in an improved cell or a more energy-efficient operation. To understand the impact of component and system changes on the cell performance, it is helpful to know about the Hall-Heroult process in terms of its components, operations, and interrelationships. This section explains the process by describing a typical prebake anode operation.

Typical Hall-Heroult Cell Operation. A typical modern aluminum electrolysis Hall-Heroult reduction cell (pot) is a rectangular steel shell 9 to 12 m long, 3 to 4 m wide, and 1 to 1.5 m deep (30 to 40 ft long, 10 to 13 ft wide, and 3 to 5 ft deep) (Fig.10.13). It has an inner lining of carbon, which is surrounded by refractory thermal insulation, that keeps it isolated thermally and electrically from the steel shell. Commercial cells range in capacity from 60,000 A to more than 500,000 A and can produce more than 450 to 4000 kg of aluminum per day, respectively.

A cell typically operates at 950 to 980 °C (1740 to 1800 °F) and yields molten aluminum and carbon dioxide. The molten aluminum has a higher density than the electrolyte (cryolite bath) and settles to the bottom of the cell, on top of the carbon lining. Molten aluminum at approximately 99.7% purity is periodically tapped by a vacuum siphon from the cell bottom. The tapped metal is transferred to holding furnaces where the metal is alloyed, and entrained gases and impurities are removed prior to casting. The carbon dioxide and other gases generated in the cell during the reduction process are collected and treated to meet environmental regulations.

Electric current enters the cell through the carbon anode and flows through 3 to 6 cm (1.2 to 2.4 in.) of electrolyte (bath) to the aluminum pad and carbon lining cathode. The aluminum pad is in intimate contact with the carbon lining and serves as the charged surface of the cathode. Steel collector bars are set near the bottom of the

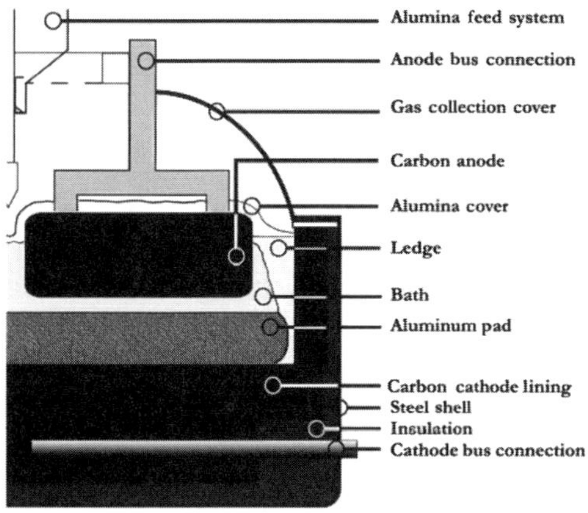

Fig. 10.13 Typical Hall-Heroult cell—half-section view

carbon lining to conduct the current to the anode of the next cell.

The 950 to 980 °C (1740 to 1800 °F) molten cryolite and aluminum used in a typical reduction cell are corrosive. Molten cryolite has low viscosity and interfacial tension that allows it to easily penetrate any porosity in the cell lining. To protect the carbon lining, the thermal insulation is adjusted to provide sufficient heat loss to freeze a protective coating of the electrolyte, known as a ledge, on the inner walls. The molten aluminum pad protects the carbon bottom of the cell. The cell is never tapped completely dry of molten aluminum. It is essential that no alumina or frozen ledge form under the metal pad. The carbon cathode must remain bare for good electrical contact with the aluminum pad and to minimize wear of the carbon surface.

The reduction reaction is continuous, and alumina must be supplied to the bath at a controlled rate to maintain constant conditions. This is accomplished with automatic feeders that break the surface crust and deposit alumina into the molten bath, where it is dissolved for reaction. Alumina is also used to cover the carbon anodes and the frozen bath surface. The alumina covering serves as thermal insulation and as a protective cover to reduce air burning of the anode.

The electrolytic reaction in a Hall-Heroult cell consumes the carbon anode. Approximately 0.45 kg of the carbon anode are consumed for each kilogram of aluminum produced. The carbon anodes provide a necessary part of the energy required to operate a cell. The distance between the carbon anode and the metal pad is kept constant by adjusting the anode as it is consumed and as the aluminum is tapped. The consumable carbon anodes must be replaced periodically, typically about every four weeks in a modern plant. The frequency of anode changing depends on the anode design and the cell operation. Anode changing represents the most frequent cell and productivity disruption. The removed portion of an anode (known as a butt) is recycled or sold as a fuel. The pot cover, which is part of the gas collection system, must be removed, the used anode must be pulled from the frozen surface crust, and the new anode must be inserted into the space of the consumed anode. This must be accomplished without significant pot crust breakage or alumina falling into the bath. Anode changing is the single largest thermal, current, and magnetic disturbance in cell operation.

Cells are arranged in long rows called potlines. They are placed as close as possible to each other while maintaining sufficient room for anode changing, alumina feeding, and reasonably low electromagnetic interference. The cells are connected electrically in series. Rectifiers, which convert alternating current (ac) to direct current (dc), are chosen to minimize capital investment and typically provide approximately 700 V (dc). Typically, a reduction cell design requires approximately 4.6 V (dc), so that a potline of roughly 150 to 180 cells would be used.

Voltage Requirements. The energy consumed in an electrolytic reaction is a function of the voltage used and the current efficiency of the operating cell (the minimum current is fixed by Faraday's law; see the section "Theoretical Minimum Energy Requirement for Reduction" in this chapter). Modern Hall-Heroult cells operate at high (+96%) current efficiencies. The approximate voltage components of a conventional cell are shown in Fig. 10.14.

The electric current flows through the cell, and the cell voltage components can be described as a set of resistors in series:

$$E = \text{Cell reaction} + \text{Overvoltage} + \text{Bath} + \text{Cathode} + \text{Anode} + \text{Connectors}$$

The cell reaction voltage is a function of temperature and, at 960 °C (1760 °F), is fixed at 1.2 V (dc) (Ref 10.20). This is the theoretical minimum voltage required for the reduction reaction to take place, and no cell can operate at 960 °C (1760 °F) below this voltage. The total cell operating voltage includes the addition of voltages required to overcome the ohmic resistance of the

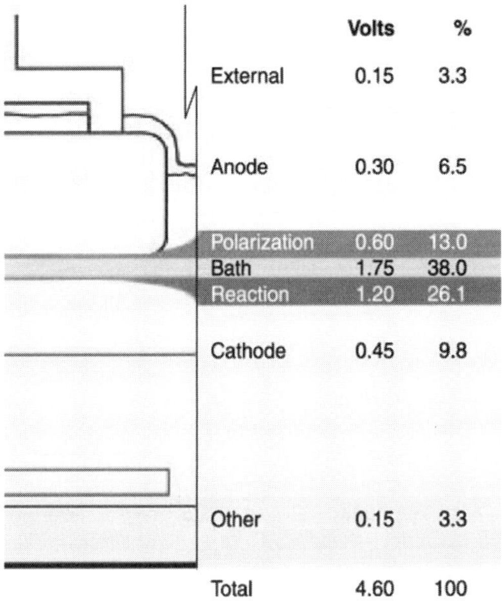

Fig. 10.14 Voltage distribution in a Hall-Heroult cell

other cell components. These are described in the following section.

Cell subsystems and variables include:

- Busbars and pot connectors
- Electrolyte
- Anode-cathode distance
- Aluminum pad
- Cathode
- Current density
- Cell polarization/overvoltage
- Anode effects
- Alumina feed
- Cell operating temperature
- Heat balance

Busbars and pot connectors electrically connect in series all the cells of a single potline, which typically contains more than 150 pots or cells. They are fabricated from highly conductive aluminum alloy and are sized for minimum overall system cost. Any voltage drop in the busbar and connector system results in energy loss.

The electrolyte or bath used in Hall-Heroult is cryolite (Na_3AlF_6). This is modified with the-addition of aluminum fluoride (AlF_3), calcium fluoride (CaF_2), and other additives to control the operating temperature, solubilities, activities of ionic species, conductivity, viscosity, interfacial tension, bath density, vapor pressure, hardness of the crust, and other factors. Bath chemistry, physical properties, and thermodynamics are very complex. The bulk electrical conductivity of the bath is influenced not only by its composition and temperature but also by the presence of anode gas bubbles and carbon dust.

Aluminum fluoride (AlF_3) is the most common bath additive. It lowers the operating temperature, the solubility of the reduced aluminum, surface tension, viscosity, and density. However, it has the undesirable effect of decreasing alumina solubility and electrical conductivity. The weight ratio of NaF to AlF_3 is referred to as the bath ratio. Controlling this ratio is important for efficient cell operation. (Note: Outside the United States, many countries use the cryolite or molar ratio, which is twice the bath ratio.)

The fluid bath circulates within the cell. As gas molecules are formed at the anode, they accumulate and coalesce into fine bubbles that aggregate into larger bubbles. These bubbles collect and move across the anode surface to escape around the edges of the anode. The buoyancy of the gas creates movement, which contributes to the motion of the bath and pad. The bath movement results from, in decreasing order of magnitude, gas bubble drag and electrolyte density difference caused by the bubbles generated at the anode, electromagnetic forces on the molten metal pad, and temperature gradients. This motion influences the concentration gradients of dissolved alumina and affects current efficiency. The motion also influences the heat transfer from the bath to the protective frozen ledge.

The anode-cathode distance (ACD) is the distance between electrode surfaces. In the Hall-Heroult cell, it is the distance from the lower face of the carbon anode to the top surface of the aluminum pad. This distance is typically approximately 4 to 5 cm (1.6 to 2.0 in.). The electrolytic bath occupies the space between the carbon anode and the aluminum pad. The voltage required for current to pass through the bath is related to the bath conductivity and the distance between the anode and the cathode. Decreasing the ACD lowers the voltage and energy requirements of the cell. The operating ACD is a compromise between keeping a low value of bath resistance, while at the same time enabling electrolyte rich in alumina ionic species to reach the charged surfaces and allowing reactant gas bubbles to escape. The ACD also must be large enough to ensure that the liquid metal does not contact the anode and short-circuit the cell.

The heat required to keep the bath molten is in part supplied by the electrical resistance of the bath as current passes through it. The amount of heat developed depends on the current path or the ACD. Changing the ACD is one method of controlling the desired bath operating temperature.

The molten aluminum pad that forms at the bottom of the cell is the cathodically-charged surface for the reduction reaction. The large amperage flowing through the cell creates electromagnetic forces and torques that cause the metallic pad to rotate. These magnetohydrodynamic (MHD) forces create this motion, which deforms the molten aluminum/cryolite interface. The MHD forces cause local metal velocities of approximately 5 cm/s (2 in./s) to occur. Control of MHD effects is one of the key factors for successful cell operation with high current efficiency and low energy consumption. Cells are designed to minimize these forces, and today (2007), velocities are typically one-third of what was common 50 years ago. Movement of the aluminum pad is mainly caused by electromagnetic forces in the cell and, to a smaller extent, by the interfacial drag of the bath fluid. Joint discontinuities in the carbon blocks create additional flow disturbances in the moving pad. The combination of all these forces causes the pad to undulate or roll, which could result in waves forming on the surface of the pad. These waves can approach the anode and result in an electrical short-circuit. The current that flows during this shorting produces no aluminum and results in a major loss of power and productivity. The motion of the aluminum pad also produces erosion of the carbon lining and shortens cell life. Because the distance between the anode and cathode is constantly changing as a result of the undulating pad, the ACD is kept large enough to avoid contact between the anode and the pad. This requires that the anode be backed away from its optimal position. Designing systems to minimize movements of the metal pad is a key factor in the efficient operation of a cell. A stable pad surface will allow the ACD to be decreased.

The newer concept of using a drained cathode cell is an approach to circumvent the difficulties associated with keeping the metal pad stable. Essentially, the bulk of the metal is drained to a sump, and the cathode is left wetted only by a thin metal sheet (see the section "Advanced Hall-Heroult Cells" in this chapter).

The bottom lining of the cell also serves as a cathode and carries current from the molten metal pad. Because 10% of the total cell voltage drop is in the carbon cathode blocks, it is important that they have the highest density and electrical conductivity possible.

In an attempt to reduce the resistance of the cathode, some have experimented with cathode blocks containing a higher content of graphitic carbon. However, while less resistive, graphite is also less wear resistant, and this compromises the life of the cathode. The cathode life generally determines cell life, because cathode replacement requires the complete dismantling of a cell. The advent of hard titanium diboride (TiB_2) coatings may offer an opportunity to increase the graphitic content of cathode blocks, and lower cell resistance, without reducing the cell life. Under optimal conditions, the life of a cathode or cell is in the range of 7 to 10 years.

Current density is a measure of the productivity of a cell. It is calculated by dividing the amperage supplied to an anode by the geometric face area of the anode. It is generally expressed in amperes per square centimeter (A/cm^2). Most potlines operate in the range of 0.8 to 1.0 A/cm^2. The quantity of aluminum produced per cell increases with increasing current density. The tradeoff is that as current density and productivity increase, current efficiency decreases, which results in a higher energy consumption per unit of metal produced. Lower current densities are more energy efficient but increase capital and labor costs per unit of output.

The reactions occurring at the anode and the cathode create localized conditions that are different from the bulk of the bath. The reactions deplete the supply of reactants and increase the quantity of products. This creates concentration gradients, which in turn cause concentration polarization. Additionally, the gas generated at the anode forms bubbles, which lower the effective bath conductivity. Localized conditions at the anode and cathode are unavoidable and require a voltage higher than the minimum reaction voltage to be applied to the cell.

No free aluminum (Al^{3+}) or oxygen (O^{2-}) ions are present in the bath. Alumina dissolves and dissociates into salt complexes in the bath. The dynamics at the anode are complicated by the release of gas bubbles. Oxygen containing ionic species are transported through the bath and discharged on the carbon anode. These anode reactions are:

$$Al_2O_2F_6^{2-} + 2F^- + C \rightarrow CO_2 + 2AlF_4^- + 4e^- \text{ and}$$

$$Al_2O_2F_4^{2-} + 4F^- + C \rightarrow CO_2 + 2AlF_4^- + 4e^-$$

At least three phenomena have been identified as contributing to the anode overvoltage:

- An increase in the local current density due to the presence of gas bubbles adjacent to the anode surface, which displace the electrolyte bath
- An ohmic component from an increase in the resistance of the electrolyte due to the presence of the bubbles
- The concentration polarization overvoltage (Ref 10.21)

Polarization effects at the cathode contribute much less to overvoltage than at the anode. The cathode reactions are $(AlF_6^{3-} + 3e^- \rightarrow Al + 6F^-)$ and $(AlF_4^- + 3e^- \rightarrow Al + 4F^-)$. The aluminum ion complexes, AlF_6^{3-} and AlF_4^- have higher ionic mobility than their anodic counterparts, which lowers the concentration polarization effect. In addition, no gas bubbles, which influence both resistance and concentration polarization, are produced at the cathode.

Control of the quantity of alumina dissolved in the bath is important for proper operation of a cell. Alumina saturation is reached at approximately 7% alumina dissolved in a typical bath. The normal operating level is approximately 3% alumina. If the level goes above 4%, some of the added alumina may not dissolve rapidly and can settle to form a sludge on the cell bottom, thereby reducing the cell conductivity. If it falls below 1% alumina, the cell is starved of the reactant, and an anode effect ensues. When this occurs, the production of metal is interrupted, and fluorine (F_2), hydrogen fluoride (HF), and two perfluorocarbon gases—tetrafluoromethane (CF_4) and hexafluoroethane (C_2F_6)—are discharged instead of carbon dioxide. These perfluorocarbon gases have 6000 to 9000 times the global-warming potential of carbon dioxide and require gas scrubbing to meet environmental regulations.

Significantly, alumina is a good absorbent of fluorine and hydrogen fluoride but not perfluorocarbon gases. This allows primary production facilities to use their alumina raw material as an absorbent in dry gas scrubbing systems. The fluorides absorbed on the alumina in the scrubbers are recycled into the potfeeding systems, so that both the alumina and fluorides can be reused in the process.

Alumina is ideally added to the cell at a rate that exactly replaces the alumina that has been reduced. If alumina is fed too fast or in large increments, it may not dissolve and can form sludge. Sludge affects fluid flows within the cell and contributes to erosion of the cathode block surface. Underfeeding the cell results in an anode effect. There is no current technology for in situ, real-time bath analysis to provide precise control of alumina concentration in the bath and alumina feed rate.

Alumina is fed by automatic handling and conveyor systems. The specifications for alumina are complex, with numerous tradeoffs. It must dissolve rapidly; it should contain few impurities and have a high surface area, yet be a relatively large particle (this apparent inconsistency is overcome by the particle having significant porosity). The particle must also be robust, create little dust, and resist breakage during handling. The introduction of dry scrubbing systems for the cell offgases, wherein the alumina is used as an adsorbent for fluoride emissions, further complicates the alumina specifications.

Alumina is side-worked in older prebake cell designs and Soderberg cells. Side-worked cells introduce alumina into the bath using automated crust-breaking and feeding machines that move along the length of the cell. Side-working is time-consuming and can take 1 to 4 h before the machine feeds the same section of the cell again. Newer prebaked cell designs use the spaces between the anodes to feed alumina into the cells. Point feeders pierce the crust and dispense small quantities of alumina at numerous points, typically in the center of the cell. One-minute intervals between point feeding are common. This frequent addition of small quantities of alumina takes advantage of the motion of the bath and provides significantly better control of local alumina concentration. This provides for better current efficiency, fewer anode effects, and less erosion caused by solid alumina. Point feeders have proven to be significantly better and have replaced most feeding systems in both prebaked and Soderberg cells.

Bath chemistry controls the operating temperature of the Hall-Heroult process. Most commercial cells operate near 960 °C (1233 K). Reducing the cell operating temperature is an obvious approach to saving energy and reducing capital costs by lowering insulation requirements. However, controlling the operating variables of a cell becomes more critical as temperature is lowered. Dissolved alumina must be available for reduction in the bath to have high cell productivity and minimal energy-consuming concentration polarization and anode effects. Lowering the cell temperature lowers the solubility range for alumina in a cryolite bath. The solubility or phase diagram for a cryolite alumina system is V-shaped with temperature on the y-axis and alumina concentration on the x-axis, as shown in Fig. 10.15. Cells must operate within the "V"; operations to the left will precipitate cryolite, and those to the right will precipitate alumina. Decreasing the temperature narrows the cell-operating range with respect to alumina solubility. In practice, bath chemistry is modified with the addition of aluminum fluoride, calcium fluoride, and other salts. These modifications lower the liquid-phase temperature. However, the basic V-shape of the diagram is retained.

Controlling the thermal balance of the cell is of prime importance for efficient operation and long cell life. There is no commercially available material that can retain its insulating value and resist penetration and chemical attack by cryolite at cell-operating temperatures. Accordingly, the method used to protect the side walls is to allow some heat loss so that the temperature of the exposed surface of the cell lining is below the freezing point of the bath. This creates a frozen layer or ledge of bath that protects the linings. Side-wall heat losses can account for 35 to 45% of the total heat loss in a cell (Ref 10.3). The frozen ledge, due to phase relationships, differs in composition from the bath. If the temperature rises above steady state, the ledge begins to dissolve, and the bath ratio (sodium fluoride to aluminum fluoride) changes. If the temperature is allowed to fall too much, ledge formation is excessive, anodes are changed with great difficulty, alumina does not dissolve as readily, and the bath ratio is affected. The frozen electrolyte ledge also provides the electrical insulation of the side walls. The thermal conductivity of frozen cryolite is an order of magnitude lower than that of the molten cryolite.

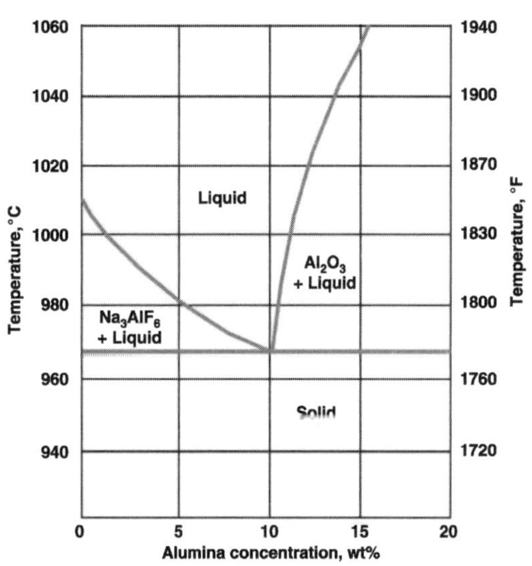

Fig. 10.15 Cryolite-alumina phase diagram

Large cell designs require less energy to maintain operating temperatures because of the lower ratio of cell surface area to volume. This is one factor in the trend to use larger cells in newer smelting plants.

Environmental Considerations

The *Energy and Environmental Profile of the U.S. Aluminum Industry* (Ref 10.8), compiled by the U.S. Department of Energy, Office of Industrial Technologies Program, and *Life Cycle Inventory Report for the North American Aluminum Industry* (Ref 10.7), published by The Aluminum Association, provide detailed environmental information on the overall aluminum production process. The by-products generated in the Hall-Heroult process that are of environmental concern can be grouped into three areas: electrolysis, anode production, and cell waste products.

Electrolysis. Electrolysis green house gas (GHG) emissions in the Hall-Heroult process can be split into three groups: reduction reaction emissions, carbon dioxide (CO_2) and carbon monoxide (CO); process upset perfluorocarbons emissions; and hydrogen fluoride (HF) formed from the inclusion of moisture (H_2O) in the raw materials. Hydrogen fluoride gas is almost completely captured and returned to the cells by the alumina dry scrubbing system used in modern facilities.

The carbon-base emissions associated with the reduction reaction come from three sources:

- *Reaction products:* The reaction produces oxygen that reacts with the carbon anode to produce CO_2 and small quantities of CO; this reaction produces 1.22 kg of carbon dioxide equivalents for each kilogram of aluminum produced (Appendix D, Table D.4).
- *Air burning:* The carbon anode loses mass to oxidation with the atmosphere. This produces 0.30 kg of CO_2 for each kilogram of aluminum produced (Appendix D, Table D.4).
- *Electricity generation and transmission:* The emissions related to the fuels used in electricity generation for U.S. primary facilities are 5.38 kg of carbon dioxide for each kilogram of aluminum produced (Appendix D, Table D.2).

A total of 6.90 kg of CO_2 gas is generated from the reduction process for each kilogram of aluminum produced in an average U.S. primary facility. It should be noted that electricity-related emissions for specific potlines vary widely. Potlines operating on electricity obtained from coal-fired power plants produce 16.0 kg of CO_2 gas for each kilogram of aluminum produced, while potlines using electricity from hydropower plants produce close to zero CO_2 gas emissions.

The perfluorocarbon emissions are related to the anode effect. If the concentration of alumina in the bath becomes too low, other reactions between the carbon anode and the bath occur, and tetrafluoromethane (CF_4) and hexafluoroethane (C_2F_6) are generated. These gases have a high global-warming potential (GWP). The GWP of a GHG is a ratio developed to compare the ability of each GHG to trap heat in the atmosphere relative to CO_2 gas. The GWP of CF_4 and C_2F_6 is 6500 and 9200, respectively. In other words, 1 kg of CF_4 released to the atmosphere is equivalent in its warming potential to 6500 kg of CO_2.

Aluminum smelting is the principal quantifiable source of perfluorocarbon in the United States. The U.S. Environmental Protection Agency (EPA) estimates U.S. emissions from aluminum production at 500 metric tons of CF_4 gas and 50 metric tons of C_2F_6 gas in 2003 (Ref 10.22). In 1995, the aluminum industry entered into a Voluntary Aluminum Industry Partnership (VAIP) with the EPA to reduce perfluorocarbon emissions by 46% during the next decade. Reductions in primary aluminum production and efficiency improvements to reduce anode effects have reduced emissions of CF_4 and C_2F_6 gases since 1990 by 71% each. The U.S. aluminum industry and EPA are continuing the VAIP to seek further GHG reductions beyond the original achievements. The total 2002 U.S. aluminum perfluorocarbon emissions are 2.0 CO_2 equivalent metric tons per ton of aluminum.

The emissions of perfluorocarbon for older side-fed cells are 1 order of magnitude higher than the emissions for cells with point feeders. The industry continues to improve point feed systems. Ultimately, with the entire industry on point feeders and advanced cell control systems, it should be possible to virtually eliminate anode effects and hence perfluorocarbon generation.

Anode Production. The CO_2 emissions for the carbon anode manufacturing amount to approximately 0.13 kg of CO_2 equivalents per kilogram of aluminum. In this report, the feedstock portion of the CO_2 emissions for carbon anode manufacturing is included in the electrolysis. More specific information on this topic can be found in the comprehensive life-cycle information published by The Aluminum Association (Ref 10.7).

Cell Waste Products. Aluminum electrolysis carbon slime and spent potlining (SPL) are unavoidable by-products of the aluminum smelting process and are listed in the United States as hazardous wastes. Development work is underway to mitigate problems associated with SPLs. Most development efforts attempt to combust the carbon linings to destroy any remaining toxic chemicals, to recover the valuable fluoride as AlF_3, and to render the remaining material inert through vitrification. In some areas of the world, SPL is destroyed by combustion in the production of cement.

Technological Change in the Next Decade

The Hall-Heroult electrolysis process, using a carbon anode and cryolite bath, is a mature technology. However, gradual improvements in both productivity and environmental performance are still possible. The typical Hall-Heroult cell life ranges from 7 to 10 years. Adoption rates of new technology and systems are governed to some degree by the cell life. There is a slow, autonomous efficiency improvement in the Hall-Heroult process because of the continued adoption of advanced cell designs, feeding systems, bath composition, control systems, and other practices. This trend has resulted in a gradual decline in energy consumption in the range of 0.2 to 0.5% per year. Energy savings are actively pursued by aluminum producers because electricity costs constitute a high percentage (20 to 30%) of total production costs. Because current efficiencies are already over 95%, the goal is generally to reduce the overvoltage in the aluminum cells to increase the overall electric efficiency.

A number of technological and engineering improvement options exist and are being adopted by the aluminum industry. These include:

- *Point Feeders:* Point feeders enable more precise, incremental alumina feeding for better cell operation. Point feeders are generally located in the center of the cell and thereby cut down on the diffusion required to move dissolved alumina to the anodic reaction sites. The controlled addition of discrete amounts of alumina enhances the dissolution process, which aids in improving cell stability and control, minimizing anode effects, and decreasing the formation of undissolved sludge on the cathode. In the jargon of modern commerce, point feeders enable "just-in-time alumina supply" to permit optimal cell operation. Point feeder improvements continue to be made as more accurate cell controllers become available.

- *Improved process controls:* Advanced process controllers reduce the frequency of anode effects and control operational variables, particularly bath chemistry and alumina saturation, so that cells remain at their optimal conditions.
- *Slotted anodes:* The use of transversally slotted anodes helps to facilitate the removal of anode gases from underneath the anode. This consequently lowers the overall anode polarization, allowing for incremental improvements in productivity and cell amperage.

Changes to the Hall-Heroult Cell and Alternative Technologies. Two innovative technological changes to the Hall-Heroult process, the wetted drained cathode and the inert anode, are on the near-term horizon for improving energy efficiency. These technologies can, with cell modifications, be retrofit into existing potlines and supporting infrastructure. Wetted cathodes are anticipated to lower energy consumption of a Hall-Heroult cell by $15^{tf}\%$ when compared to a modern Hall-Heroult cell. This report defines a modern cell as one that operates at 4.6 V (dc) and 95% current efficiency with the voltage distribution shown in Fig. 10.14. The combination of an inert anode with a wetted cathode could provide a $18^{tf}\%$ reduction in energy consumption and the elimination of cell CO_2 emissions. These technologies are described in the section "Advanced Hall-Heroult Cells" in this chapter, and the energy impacts are calculated in Appendix H. Multipolar cells using Hall-Heroult chemistry require the use of inert anode and wetted cathode technologies. The multipolar design allows for a more compact, more productive cell with significant thermal energy savings. The section "Advanced Hall-Heroult Cells" also describes multipolar electrolytic cells.

Two alternative technologies to the Hall-Heroult process, carbothermic reduction and kaolinite reduction, have been studied by several groups for many years. These alternative technologies could displace existing Hall-Heroult cells in the future. They are described in the section "Alternative Primary Aluminum Processes" in this chapter. Both these processes could potentially change where and how the aluminum industry operates, while lowering energy consumption. These alternatives consume more carbon and have higher on-site carbon emissions than the Hall-Heroult process. However, their electrical demands are lower, which results in lower overall (utility-to-metal) CO_2 emissions. The carbothermic process is anticipated to save $21^{tf}\%$ in energy and be economical at a much smaller scale ($\frac{1}{10}$) than the Hall-Heroult facilities. The kaolinite reduction process is anticipated to save approximately $15^{tf}\%$ of the energy required for a modern Hall-Heroult system. This value is impacted by the need to prepare additional ore mass and carbon for the process. However, the kaolinite reduction process is commercially interesting because of its lower on-site energy demands, domestically available ore, and lower-cost raw materials.

Table 10.11 summarizes the estimated energy performance of these near- and midterm technologies (Appendix H, Table H.4). The on-site and tacit energy values in the table allow the processes to be compared on a reaction, raw material, or a complete ore-to-metal basis. The table provides the energy associated with anode production and feedstock energies. The energy performance of the near-term technologies, wetted cathode and inert anode, are based on voltage changes in the electrolysis cell (Appendix H, Tables H.1 and H.2). These voltage changes are supported by theory and reported experiments and provide a good estimate of energy use. The energy values reported for midterm technologies, carbothermic reduction and kaolinite reduction, are approximations based on the theoretical energy requirements and assumed reactor inefficiencies. Both midterm technologies involve multiple reaction and separation zones. To date, no fully integrated reactor systems have been built. These midterm energy approximations assume that there is significant heat integration (recovery) within a facility.

Many technical hurdles remain to be solved in these new processes before they become commercially viable. Wetted cathodes and inert anodes will be adopted as their performances are proven and existing cells need rebuilding.

Industry will require significant demonstration time before adopting any alternative to Hall-Heroult technology.

Advanced Hall-Heroult Cells

The Hall-Heroult cell system efficiency can be improved with the adoption of new cell technologies, described subsequently. Wetted drained cathode technology offers the most significant improvement in energy efficiency, but its adoption will be gradual. Typical cell life is seven to ten years. The industry will optimize the capital invested in existing cells before retrofitting with new technology. Inert anodes offer significant environmental benefits, lower maintenance costs, and can be retrofit into the existing cells. Inert anodes could be adopted more quickly by industry because the carbon anodes are replaced approximately every four weeks. However, their superior performance must still be proven in industrial trials.

Wetted Drained Cathode

Wetted drained cathodes allow the anode-cathode distance to be reduced and are expected to result in a 20 (18[tf])% reduction in the electrolysis energy required to produce aluminum.

Molten aluminum does not wet the carbon lining of a Hall-Heroult cell. The aluminum pad rests on an extremely thin sheet of cryolite bath. This creates an electrical junction similar to the air gap between two metal bars and causes a small voltage drop. If the bars are clamped tighter, providing better contact, the junction voltage drop decreases. The thicker (heavier) the aluminum pad is, the thinner the cryolite sheet becomes and the lower the junction voltage drop. Modifying the cathode surface to make it more wettable would allow the same electrical contact with a decreased thickness in the pad. A thinner pad would be more hydrodynamically stable, have lower wave height, and allow a decrease in the anode-cathode distance (ACD). A decrease in the ACD results in energy savings. A cell lining that is completely wetted and inert to cryolite would be even more efficient. This combination of properties would allow the aluminum pad to be drained out of the anode-cathode spacing. Removing the unstable aluminum pad would allow the ACD to be considerably reduced and provide significant energy savings.

Titanium diboride (TiB_2) has been found to be a wettable cathodic material, and several approaches to incorporate TiB_2 into a Hall-Heroult cell are being studied (Ref 10.23). Cathodes made wettable with TiB_2 appear to increase cell life by making the cathode less susceptible to penetration by bath material. This is a considerable benefit, because cell rebuilding costs are a major contribution to primary aluminum operational costs. Longer-life cells also generate less cell lining waste material (spent potliner) per ton of aluminum produced. These benefits have to be balanced with the higher costs associated with the TiB_2 material. Recent evidence also suggests that the wetted cathodes reduce the formation of sludge (undissolved alumina) on the cell bottom and improve cell operations (Ref 10.24). Chalco is testing and has made wetted cathode blocks available to other primary aluminum producers.

Several concepts for wetted cathode and draining cells have been proposed. Figure 10.16 shows:

- A conventional cell
- A low metal pad with a wetted cathode in a conventional cell
- A hybrid cell with a metal sump
- A fully drained slanted cell

The decrease in ACD is shown for each configuration. These configurations offer attractive alternatives because they can be incorporated as a retrofit to the existing facilities. Some of these configurations are currently being evaluated in commercial cells and could soon be available (Ref 10.24). However, as ACD is decreased, less bath is available for circulation and mixing. This requires retrofit cell designs to account for the change in bath dynamics. Designs must ensure that dissolved alumina is available across the anode surface to maintain high current efficiency and productivity and to avoid anode effects. Designs also need to compensate for the heat energy lost due to the lower cell voltage operation.

Conceptually, the fully drained inclined cathode design offers the greatest potential for energy savings. By inclining the anode and cathode a few degrees, the molten metal pad (with all its complexities) is removed completely to a sump. Without the pad, the ACD is dimensionally stable and can be narrowed, which significantly reduces the total electrical resistance of the bath. The alumina feeding of the bath is not compromised, because the buoyancy of the gas

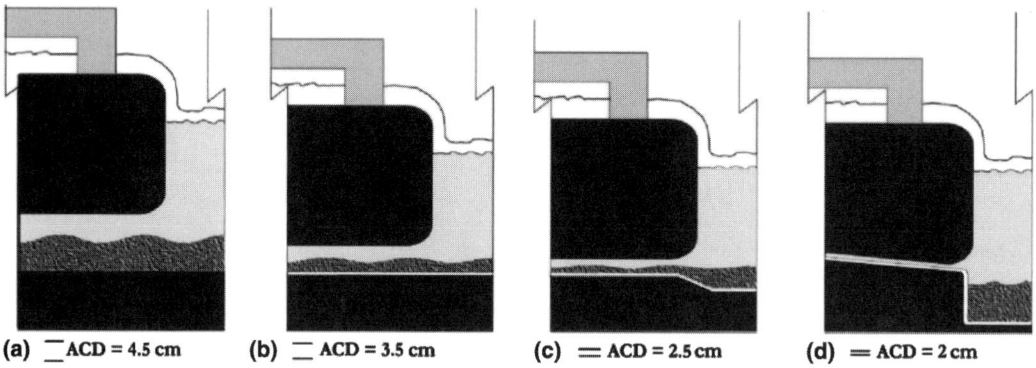

Fig. 10.16 Conceptual wetted cathode cells. (a) Traditional Hall-Heroult cell. (b) Wetted cathode with reduced metal pad. (c) Wetted drained cathode. (d) Wetted sloped drained cathode. ACD, anode-cathode distance

bubbles generated during reduction causes bath circulation, and fresh alumina is drawn into the ACD gap and across the electrode face.

Energy Savings for Wetted Cathode Technologies. Decreasing the ACD results in a proportional decrease in the voltage drop associated with the electrolytic bath. Energy is saved when the ACD reduction is matched with the ability to maintain current efficiency and heat balance. The energy savings for the different wetted designs shown in Fig. 10.16 are estimated in Appendix H, Table H.1.

The modern Hall-Heroult cell, shown in Fig. 10.16(a) has an ACD of 4.5 cm (1.8 in.) and a voltage distribution similar to that shown in Fig. 10.14. The traditional cell voltage distribution has 1.75 V (dc) associated with the ACD, which accounts for approximately 38% of the total 4.60 V (dc) drop across the cell. As discussed in the previous section, a wetted cathode provides better electrical contact between the metal pad and the cathode, allowing the ACD to be decreased. If the ACD were lowered from 4.5 to 3.5 cm (1.8 to 1.4 in.), as shown in Fig. 10.16(b), the voltage associated with the bath would be proportionally lowered to 1.36 V (dc). The total voltage would decrease to 4.21 V (dc), which would provide an 8% reduction in the electrical energy usage.

Draining the metal pad into a sump would eliminate the unevenness of the pad and permit an even smaller ACD with greater energy savings. If the ACD were lowered from 4.5 to 2.5 cm (1.8 to 1.0 in.), as shown in Fig. 10.16 (c), the voltage associated with the bath would be proportionally lowered to 0.97 V (dc). The total voltage would decrease to 3.82 V (dc), providing a 16% reduction in the electrical energy usage.

Reducing the ACD is limited by the ability to transport the reactant (dissolved alumina) to the electrode interface and to remove products (aluminum and carbon dioxide) from the electrode interface. Sloping the electrode interface slightly, as shown in Fig. 10.16(d), removes products and supplies reactants more effectively by using the buoyancy of the gas to induce bath circulation. It is estimated that under the best conditions, the ACD for a sloped configuration could be reduced to as little as 2.0 cm (0.8 in.) (Ref 10.25, 10.26), in which case, the voltage associated with the resistance of the bath would be 0.78 V (dc). The total voltage required would then decrease to 3.63 V (dc), providing nearly a 20% reduction in the electrical energy usage.

Environmental Impacts for Wetted Cathode Technologies. The by-products generated in the Hall-Heroult process that are of environmental concern are grouped into three areas: electrolysis, anode production, and cell waste products. Of these, the wetted cathode technology impacts electrolysis and cell waste products.

Electrolysis GHG emissions can be split into three groupings: reduction reaction emissions, process upset perfluorocarbons emissions, and hydrogen fluoride emissions from the bath. Wetted cathode technology does not change the emissions related to process upsets or bath emissions. The reduction reaction emissions come from three sources: the reaction products, anode air burning, and electricity generation and transmission. Wetted cathodes do not change the reaction products or anode air burning. Both wetted and traditional cathodes will produce 1.22 kg of carbon dioxide from the reduction reaction ($2Al_2O_3 + 3C \rightleftharpoons 4Al + 3CO_2$), 0.30 kg of carbon dioxide for each kilogram of aluminum produced from air burning of the carbon anode, and 0.12 kg of carbon dioxide for each kilogram of aluminum for the fuels associated with manufacturing the anodes.

The environmental benefit of wetted cathode technology is related to the emissions associated with the electricity production. A wetted cathode lowers the electrical energy requirement, which in turn reduces the emissions related to the fuels used in electricity generation and transmission. The electricity production (14.4 kWh/kg of aluminum) for a modern Hall-Heroult cell emits 5.04 kg of carbon dioxide equivalents (CO_{2e}) for each kilogram of aluminum produced. A wetted-sloped cathode cell with a 2.0 cm (0.8 in.) ACD will lower CO_{2e} associated with electricity generation and transmission by nearly 21% to 3.98 kg CO_{2e}/kg of aluminum produced. This lowers the total CO_{2e} associated with a wetted-sloped cathode cell from 10.62 to 7.90 kg CO_{2e}/kg of aluminum produced (Appendix D, Table D.4).

One of the benefits of using wetted cathodes is longer cell life. Rebuilding cells less frequently will lower the quantity of spent potliner waste per unit of aluminum produced.

Hall-Heroult Inert Anode

Carbon anodes are consumed in the Hall-Heroult process, making the continuous manufacture of new anodes and constant changing of the consumed anodes necessary. Anode changing upsets the stability, production, and energy efficiency of the cell. For more than 125 years, Charles Hall and other primary metal producers have attempted to find an inert anode that would eliminate the manufacturing and handling of consumable anodes (Ref 10.20). The material demands of an inert anode require that it be highly conductive, and be thermally and mechanically stable at 800 to 1000 °C (1470 to 1830 °F). Further, it should not react or dissolve to any significant extent in the cryolite, or react or corrode in the 800 to 1000 °C (1470 to 1830 °F) oxygen-containing atmosphere. Any undesirable reaction with the bath must occur at a very slow rate, because this will result in the anode material contaminating the aluminum product.

Few materials are truly inert under the extreme conditions of a cell. Research has focused on three classes of inert anode materials: ceramics, cermets, and metals. The challenges to finding the most efficient material are substantial. Ceramics typically have poor thermal-shock resistance, are not mechanically robust, display poor electrical conductivity, and are difficult to connect electrically. Metals have good thermal, mechanical, and electric properties but are attacked by the hot oxidizing atmosphere. Metal oxides are somewhat soluble in cryolite and can resist the hot oxygen atmosphere, but they exhibit lower electrical conductivity than metals. When in solution, they electrochemically reduce and contaminate the aluminum metal. Metal oxide solubility can be reduced dramatically in cryolite by operating with a bath saturated in aluminum oxide and alumina. However, this presents significant bath feeding challenges. Cermets combine the advantages and disadvantages of ceramics and metals. In addition to overcoming technical hurdles, the likely higher cost of manufacturing inert anodes of commercial size must be compensated with longer life, lower energy consumption, and higher productivity.

An inert anode would enable greater control of the critical ACD, which represents the largest voltage drop in the cell (Fig. 10.14). When used in conjunction with a drained cathode, it is estimated that an inert anode may save up to 21[tf]% of the energy required for reducing aluminum. Inert anodes offer a major environmental advantage and the potential of producing a valuable coproduct. Replacing the carbon anode with an inert material results in oxygen being discharged rather than carbon dioxide. If a significant market for oxygen is near the reduction facility, the oxygen produced could be collected and sold as a coproduct. Carbon credits are an unknown but potentially large economic force that could hasten the development of inert anodes.

Inert anode technology could potentially be retrofit into the existing cells with limited changes and use the existing alumina- and aluminum-handling infrastructures. Some electrical infrastructure changes would be required because the inert anode will operate at a higher voltage than the carbon anodes. Because frequent access to the cells is not required for changing anodes, cells can be sealed more effectively to provide better gas collection and treatments.

Notable progress in the production and testing of potential inert anode materials has been made in recent years. (Ref 10.27). Some companies are now conducting trials with relatively stable materials that offer the promise of inert anode performance. Using an inert anode and wetted cathode could also lead to the design of multipolar, vertical electrode cells, which would increase productivity and further reduce energy usage.

Theoretical Energy for Hall-Heroult Inert Anode. The theoretical minimum energy requirement for an ideal Hall-Heroult cell using an inert anode can be calculated from the thermodynamics of the reaction, $2Al_2O_3 \rightleftharpoons 4Al + 3O_2$ (Appendix F, Table F.2). The production of aluminum shown in Fig. 10.17 assumes that the reactant (alumina) enters the cell at 25 °C (75 °F), the oxygen by-product leaves the cell at 25 °C (75 °F), and the aluminum product leaves the cell as molten metal at 960 °C (1760 °F). The theoretical minimum energy requirement is calculated under perfect conditions: an environment with no reverse reactions, parasitic reactions, heat/energy losses external to the system, or limitations to the ionic species reacting at the electrodes. The results show that the energy required to drive the reaction forward (ΔG) is 8.16 kWh/kg, the thermal energy ($\Delta H - \Delta G$) required to maintain equilibrium is 0.48 kWh/kg, and the thermal energy (C_p) associated with the molten aluminum is 0.39 kWh/kg of aluminum. The total theoretical minimum energy requirement for an inert anode adds up to 9.03 kWh/kg of aluminum. (Note: If the O_2 gas emission at 960 °C, or 1760 °F, is included, the total theoretical minimum energy requirement is 9.30 kWh/kg of aluminum.)

In actual cell operations, current efficiencies of less than 100% can result from reverse oxidation reactions between part of the metallic aluminum and the oxygen gas produced by the reaction. Inert anode cell designs must ensure that oxygen bubbles are discharged from the system without contacting the molten aluminum being produced.

Comparative Energy Requirements for Carbon and Inert Anode Hall-Heroult Cells. The theoretical minimum energy requirement for an inert anode reaction is 51% higher than the carbon anode requirement. Both the Hall-Heroult carbon and inert anode processes use the same starting raw material, alumina, and require the same total theoretical energy, 9.03 kWh/kg of aluminum. The Hall-Heroult carbon anode is consumed as part of the reduction reaction and provides part of the energy to the system. The energy provided by oxidizing the carbon (3.04 kWh/kg of aluminum) lowers the carbon anode system theoretical energy requirement to 5.99 kWh/kg of aluminum. The inert anode must supply the full minimum energy requirement for alumina reduction. The inert anode system requires 2.2 V (dc) compared to the Hall-Heroult reaction requirement of 1.2 V (dc) (Ref 10.28). This 83% increase in reaction voltage supplies the additional 3.04 kWh/kg of aluminum minimum energy required by the inert anode system to reduce alumina.

Although the inert anode reaction voltage is 1 V (dc) higher than a carbon anode Hall-Heroult cell, three factors are expected to provide the inert anode with an overall improved operational energy performance than the carbon anodes:

- Elimination of carbon anode manufacturing
- Reduction of anode polarization overvoltage
- Reduction in ACD that results from a stable anode surface

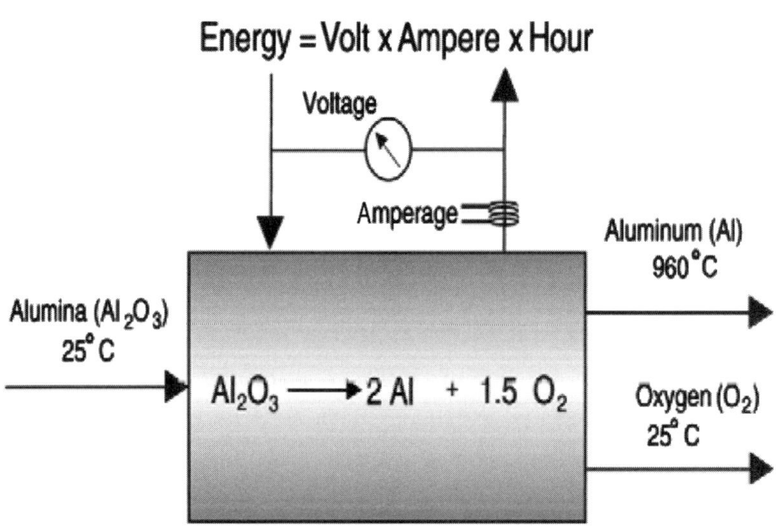

Fig. 10.17 Theoretical energy schematic for an inert anode

The impacts of these three factors are shown in Table 10.12, and calculations are presented in Appendix H, Table H.2.

The energy inherent in the carbon and required for carbon anode manufacturing exceeds the additional theoretical energy need for the inert anode system. A complete account of the energy associated with the carbon anode production shows that 6.02^{tf} kWh/kg of aluminum produced is associated with the anode manufacture and fuel value. This value is higher than the 3.04 kWh/kg of aluminum that the carbon anode supplies in the reaction for theoretical reduction. Because the actual values for manufacturing inert anodes are unknown at this time, a conservative estimate is made in Appendix H, Table H.3. This estimate, approximately 0.77 kWh/kg of aluminum, provides a basis for comparing the two processes. The total energy associated with an inert anode operating cell is 0.77^{tf} kWh/kg for anode manufacturing, 3.14 kWh/kg for the on-site additional voltage requirement, and 3.86^{tf} kWh/kg for the generation and transmission losses associated with the additional anode voltage. The inert anode requires 7.77^{tf} kWh/kg of aluminum produced, 29^{tf}% more than the tacit energy associated with the Hall-Heroult carbon anode requirement.

The total energy consumption associated with the cell operation of a simple inert anode is 18.3 (51.7^{tf}) kWh/kg of aluminum. This includes a 5.25^{tf} kWh/kg energy savings for anode manufacturing, as well as added energy consumption of 3.14 kWh/kg for the additional on-site voltage requirement, and 3.86^{tf} kWh/kg for the generation and transmission losses associated with the additional anode voltage requirement. Overall, the inert anode cell consumes an additional 3.14 (9.10^{tf}) kWh/kg of aluminum produced over the traditional Hall-Hèroult carbon anode requirement, representing an increase in energy consumption of 22 (8^{tf})%.

As gas molecules evolve on the anode surface, they coalesce and form bubbles, which creates a layer of bubbles adjacent to the horizontal electrode. These bubbles jostle and move along as a result of their density difference with the bath and by the motion imparted from the aluminum pad. Experimental evidence shows that the oxygen gas evolved in the bath has a different froth/foam dynamic than carbon dioxide. The molar volumes of gases that result from the reduction reaction are equal. However, because oxygen is 27% lighter than carbon dioxide, less mass moves through the electrode area. The physical properties, in practice, contribute to a

Table 10.12 Energy impact of inert anode technology

Energy input kWh/kg Al		Modern prebaked Hall-Heroult ACD = 4.5	*Inert anode* 1) Elimination of carbon anode ACD = 4.5	*Inert anode* 1) Elimination of carbon anode 2) Low polarization ACD = 4.5	*Inert anode* 1) Elimination of carbon anode 2) Low polarization 3) Wetted cathode ACD = 2.0
On-site energy demands					
Raw materials					
Bauxite-alumina		7.59	7.59	7.59	7.59
Anode materials		0.61	0.76	0.76	0.76
Total		8.20	8.20	8.20	8.20
Reaction energy					
Reaction electrolysis		3.76	6.90	6.90	6.90
Cell ohmic		10.67	10.67	9.25	6.20
Total reaction		14.43	17.57	16.15	13.11
Total on-site		22.63	25.76	24.35	21.30
Percent energy savings	Reactions		−22%	−12%	9%
	Reactions and anode		−22%	−13%	8%
	Reactions, anodes, and ore		−14%	−8%	6%
Tacit energy demands					
Raw materials					
Bauxite-alumina		8.21	8.21	8.21	8.21
Anode materials		6.01	0.76	0.76	0.76
Total		14.22	8.98	8.98	8.98
Reaction energy					
Reaction electrolysis		10.91	20.00	20.00	20.00
Cell ohmic		30.91	30.91	26.82	17.98
Total reaction		41.83	50.92	46.83	37.99
Total tacit		56.05	59.89	55.80	46.96
Percent energy savings	Reactions		−22%	−12%	9%
	Reactions and anode		−8%	1%	19%
	Reactions, anodes, and ore		−7%	0%	16%

ACD, anode-cathode distance

lower anode overvoltage (Ref 10.29). The inert anode can also be shaped to allow for better release of the oxygen generated (Fig. 10.18). The lower overvoltage of the inert anode, compared to a carbon anode, reduces the energy consumed. The overvoltage reduction, combined with the elimination of carbon anode manufacturing, is estimated to provide a 5% energy improvement.

A more stable inert anode surface will allow the ACD to be reduced and will be simpler to control. Carbon anodes must be lowered as they are consumed to maintain an optimal ACD. Combined with a wetted cathode, the inert anode probably offers the greatest potential for bath voltage reduction. Inert anodes will also eliminate cell disruptions that occur with carbon anode changing.

Overall, inert anodes, when combined with a wetted cathode and compared to the traditional Hall-Heroult cells, are expected to provide:

- A 10% reduction in operating costs (elimination of carbon anode plant and labor costs associated with replacing anodes) (Ref 10)
- A 5% increase in cell productivity (Ref 10.30)
- A 32% reduction in GHG emissions (Appendix D, Table D.4)

These practical scenarios provide the inert anode cell with an overall lower energy requirement than the state-of-the-art Hall-Heroult cell. However, the challenging engineering designs for the inert anode cell systems must incorporate effective approaches for minimizing thermal losses from the reduction cells, the current-carrying bus systems, and the connectors external to the cell.

Multipolar Cells

Present Hall-Heroult industrial cells consist of a single cathode surface and essentially a single anode surface immersed horizontally, one over the other, in a bath. This single-pole arrangement makes aluminum reduction a capital-intensive process. Use of multipolar electrolytic cells would greatly increase productivity-per-unit-reactor volume by placing multiple electrodes in a single reactor. This arrangement would also provide better control of heat losses. For practical engineering reasons, a multipolar cell requires that the ACD be stable, which in turn requires an inert/stable anode. To date, the lack of suitable materials of construction have ruled out multipolar cells using molten fluoride electrolytes.

Two variations of electrochemical multipolar cells are possible, as shown conceptually in Fig. 10.19. One system uses a cell with an anodic surface on one side, multiple bipolar electrodes in the center portion, and a cathodic surface on the opposite side, as in Fig. 10.19(a). One surface of a bipolar electrode plate acts anodically, while the opposite surface acts cathodically. Bipolar electrodes must be electrically isolated and sealed to the cell walls to avoid bypass current efficiency losses. The second system uses independent pairs of electrodes immersed in the same cell, as shown in Fig. 10.19(b). Multipolar cell designs are complicated by the need to remove liquid metal and collect the gaseous reaction products.

The bipolar-multipolar concept was successfully demonstrated for electrolytic aluminum chloride reduction (Ref 10.31). In 1976, Alcoa

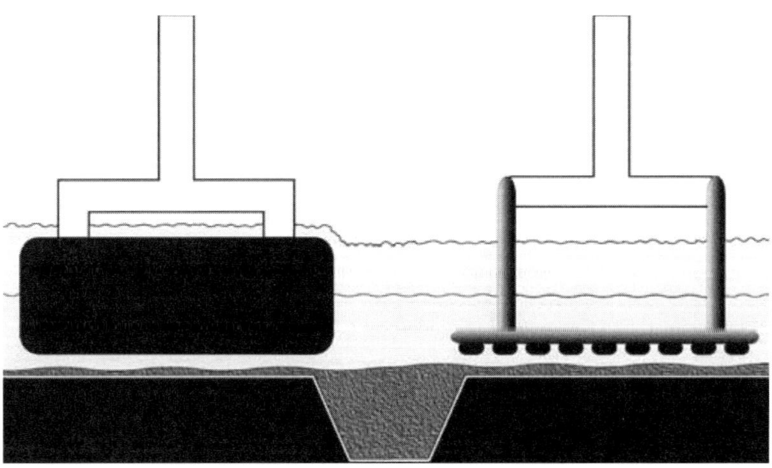

Fig. 10.18 Inert anode schematic (right) compared with conventional carbon anode (left)

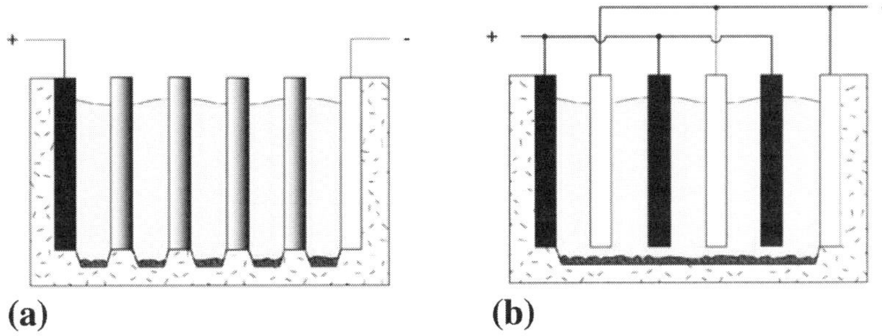

Fig. 10.19 Multipolar cells

Inc. began operating a multipolar, prototype plant with a capacity of over 18,000 kg of aluminum per day. Alcoa obtained 90% current efficiency at 9.26 kWh/kg of aluminum. The Alcoa multipolar cell was electrically 40% more efficient than a present-day Hall-Heroult cell. However, the prototype plant was eventually shut down because of the combination of the costs to produce anhydrous aluminum chloride feed, the failure to reach full design capacity, the need to remove and destroy trace amounts of chlorinated biphenal by-products, the reactor capital costs, and general maintenance costs (Ref 10.32).

The Alcoa cell had distinct technical advantages over the classic Hall-Heroult cell: it worked at substantially lower temperatures, it had relatively high current densities, its carbon anodes were not consumed, it had no fluoride emissions, and the plant had a smaller footprint. The improvement in electrical energy efficiency was a result of the higher electrical conductivity of the electrolyte and the small interpolar distance. The cell feed consisted of 3 to 10% of purified aluminum chloride ($AlCl_3$), obtained by carbochlorination. The electrolyte consisted mainly of sodium chloride and potassium chloride, or lithium chloride. Electrolysis was performed in a sealed cell consisting of 12 to 30 bipolar carbon electrodes stacked vertically at an interpolar distance of 1 cm (0.4 in.). The electrodes remained immersed in the electrolyte at an operating temperature of 700 ± 30 °C (1290 ± 55 °F). A current density of 0.8 to 2.3 A/cm^2 and a single-cell voltage of 2.7 V (dc) were typical operating conditions. An operating life of nearly three years was claimed for the electrodes.

A useful feature of the Alcoa cell was that the buoyant chlorine created a flow in the narrow gap between the bipolar electrodes. This aided in sweeping the aluminum from the cathodes and enhancing the formation of aluminum droplets. The gas also helped maintain a continuous flow of electrolyte across the cell. Chlorine was collected at the top of the cell, and molten aluminum was collected in the bottom. The chlorine gas generated by the reduction reaction was completely recycled to produce new aluminum chloride feed for the cell.

A Hall-Heroult multipolar design, involving multiple inert anodes and wettable cathodes arrayed vertically in a fluoride electrolyte cell, has been explored by Northwest Aluminum (Ref 10.33). In this concept, the operating temperature was 750 °C (1380 °F), much lower than Hall-Heroult technology. Alumina saturation at the lower temperature of the electrolyte bath was maintained by controlling a suspension of fine alumina particles in the bath. This technology offered all the benefits of reduced energy consumption (low ACD), elimination of carbon anodes and associated emissions, as well as a significant increase in productivity per cell, as a result of the multipolar design. This work came to an unfortunate halt with the closure of Northwest Aluminum.

Another multipolar cell design using a modified bath composition has been under investigation at Argonne National Laboratory (Ref 10.34). The modified bath allows for the cell-operating temperature to be lowered to 700 °C (1290 °F). This temperature enables the use of commercially available metal anodes (e.g., aluminum bronzes) with wetted cathodes in a vertical (nonhorizontal) arrangement to produce high-purity aluminum metal product and oxygen gas.

Alternative Primary Aluminum Processes

Carbothermic reduction of alumina and chloride reduction of kaolinite clay have been under study for more than 45 years. These technologies have the potential to be alternative processes to the Hall-Heroult process. The nonelectrolytic, carbothermic reduction of alumina is estimated to produce aluminum with significantly less energy consumption, reduced emissions, and lower investment costs. Multipolar-reduction technology to produce aluminum using aluminum chloride obtained from low-grade domestically available clays is also estimated to reduce energy consumption, capital costs, and emissions. The aluminum industry has yet to adopt these processes for producing aluminum, mainly because of the uncertain capital and operating economics associated with large-scale plants.

Carbothermic Technology

Carbothermic reduction of alumina is the only nonelectrochemical process that has shown potential for aluminum production. This technology has been an industry objective and the subject of extensive research for more than 45 years (Ref 10.35). Carbothermic technology is projected to produce aluminum from ore at 36.7[tf] kWh/kg, or 34% below a modern Hall-Heroult carbon anode technology (Table 10.13).

Carbothermic reduction produces aluminum using a chemical reaction that takes place within the volume of a reactor. This volumetric reaction requires much less physical space than the area reaction of Hall-Heroult. Specifically, with this distinction of volumetric versus area processing, the Hall-Heroult reduction electrolytic reaction occurs on the surface of an electrode in a reactor (cell). Capacity is doubled simply by doubling the electrode surface area.

Table 10.13 Comparison of Hall-Heroult and carbothermic reduction

Energy input kWh/kg Al		Modern prebaked Hall-Heroult	Carbothermic reduction
On-site energy demands			
Raw materials			
Bauxite-alumina		7.59	7.59
Kaolinite		...	...
Anode materials		0.61	...
Reaction carbon		...	0
Total		8.20	7.59
Reaction energy			
Reaction thermal			7.71
Furnace losses		...	1.36
Reaction		3.76	...
Cell ohmic		10.67	...
Total reaction		14.43	9.07
Total on-site		22.63	16.66
Percent energy savings	Reactions		37%
	Reactions and anode		40%
	Reactions, anodes, and ore		26%

Nonelectrolytic reduction chemistry occurs within a volume of fluid. A spherical volumetric reactor can be doubled by increasing its radius 1.4 times. This difference between volume and surface area reactions provides significant savings in capital cost for a reactor.

This distinction gives carbothermic technology a smaller footprint ($\frac{1}{10}$ the size of Hall-Heroult) and makes it much less dependent on the economies of scale (large and long potlines) required for an economically efficient Hall-Heroult facility. If successful, this technology could transform the structure of the aluminum industry, allowing industry the freedom to relocate from regions of inexpensive power to centers of manufacturing. Aluminum production "mini-mills" could be placed adjacent to or within aluminum casting facilities. These mini-mill installations could provide molten aluminum and/or specifically alloyed metal directly to the casting operations. This would provide additional energy, economic, and environmental benefits to the industry (Ref 10.36, 10.37).

Carbothermic reduction of alumina to aluminum is a multistep chemical reaction process. The thermodynamic optimization for the reactions requires a multizone furnace operating at very high temperatures (Fig. 10.20). In the first-stage net reaction, alumina and carbon form an alumina/aluminum carbide slag at ~1900 °C (3450 °F) ($2Al_2O_3 + 9C \rightleftharpoons Al_4C_3 + 6CO$). In the next-stage net reaction, aluminum carbide is reduced by alumina to form aluminum metal at ~2000 °C (3630 °F) ($Al_4C_3 + Al_2O_3 \rightleftharpoons 6Al + 3CO$). The thermodynamics (temperatures and chemical equilibria) of these reactions are very complex. A significant portion of aluminum evolves as gas-phase components (aluminum and Al_2O) at these operating temperatures. Careful process control is necessary to minimize the generation of volatiles. Recovery of these components in the form of Al_4C_3 in a vapor recovery system is required for the process to be economically viable. If the aluminum and Al_2O back-react with CO to form Al_2O_3, then the productivity of the process is decreased, and the Al_4C_3 required to satisfy the process stoichiometry is deficient.

The complex thermodynamic controls, sophisticated equipment, and construction materials required to successfully develop an economical commercial system have eluded the industry so far. Current R&D efforts are reevaluating carbothermic technology in hopes of capitalizing on new, advanced, high-intensity electric-arc

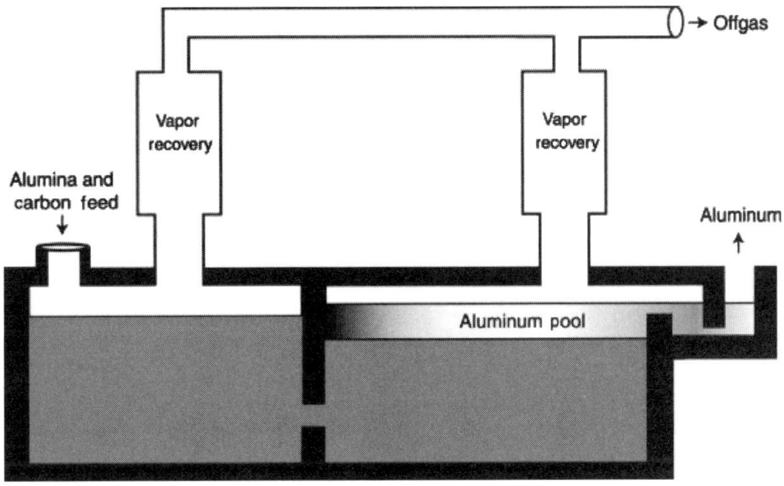

Fig. 10.20 Carbothermic reactor

furnace technology, advanced thermodynamic and system modeling techniques, and an improved understanding of the process dynamics (Ref 10.38).

At a minimum, the carbothermic process will have at least twice the inorganic metallic impurities (and possibly more, because petroleum coke is a waste product of overall refining operations), which will end up in the metal. In addition, any erosion or corrosion of furnace liners or plant hardware, including wear of an auger that may be used to force aluminum carbide from the vapor recovery stage back into the stage 1 and 2 reactors, will likewise end up contaminating the metal. Probably, it will have to be experimentally determined if the dilution of the metal with scrap or virgin metal additions during the decarbonation stage is sufficient to meet the required metal purity standards without an additional impurity removal or filtration step, which would adversely impact the process economics.

Theoretical Energy for Carbothermic Reduction of Alumina. The theoretical minimum energy requirement for producing aluminum by the carbothermic reduction of alumina is 7.32 kWh/kg.

Carbon is a reactant in the carbothermic reduction reaction process and supplies part of the energy necessary to drive the reaction forward. This gives the carbothermic reduction process a lower theoretical energy requirement than the direct electrolytic reduction of alumina to aluminum.

The net reaction for the carbothermic reduction is $Al_2O_3 + 3C \rightleftharpoons 2Al + 3CO$. It is assumed that the reactants (alumina and carbon) enter the reactor system boundaries at 25 °C (75 °F), the carbon monoxide byproduct leaves at 25 °C (75 °F), and the aluminum product leaves the reactor system as molten metal at 960 °C (1760 °F) (Fig. 10.21). The assumed molten metal temperature is lower than the actual reactor metal discharge temperature. Theoretically and practically, the energy in the higher-temperature reactor discharge can be recovered efficiently (e.g., by mixing high-temperature reactor aluminum with solid metal scrap to recover the heat energy and lower the temperature). To compare the theoretical limits of the various aluminum production processes, the same molten product temperature of 960 °C (1760 °F) is used throughout this chapter. The theoretical reaction occurs under perfect conditions when there are no reverse reactions, parasitic reactions, or heat/energy losses external to the system.

The calculation of theoretical minimum energy requirement for the carbothermic reaction is detailed in Table F.3 of Appendix F. The results show that the energy required to drive the reaction forward (ΔG) is 6.03 kWh/kg, and this energy is supplied as thermal energy versus the electrical energy used in a Hall-Heroult cell. The thermal energy ($\Delta H - \Delta G$) required to maintain equilibrium is 0.90 kWh/kg, and the thermal energy (C_p) associated with the molten aluminum is 0.39 kWh/kg of aluminum. The theoretical minimum energy requirement under these conditions adds up to 7.32 kWh/kg of aluminum. (Note: If the CO gas emission at 960 °C, or 1760 °F, is included, the total theoretical

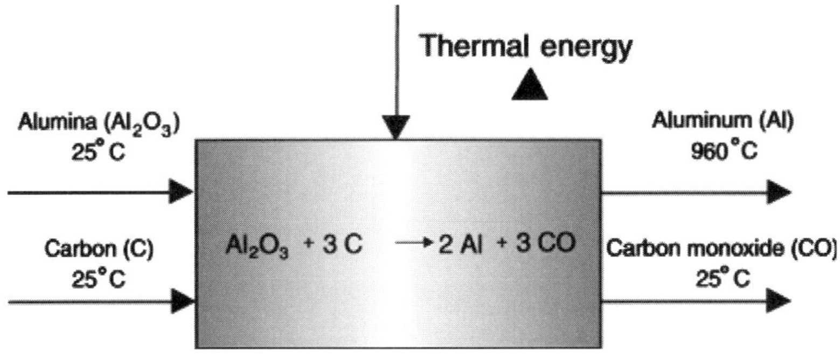

Fig. 10.21 Carbothermic theoretical minimum energy

minimum energy requirement is 7.51 kWh/kg of aluminum.)

The carbon monoxide by-product has a fuel value and would likely be captured and used to supply thermal energy to the carbothermic facility.

Comparative Benefits for Carbothermic Reactors and Hall-Heroult Cells. Electric furnace technology provides 90% thermal efficiency in many applications. Heat losses are limited to conduction and radiation losses from the furnace shell. Flue gas heat losses are eliminated, as are any undesired reactions between flue gases and molten aluminum metal. It is reasonable to assume that the carbothermic furnace/reactor can be designed with more than 85% thermal efficiency. If the thermodynamics of the reaction and offgas recovery can be controlled within 95% of the theoretical requirements, the electrical energy required for an operating carbothermic reactor would be (7.32 kWh/kg ÷ [0.85 × 0.95]) or 9.07 kWh/kg of aluminum produced. This represents a 37% reduction in energy use when compared to the 14.4 kWh/kg of aluminum from a modern Hall-Heroult cell. Table 10.13 shows the on-site and tacit energy savings potential of carbothermic technology.

In addition to electrical energy savings, carbothermic technology is expected to provide other benefits, such as:

- A reduction in capital costs by 50% or more as a result of volumetric processing through high-intensity smelting
- A reduction in production costs by 25% through lower electrical demand and elimination of carbon anode manufacturing and handling
- The production of a high-purity CO/CO_2 stream for coproduct sale or energy integration
- The potential to use small blocks of electrical power due to a high turndown ratio
- The potential of widely locating mini-mills with integral, captive smelters delivering molten metal

Environmental Impacts of Carbothermic Technology. The carbothermic process, when compared to Hall-Heroult technology, results in significantly reduced electrical consumption and the elimination of perfluorocarbon emissions that result from carbon anode effects, hazardous spent potliners, and hydrocarbon emissions associated with the baking of consumable carbon anodes.

The total carbon dioxide emissions from carbothermic reduction, as with Hall-Heroult, depend on the source of electricity. Hydroelectric power generation emits almost no carbon dioxide, whereas the carbon dioxide emissions associated with the average U.S. grid electricity are 0.49 kg CO_{2e}/kWh (Appendix D, Table D.1). Because 39% of the U.S. primary industry operates on hydroelectric power, the aluminum industry's average electrical generation emission rate is 0.35 kg CO_{2e}/kWh. The carbothermic reaction results in the generation of carbon-base GHGs, mainly carbon monoxide (CO), at twice the rate of the Hall-Heroult reaction. However, the carbothermic process only requires electricity for heating and not for the reduction reaction. Assuming carbothermic technology is used in mini-mills operating off the average U.S. electric grid, the total GHG emissions from "utility to metal" for the carbothermic process are reduced relative to the average U.S. Hall-Heroult system, because of carbothermic lower electrical intensity. Table 10.14 (Appendix D, Table D.4) shows the lower electrical demand of carbothermic technology resulting in lower total carbon

Table 10.14 Carbon dioxide equivalent comparison of Hall-Heroult and carbothermic reduction

Emission sources	Modern Hall-Heroult, kg CO_{2e}/kg Al	Carbothermic, kg CO_{2e}/kg Al
Carbon anode	1.68	2.45
Reaction energy requirements:		
kWh/kg Al	14.43	9.07
kg Co_{2e}/kWh	0.492	0.492
	7.10	4.46
Process	2.20	0.0
Total	**10.98**	**6.91**

dioxide equivalent emissions (CO_{2e}) than for a Hall-Heroult system. Other studies (Ref 10.39) have also shown a reduction in CO_{2e}.

Kaolinite Reduction Technology

Production of pure aluminum by reduction of aluminum chloride was discovered before the Hall-Heroult process, in 1825. Alumina conversion to aluminum chloride and reduction to aluminum using bipolar technology was demonstrated in the late 1970s, but it was not commercialized because of problems with the product purity and projected high capital and operating costs. New construction materials, improved thermodynamic understanding, and the potential to use low-cost alumina containing clays have maintained the interest in chloride reduction technology for producing aluminum.

Compared to the current Bayer refining and Hall-Heroult fluoride-base smelting of alumina, the chlorination of alumina-containing clays promises many potential advantages:

- Raw materials are widely available, inexpensive, and indigenous to the United States.
- Thermodynamics provide high-speed, high-conversion reactions with lower electrical demand.
- No bauxite residue is produced, although there is waste from the clay chlorination.
- Conventional materials-of-construction (i.e., mild steel) can be used.

Industry has spent more than 35 years developing process technologies for the chlorination of widely available, low-grade kaolin clays. These clays contain kaolinite (hydrated alumina silicate, i.e., $Al_2O_3{\cdot}2SiO_2{\cdot}2H_2O$), significant amounts of titanium dioxide (1 to 5%), and other materials. Titanium tetrachloride and other metal chloride by-products are also produced when processing kaolin clays.

The basic steps of the clay-to-aluminum process are shown in Fig. 10.22 (Ref 10.40).

First, the kaolin clay and process coke are dried. All feed materials to the clay chlorination units must be dried to minimize chlorine absorption in any water and to reduce the corrosive effects of moisture in the chlorination offgas stream. The drying process involves controlled, catalyzed heating of the finely ground clay at ~800 °C (1470 °F) in a fluidized reactor, using coke and air to provide the heating energy. The hot reactor produces a dehydrated calcined clay, which is sent to carbochlorination.

The carbochlorination step is a high-speed, catalyzed exothermic reaction of calcined clay with chlorine and coke:

$$3(Al_2O_3{\cdot}2SiO_2) + 14C + 21Cl_2 \rightleftharpoons 6AlCl_3 + 6SiCl_4 + 7CO + 7CO_2$$

Chlorine is injected into the bottom of the fluidized reactor with the clay oxides and coke. The metallic oxides, principally those of aluminum and silicon, are converted to their

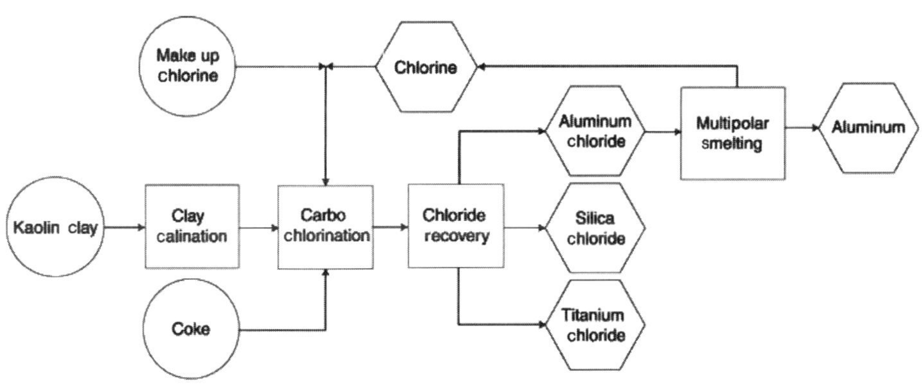

Fig. 10.22 Clay-to-aluminum process schematic

chlorides. The chlorides exist as vapors at the reaction temperatures. The next process step is designed to suppress the formation of silicon tetrachloride. The offgases of the chlorination reactor are treated with a small amount of catalyst vapor and reacted in a second fluid bed reactor with additional calcined clay. The alumina portion of the clay is converted to aluminum chloride, and silicon tetrachloride is converted to solid silica and discharged ($3SiCl_4 + 2Al_2O_3 \rightleftharpoons 4AlCl_3 + 3SiO_2$). The net effect of this step is the preferential chlorination of alumina relative to silica. The hot vapors from the second reactor are cooled, and crude aluminum chloride is recovered. Several by-products are produced as an extension of the chlorination step. Titanium tetrachloride, silicon tetrachloride, and boron trichloride are not condensed in the cooler and can be recovered in subsequent processing. Silicon tetrachloride can be reacted with oxygen to recover chlorine, which is recycled back to the chlorination step, as follows:

$$4SiCl_4 + 5O_2 \rightleftharpoons 2SiO_2 + 3O_2 + 8Cl_2$$

The crude aluminum chloride must be purified for ease of operation of the electrolytic reduction cells and for the final aluminum quality. The main impurity is iron, which is present as ferric chloride at levels as high as ~50,000 ppm. Impurities are removed by chemical treatment.

The electrolysis of aluminum chloride to aluminum takes place in an aluminum chloride smelting cell, which comprises a stack of horizontal bipolar graphite electrodes, between which the aluminum chloride is converted into high-grade aluminum and chlorine gas ($2AlCl_3 \rightleftharpoons 2Al + 3Cl_2$). The electrodes are immersed in a chloride bath, which is contained in fully enclosed, thermally lined vessels. The bipolar electrodes are supported and separated by inert spacers resting on the electrode below it, and the entire stack, consisting of bottom cathode, bipolar electrodes, and top anode, is supported by the walls of the cell.

Theoretical Energy for Kaolinite Reduction of Alumina. The theoretical minimum energy requirement for producing aluminum from kaolinite is 5.76 kWh/kg of aluminum produced. The theoretical minimum for the chloride reduction step of aluminum chloride is 7.66 kWh/kg of aluminum.

The theoretical minimum energy requirement can be calculated from the net chemical reaction in the kaolinite-to-aluminum process:

$$7(Al_2O_3 \cdot 2SiO_2) + 14C \rightleftharpoons 14Al + 14SiO_2 + 7CO + 7CO_2$$

It is assumed that the reactants (kaolinite and carbon) enter the system boundaries at 25 °C (75 °F), the carbon monoxide and carbon dioxide by-products leave at 25 °C (75 °F), and the aluminum product leaves the system as molten metal at 960 °C (1760 °F). The theoretical net reaction occurs under perfect conditions when there are no reverse reactions, parasitic reactions, or heat/energy losses external to the system. These assumptions yield a theoretical minimum energy requirement of aluminum production from kaolinite as 5.76 kWh/kg of aluminum (Appendix F, Table F.5). This assumes a pure kaolinite feedstock.

The kaolinite process is accomplished in two steps: carbochlorination and aluminum chloride reduction. The electrolytic aluminum chloride reduction process used in the kaolinite process requires the exact same minimum amperage (2980 Ah/kg of aluminum) as any electrolytic process for reducing aluminum. The theoretical chloride reduction reaction occurs under perfect conditions (where there are no reverse reactions, parasitic reactions, or heat/energy losses external to the system) and requires 7.66 kWh/kg of aluminum produced. The theoretical minimum energy requirement used in this chapter is calculated at 960 °C (1760 °F) to provide comparison with the other processes. The proposed multipolar system operates at approximately 700 °C (1290 °F), which would lower the theoretical energy demand by approximately 0.09 kWh/kg of aluminum. The exothermic nature of the carbochlorination reaction (−1.90 kWh/kg of aluminum) results in the overall kaolinite-to-aluminum theoretical energy requirement being lesser than the minimum energy requirement for the aluminum chloride reduction.

Comparative Benefits for Kaolinite Reduction and Hall-Heroult Cells. The kaolinite reduction process offers potential advantages when compared to the Hall-Heroult process. Table 10.15 shows the on-site and tacit energy comparison for the Hall-Heroult and kaolinite processes.

The combined clay carbochlorination and bipolar aluminum chloride smelting process is

Table 10.15 Comparison of Hall-Heroult and kaolinite reduction

Energy input kWh/kg Al	Modern prebaked Hall-Heroult	Kaolinite ($AlCl_3$) reduction
On-site energy demands		
Raw materials		
Bauxite-alumina	7.59	...
Kaolinite	...	8.14
Anode materials	0.61	0.76
Reaction carbon	...	0
Total	8.20	8.91
Reaction energy		
Reaction thermal	...	–1.90
Furnace losses	...	0.40
Reaction	3.76	6.48
Cell ohmic	10.67	2.93
Total reaction	14.43	7.91
Total onsite	22.63	16.82
Percent energy savings:	Reactions	45%
	Reactions and anode	42%
	Reactions, anodes, and ore	26%
Tacit energy demands		
Raw materials		
Bauxite-alumina	8.21	...
Kaolinite	...	8.81
Anode materials	6.01	0.76
Reaction carbon	...	11.34
Total	14.22	20.92
Reaction energy		
Reaction thermal	...	–1.90
Furnace losses	...	0.60
Reaction	10.91	18.79
Cell ohmic	30.91	8.48
Total	41.83	25.98
Total tacit	56.05	46.90
Percent energy savings:		
	Reactions	38%
	Reactions and anode	44%
	Reactions, anodes, and ore	16%

estimated to be 16[tf]% more tacit energy efficient than the Hall-Heroult, has a smaller plant footprint, can be more flexible in regard to use of off-peak power and power fluctuations, and produces fewer emissions and process wastes. On-site energy use is approximately 42% lower than a modern Hall-Heroult cell. The large tacit energy improvement results from a decrease in electrical use. Compared to the current Hall-Heroult smelting technology, the bipolar aluminum chloride smelting process consumes less electricity, yielding 60% more metal for the same electrical input. Bipolar cell designs provide significantly lower reactor volume per unit of product output. This lower volume allows the cell to idle and hold temperature much more efficiently than a single electrode cell. These properties allow multipolar cells to take better advantage of off-peak electrical costs. Chloride cells also operate at lower temperatures, providing additional savings.

Raw materials used in the kaolinite process require approximately 47% more energy compared to those in a modern Hall-Heroult process (Table 10.15). Kaolinite requires more energy for extraction on an aluminum basis, and the carbochlorination step requires 2.5 times the carbon of a Hall-Heroult carbon anode system. The raw material energy expenses may be compensated for by the lower-cost domestic supply of kaolinite and the ability to use lower-cost carbon than Hall-Heroult systems.

It is important to emphasize that while significant elements of the kaolinite process have been studied and developed, no integrated production of aluminum from kaolinite clays has yet been attempted. Significant work in developing an integrated production process has been reported.

Environmental Impacts of Kaolinite Technology. Table D.4 in Appendix D tabulates CO_{2e} for the kaolinite-to-aluminum process. Operating off the average U.S. electrical grid, this process would produce 29% fewer CO_{2e} than a typical Hall-Heroult facility.

Secondary Aluminum (Recycling)

The production of primary aluminum ingots from bauxite ore requires approximately 23.8 (59.1^{tf}) kWh/kg of aluminum. Recovering aluminum from scrap to produce secondary aluminum ingot consumes approximately 5% of the energy required to produce primary aluminum (Ref 10.11). This significant energy difference drives the emphasis placed on aluminum recycling in today's (2007) society and in the aluminum industry.

Recycling in the United States saved more than 172 × 10^9 kWh (0.59 quad) of energy in 2005, the equivalent of 19,700 Mw. Each kilogram of aluminum that is recovered by recycling saves 57.3^{tf} kWh of the 60.5^{tf} kWh of energy consumed in producing a primary aluminum ingot from bauxite ore (Appendix E, Table E.6). Any process that improves the recovery of scrap aluminum is effectively making an order of magnitude change in the energy associated with aluminum production.

The growth of aluminum recycling represents the greatest change in the structure of the industry and in the energy associated with aluminum manufacturing. A common practice since the early 1900s, recycling was a low-profile activity until 1968, when aluminum beverage can recycling vaulted the industry into public consciousness. In 1960, recycled aluminum accounted for 18% of the nation's total aluminum supply (401,000 metric tons). Over the next 45 years, production of recycled aluminum rose by almost 746% to 2,990,000 metric tons. During those same 45 years, the total U.S. aluminum metal supply increased 300% (Table 10.3). In 2005, 75 plants in the United States produced secondary ingots. Over half—54.7%—of the aluminum metal produced in the United States in 2005 was from recycled material (Ref 10.9).

The growth of the market for recycled aluminum is due in large measure to economics. It is cheaper, faster, and more energy-efficient to recycle aluminum than to manufacture it from ore. Recovered aluminum is easily melted at relatively low temperatures (aluminum alloys typically melt at temperatures below 660 °C, or 1220 °F). Producing a recycled aluminum ingot consumes only approximately 5% of the energy required to produce a primary aluminum ingot from bauxite ore. In addition, to achieve a given output of ingot, recycled aluminum requires only approximately 10% of the capital equipment costs compared with those required for the production of primary aluminum.

Aluminum products are corrosion resistant, which allows them to be easily and repeatedly recycled into new products. The corrosion resistance is due to the metal properties. When the surface of aluminum is exposed to air, it rapidly forms a tenacious, self-limiting, protective oxide layer. Other surface treatments can be applied to further enhance the corrosion resistance of aluminum.

Figure 10.23 shows the annual growth rates of the three sources of metal supply between the 1995 and 2005 period, together with the growth in the major aluminum product markets. Aluminum imports are growing at a rate of 6.9% per year, while both U.S. primary and secondary productions are in decline (Table 10.3).

Aluminum scrap is categorized as new or old. New scrap is generated when aluminum products are manufactured. It includes defective products; scalping chips; edge and end trim from rolling processes; skeleton scrap from stamping and blanking operations; flash, gates, and risers; extrusion butts and ends; and turnings and borings. Old scrap (post-consumer or obsolete products) comes from discarded, used, worn-out, or out-of-date products that include automotive parts, white-good parts, containers such as used beverage cans and closures, wires, cables, and building materials. Runaround scrap is new scrap that is recycled by the same company that generated it. Because runaround is usually not sold or marketed, it is not reported in the U.S. recycling statistics.

Secondary Aluminum Production. Secondary aluminum producers represent a separate and vital segment of the aluminum industry whose principal activities are converting purchased scrap and metal recovered from skim and dross generated in molten metal operations into usable aluminum alloy products. Recycling in primary aluminum operations is typically confined to in-house or runaround scrap and manufacturing scrap returned directly from their customers. Secondary aluminum producers specialize in melting and processing a wide range of new and old, segregated and mixed, high- and low-quality scrap.

The most desirable form of recycling is closed-loop, in which scrap from specific product applications are returned for remanufacturing of the same products. Rigid container stock used in beverage cans is an example of closed-loop recycling, because used beverage cans are

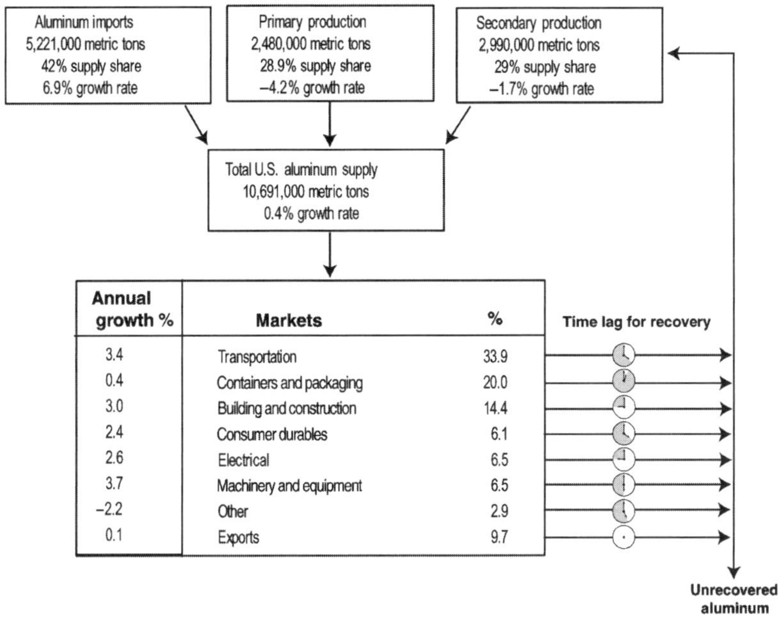

Fig. 10.23 U.S. aluminum markets and growth over the period 1995–2005

processed exclusively to remanufacture can sheet. Secondary aluminum producers are involved in closed-loop recycling as contractors to primary producers for this and other products.

Scrap segregated by alloy has greater value than mixed scrap because it can be used predictably and most efficiently in the production of the highest value-added compatible compositions. Mixed scrap presents the greatest challenge, generally requiring added steps in melting, composition identification, and often casting into ingot form before consumption.

Casting alloys are the largest and most important but not the only market for secondary aluminum. Scrap is used, with primary metal, in the production of extrusion billet and fabricating ingot. Reclaimed smelter ingot (RSI) is produced from scrap and dross, often on a toll conversion basis.

Alloys used by aluminum foundries in the production of shape or engineered castings include compositions designed to facilitate production from scrap. These alloys, which have been historically prominent, typically specify broader element ranges and higher impurity limits than the alloys developed for more specialized purposes and whose compositions require production from primary metal sources. The increased use of aluminum in structural applications in the ground transportation sector typically requires primary compositions, but the expanded use of aluminum in engine blocks, cylinder heads, and other power train parts relies on castings in secondary alloys. The emphasis on light-weighting for improved fuel efficiency and reduced environmental degradation is encouraging the adoption of various wrought aluminum, products, such as auto body sheet, fuel tanks, seat backs, extruded drive shafts and stringers, and forged connecting rods. As the proportion of aluminum increases in ground transportation designs, segregating and blending scrap into saleable alloy compositions will become common practice.

New scrap-sorting technology is developing, including those related to the recovery of aluminum from the large transportation and white-goods markets (Ref 10.41). New technologies with computer screening are using color segregation and laser-induced breakdown spectroscopy to sort wrought from cast and, in some instances, one alloy from another (see Chapter 8). Aluminum alloy separations will allow scrap to be segregated into more specific alloy groupings with higher economic value.

Production, Capacity and Growth. Aluminum scrap is widely recycled and supports a large secondary aluminum industry. In 2005, the United States produced 2,990,000 metric tons of secondary aluminum, amounting to nearly one-third of its total supply of aluminum. This market had an annual growth rate of

–1.7% over the last ten years (Fig. 10.24). The supply of scrap metal has become more precious, and U.S. secondary producers must compete in a global market to obtain scrap.

Market demand for recycled aluminum will remain strong due to its inherent low energy cost relative to primary metal. Aluminum use in consumer products has become widespread in a trend described as the "urban mine." Challenges in scrap recovery, alloy sorting, and impurity removal can be addressed with the current technologies. The limitation to secondary metal market growth is the economic supply of scrap.

The urban mine has been accumulating scrap products for the past 100 years. There is a lag time from when aluminum leaves the shipping dock to when it becomes available for recycling. This lag time, or use-phase life, varies significantly among aluminum products. For packaging products such as beverage cans, this lag is only approximately 65 days. Long-life products, such as building frames, can have a lag time of 40 to 50 years. Automobile lag time is 12 to 15 years. Aluminum in automobiles accounted for 5 to 10% of scrapped automobile weight in 2000 but represents 35 to 50% of its scrap value. This source will grow rapidly as new automobile designs continue to use a greater proportion of aluminum. Studies of the future supply of scrap in the United States show the potential for significant shortfalls (Ref 10.42).

Considering the large amounts of aluminum that are stored in long-life products (accumulated within the urban mine) and the continued growth in aluminum demand, recycling will continue to increase and be a significant contributor to the U.S. metal supply. In fact, recycling has overtaken primary production as the main source of domestically produced aluminum in the United States.

Recycling Processes. The objectives of the recycling process are maximum metal recovery, minimum contamination, and lowest conversion cost. Safety is of prime importance because components mixed with scrap can present the risk of explosive reactions and other concerns. Because moisture is a safety concern, remelt ingot and RSI are routinely preheated before charging to the furnace hearth. Preheating methods vary from heating on the charging doorsill to using dedicated preheating ovens. Preheating standards from times of exposure and temperature also vary. In either case, energy is consumed for the promotion of operational safety.

Scrap for recycling is available in many forms. Light scrap is typically baled or briquetted to reduce transportation costs. Bales and briquettes are typically split for inspection of integrity and contamination and/or are crushed, shredded, or "shripped" (sheared and ripped) to controlled flowable particle sizes. Conveyor systems segregate particle fines for separate processing, provide magnetic separation, and allow for visual or automated inspection.

Large volumes of aluminum scrap contain paint, enamel, lacquer, or porcelain coatings. These coatings, which contain oxidizing compounds, would significantly reduce metal recovery if not removed before melting. Conveyor furnaces or rotary kilns operating at temperatures near the melting point are required for their removal. Turnings, borings, and other fine scrap may contain oil from cutting fluids that must be removed for satisfactory metal recovery by swarf (metal fines and chips) drying, in which the oil contributes to the energy required,

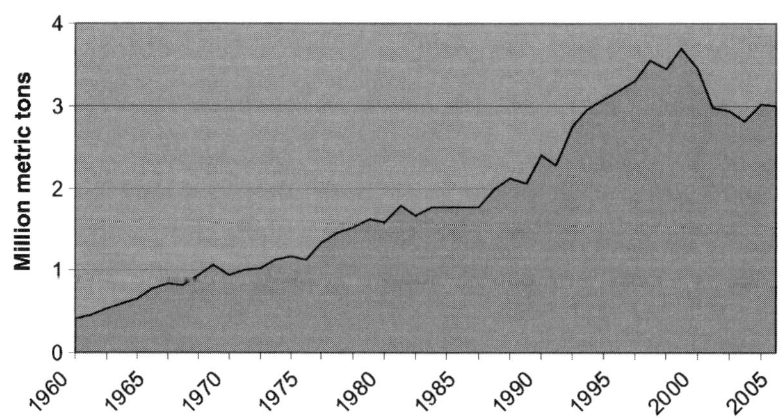

Fig. 10.24 U.S. production of secondary aluminum from 1960 to 2005

or by elevated-temperature treatment in rotating kilns. Fine scrap is conveyed to external charging wells for submergence by mechanical methods, including the use of recirculating molten metal pumps. Light scrap is charged directly to the furnace hearth and is covered by additional heavier charge components.

Melting is typically accomplished in gas-fired furnaces ranging in size from 75,000 to more than 250,000 lb holding capacity. Coreless induction furnaces are in common use for rapidly melting fine scrap. Molten metal from these furnaces is transported to the main furnaces for further processing. After iterative alloying steps, molten metal is held and processed before casting. Furnaces have thermal efficiencies ranging from approximately 20 to 45%. Rotary kilns and conveyor furnaces operate in essentially the same efficiency range. Induction melting is more efficient, at approximately 90% but furnace capacities are limited, and additional steps are required to address oxide concentrations created by electromagnetic stirring. In-furnace and in-line molten metal treatments are employed to remove dissolved hydrogen and entrained oxides and other nonmetallics before casting. In-line systems are usually internally heated using gas-air or electric resistance immersion elements. Troughing and filter basins are preheated using air-gas torches or resistance elements. Open molds for casting remelt ingot and sow are routinely preheated before use.

The aluminum, recoverable from skim and dross obtained from foundries and primary and secondary molten metal processing units, is typically extracted in secondary producers. While other dross treatment processes have been developed, rotary furnaces in which dross and salts are mixed and heated remain the most commonly used. The products of dross recovery treatment by this process are aluminum and black dross, a mixture of unrecovered aluminum, metallic oxides, and salt. Tertiary processes have been developed for separating the components of black dross into saleable salt fluxes, metal, and value-added nonmetallic derivatives, such as calcium aluminate and additives for low-density cement. The economics of tertiary processing are strongly and adversely influenced by the current low-cost alternative of landfill disposal.

In summary, the energy-intense operations common in recycling are:

- Swarf drying for removal of combustible contaminants
- Kiln or conveyor delacquering of coatings
- Preheating charge components
- Melting
- Holding and melt processing
- Rotary furnace operation

Theoretical Energy for Secondary Aluminum. The theoretical minimum energy required to produce secondary aluminum at 960 °C (1760 °F) is 0.39 kWh/kg of aluminum. On a theoretical and a practical basis, the energy required to produce secondary aluminum is less than 6.5% of the energy required to produce primary metal.

If the system boundaries are drawn around a secondary aluminum facility, the material entering and leaving is aluminum metal. Because no chemical change has occurred, the theoretical minimum energy requirement for scrap conversion is only the energy required to melt and raise the metal temperature to that required for casting. Molten metal furnace temperatures vary depending on the furnace, alloys, and other proprietary factors. A molten metal temperature of 960 °C (1760 °F), the same production temperature as for primary aluminum, is used in this chapter for secondary aluminum to easily compare aluminum process technologies. In reality, normal pouring temperatures are much lower, in the range of 650 to 750 °C (1200 to 1380 °F).

The theoretical minimum energy requirement to bring room-temperature (25 °C, or 75 °F) aluminum to its molten form at 960 °C (1760 °F) is 0.39 kWh/kg, which is less than 6.5% of the theoretical energy requirement for the primary production of aluminum. The theoretical energy required to heat aluminum from room temperature to its melting point, melt it, and raise the molten aluminum to a higher temperature is calculated and explained in Appendix G. Pure aluminum melts at 660 °C (1220 °F) and requires 0.30 kWh/kg to melt, 23% lower than the value used in this chapter.

The energy efficiency of the entire aluminum industry can be further increased by capturing a greater percentage of material for recycling and by improving technology for scrap handling and melting. Nontechnological and nonmarket factors are also important for the continued growth of recycling. Two of these factors, consumer awareness and incentives, can contribute significantly to the recycling volume. Consumer awareness requires continuing public education about the energy and environmental benefits of recycling. Offering incentives will aid in the return of aluminum scrap to the manufacturing base.

Recycling energy efficiency will be enhanced by developing technologies that minimize oxidation and improve thermal inefficiencies in scrap processing and melting. Improved collection systems and separation devices (e.g., eddy current, color sorting, laser sorting) can increase aluminum scrap recovery by 20 to 30%. Improved technology can be applied to increase scrap recovery rates, especially with regard to aluminum in municipal solid waste. Incremental improvements in existing furnaces can further reduce the recycling energy requirements. They can be achieved by recuperating stack gas energy for preheating combustion air and metal feedstock, by modifying burner and furnace designs, and by controlling furnace practice and operating conditions (Ref 10.43, 10.44). New technologies must also be developed to ensure more significant progress.

Aluminum Processing

The aluminum industry can be divided into metal- and product-producing sectors. In 2005, the metal-producing sector manufactured approximately 2,480,000 metric tons of primary metal and 2,990,000 metric tons of secondary metal. The product-producing sector processes these domestically produced and imported metals into approximately 5,497,000 metric tons of rolled products, 1,719,000 metric tons of extrusions, and 2,513,000 metric tons of shape castings (Appendix E, Table E.4) (Ref 10.43). Wire rod and bar, forgings, impacts, and powder products comprise less than 8% of the aluminum product market and have not been studied as a part of this chapter.

Recoveries (yields) in each product-processing sector are less than 100%. Handled, melted, and worked tonnage exceeds that implied by the product statistics. The manufacturing of circles and blanks from aluminum sheet, for example, may recover as little as 35% of the original ingot weight as shipped product. The gross-to-net weight ratio in gravity castings is often 2:1; in other words, 50% of the original cast weight is automatically designated for remelting. Furthermore, many products have a genealogy of repeated melting and casting steps. Ingot cast at a primary smelter is remelted at a secondary producer for casting remelt ingot, which is again remelted at an aluminum foundry for casting production. It is estimated that more than 11,000,000 metric tons of aluminum alloys are melted each year to support aluminum industry shipments. Some casting operations are located sufficiently close to primary smelters and secondary aluminum producers to allow the molten metal to be shipped in insulated and protected crucibles, saving the energy of remelting.

Minimizing planned and unplanned scrap represents a large opportunity for energy- and cost-savings. Each kilogram of metal that does not go into a final product must be remelted, recast, and reworked. Remelting also leads to the loss of a percentage of metal to oxidation, which must be replaced with energy-intensive primary metal. Melting and melt-processing operations are the most energy-intense of all postsmelting processes.

Melting, Alloying, and Melt Treatment

In an aluminum primary smelting facility, molten metal is transferred from the smelting cells to furnaces for alloying and melt treatment prior to casting. In secondary and other casting plants, ingot, metallurgical metals, and master alloys must be melted and alloyed. The melting arrangement in most larger plants provides high-heat-input, high-melt-rate furnaces for melting, and separate holding furnaces to which molten metal is transferred for final alloying and preparation for casting. Some operations have combination melting/holding furnaces.

Furnaces and Melt Treatment. There are a wide variety of furnace types and designs for melting aluminum. Furnace choice depends on the required melt volume, melt rate, availability, cost of fuel and electrical energy, and emission standards. The most common in primary and secondary operations are natural gas-fired reverberatory furnaces reaching capacities of more than 120,000 kg. Crucible furnaces with capacities ranging from 160 to 4500 kg are more common in small- and medium-sized foundries. Other furnace types include coreless and channel induction, electrical resistance, and radiant tube furnaces.

Reverberatory furnaces are box-shaped and consist of an insulated steel shell with a refractory lining. Fuel-fired reverberatory furnaces are used when the melt rate and/or capacity are large. The fuel-fired reverberatory furnace fires natural gas, propane, or oil directly into the furnace from either the roof or, more typically, the sidewall. The heat is transferred to the surface of the molten aluminum predominantly by refractory radiation and some convection. There are a large number of reverberatory furnace design variations: charging and access doors, refractory specifications, sidewells for charging and/or recirculation, hearth or sidewall induction stirring, split hearths, dry hearths, divided zones for melting and holding, and various burner capacities and types. Recuperation concepts include charge preheating, preheating combustion air, and cogeneration.

The growth in recycling has resulted in a number of specialized processes, furnaces, and systems to improve metal recovery from scrap. Molten metal pumps have been incorporated into these designs to provide rapid ingestion of fine scrap and more rapid melting of larger scrap forms. Salt flux additions maintain system cleanliness and aid in the separation of oxides. Pump-induced flow external to the furnace may include provisions for melt treatment and the separation of oxides as well as for melting.

Natural gas or oil-fired reverberatory furnaces use approximately 0.87 to 1.96 kWh/kg of aluminum (Ref 10.44). In addition, a gas furnace increases metal losses due to oxidation. Gas furnaces have 5 to 8% metal loss compared to 0.5 to 3% loss in electric furnaces. Recent design innovations in fossil-fuel reverberatory furnaces help capture the waste heat in the stack gas to preheat incoming materials. This increases energy efficiency and reduces the time required to melt the metal. Recuperated waste heat can also be used to preheat combustion air. These technologies can reduce fuel usage to less than 0.57 kWh/kg of aluminum. Large commercial gas-fired furnaces are reported with efficiencies as high as 56% (0.30 kWh/kg) (Ref 10.45).

Crucible furnaces are more versatile with regard to alloy changes and melt quantities. Combustion occurs between an insulated steel shell and a crucible of silicon carbide, graphite, clay graphite, or other refractory material resistant to molten aluminum attack. Electrical resistance elements can be substituted for gas burners in crucible furnace designs.

In recent years, much research has been done on using electric immersion heaters as a way to remelt aluminum (Ref 10.46). Immersion heaters have very low rates of heat loss (~97% thermal efficiency) and have energy usage levels of less than 0.50 kWh/kg of aluminum. This high thermal efficiency is obtained on site but is offset by the lower efficiency (30 to 40%) for electricity generation and transmission.

Oxide naturally and rapidly forms on the surface of molten aluminum, resulting initially in a thin protective film. With increase in time and temperature, the thickness of the oxide layer increases. Turbulence and agitation accelerate oxide formation and result in the intermixing of metal and oxides. Oxidation rates are influenced by alloy content and increase with temperature, especially when magnesium is present in the alloy. The oxide layer also effectively insulates the bath from radiation heat transfer and must be periodically removed to maintain thermal efficiency in reverberatory furnaces.

If fluxes are employed to treat the skim before it is removed from the furnace, the oxides are typically dewet, and a large portion of the molten aluminum entrained in the skim layer separates to the melt. In either case, untreated (skim) or flux-treated (dross) contains entrained free metal as a result of the skimming action. Efforts are usually made to recover entrained free aluminum after skimming. While still hot, metal can be drained from the skim gravimetrically and with vibration. Alternatively, skim may be rapidly cooled by inert-gas quenching or in rotating water-cooled steel drums, after which free aluminum may be physically separated. The residue comprising unrecovered aluminum and oxides is normally further processed for its metal content.

Gross-melt losses typically range from 1 to as high as 8% in small, poorly designed furnaces. The magnitude of the loss is dependent on the type of furnace and burners, the surface-to-volume ratio, the practices that are used, and the material being melted (Ref 10.9). Melt loss has significant economic impact, because oxidized metal must be replaced in the supply chain with new primary aluminum metal.

Specific elements or combinations of elements are added to molten aluminum to produce aluminum alloys. Alloying provides the basis for a remarkable range of physical and mechanical property capabilities not displayed by unalloyed aluminum or by any other metal system. Among the elements and combinations of elements added to molten aluminum are modifiers and refiners, which provide finer grain structures and controlled microstructural features, including metallurgical phases that influence recrystallization behavior.

Molten aluminum contains dissolved hydrogen, entrained oxides, and other nonmetallic inclusions that, if not removed, would adversely affect metal acceptability and performance. Treatment with salt fluxes or active fluxing gases changes the interfacial relationship of included particles with the melt so that gravitational separation is facilitated. Fluxing with argon, nitrogen, and/or other gases results in flotation of entrained matter, while dissolved hydrogen is reduced by partial pressure diffusion. Metal treatment takes place in the melting furnace, holding furnace, or in line between the furnace and the casting unit. Rotary degassers have been developed to provide the finest dispersion and intermixing of metal and fluxes for these purposes.

Molten metal filtration for the removal of particulate contamination was introduced in the 1950s and has grown in importance and application since that time. The first and still among the most effective filtration processes are deep bed filters using tabular alumina as the filtration medium. Crushed carbon beds are capable of fine inclusion removal and a reduction in sodium content that is important for many products. Porous foamed ceramics are widely used for commercial-grade and higher-quality requirements. Fused ceramic and refractory filtration elements are also available.

Energy Requirements for Melting Aluminum. Melting is an energy-intensive process; it requires nearly the same amount of energy to raise 1 kg of aluminum to a molten 700 °C (1290 °F) state as it does to raise 1 kg of iron to 1500 °C (2730 °F). However, nearly three times the volume of aluminum is produced compared to iron, because of density differences.

The energy requirements for melting aluminum are presented in the section, "Theoretical Energy for Secondary Aluminum" in this chapter. When the system boundaries are drawn around an aluminum melting facility, the material entering and leaving is aluminum metal. Because no chemical change has occurred, the theoretical minimum energy requirement is only the energy required to melt the metal. Molten metal furnace temperatures vary depending on the furnace, alloys, and other proprietary factors. The theoretical minimum energy requirement to bring room-temperature (25 °C, or 75 °F) aluminum to a molten 960 °C (1760 °F) metal is 0.39 kWh/kg. The theoretical energy requirement for melting pure aluminum to molten metal at various temperatures is presented in Appendix G.

Basic natural gas or oil-fired reverberatory furnaces range in efficiencies from approximately 20 to 45%. The more efficient furnaces employ recuperation of stack gas heat for reduced melting energy requirements through charge preheating or for more efficient burner operation through preheating combustion air. Furnace condition and operating practices have large effects on energy performance. Because heat transfer in reverberatory furnaces takes place principally through radiation, melt surface temperatures are considerably hotter, leading to more rapid oxidation and higher melt losses.

Electric furnaces, typically used in small processing operations, do not require a flue, and their heating chambers can be made nearly airtight. A sidewell is provided for charging metal and alloying materials. The sidewell removes the need to open the furnace door and prevents a major convective heat loss. Energy losses (excluding electrical generation and transmission) in electrical furnace heating are principally due to conduction and radiation losses from the exposed furnace shell. Losses are typically 0.49 to 0.81 kWh/kg of aluminum. Induction furnaces are typically more than 90% energy efficient, while gas-fired crucibles are 15 to 28%, and electrically heated crucibles 83% efficient in terms of on-site energy use.

Technological Change in the Next Decade. The industry is constantly evaluating, adopting, and improving furnace technologies and practices. This provides not only energy and environmental benefits but also cost savings. New burner technologies and oxygen-enhanced combustion systems are being developed and evaluated to further improve efficiency and reduce the costs of melting without increasing emissions. All industry segments have transitioned to greater reliance on scrap and recycling for their metal needs. Technologies for sorting, handling, and remelting scrap in all forms with optimal metal recoveries and lowest costs are continuously being developed and refined. New technologies for melting thin-gage material to minimize oxidation losses are being developed and implemented. Also, industry is continually seeking better methods to recover the aluminum that is trapped in dross. Additional efforts are directed at the closed-loop recycling of dross-related wastes, including saltcake.

New laser technologies will speed the in situ chemical analysis of molten metal and minimize the processing time required for alloy compliance. The development of better, longer-lasting ceramic materials for furnace linings is ongoing and will reduce the time required for furnace maintenance. The aluminum industry has recently published a technology roadmap in conjunction with the advanced ceramics industry, to encourage the development of superior furnace construction materials (Ref 10.47).

Finally, new melting technologies now under development offer the prospects for revolutionary improvements in melting efficiencies that may be applicable to the scale and operational demands of much of the industry. One exciting development is that of immersion heating with high-watt-density elements for melting as well as temperature maintenance. The development of commercial immersion heaters for aluminum remelting is very likely to occur within the next few years.

Ingot Casting

Ingot casting is the solidification of molten alloys into shapes that are suitable for subsequent thermomechanical processing or for remelting. Ingots are made by controlled solidification in molds designed to produce the desired geometrical configuration and metallurgical characteristics.

Ingot casting is, by itself, not energy-intensive; however, casting is typically a batch process, and large quantities of molten metal are held in furnaces in which alloying, fluxing, and degassing are performed. Accordingly, conductive and radiant heat losses occur from these furnace operations.

Ingot for wrought product applications is almost universally cast by the semicontinuous direct-chill (DC) casting process. The process includes different means of introducing and controlling the flow of molten metal into the mold, lubrication methods, the use of insulation in mold construction, and the injection of air or imposition of an electromagnetic field for reducing or eliminating contact between molten metal and the mold. The process produces rectangular cross-sectional ingots for rolling; round loglike billets for extrusion; squares for wire, rod, and bar products; and various shapes as fabricating ingot in forging.

The DC casting process begins when aluminum flows from the furnace through troughs to the casting station. At the casting station, the aluminum flows into one or multiple water-cooled stationary molds that rest on the casting station table. The DC ingot molds are only a few inches deep and form the cross section of the ingot or the billet. The ingot is initially formed in the water-cooled mold. When perimeter solidification has begun, the casting table is gradually lowered into the casting pit, while additional molten aluminum is supplied to the top of the mold. The water-cooled mold remains at the top of the pit and continues to shape the casting. Water sprays impinging on the solid shell continue the solidification process of the molten ingot core. The casting table is lowered into a casting pit until the desired length is achieved. After casting, ingot intended for wrought fabrication may be stress relieved, scalped, cut to length, and homogenized. Cutting, shearing, forming, and other mechanical operations, as well as melting, heating, casting, heat treating, and other thermal operations are used by the product-producing sector.

Casting operations attempt to control the crystal/grain structure and composition gradient of cast products. Grain size and boundaries are important factors affecting the physical and mechanical properties of the material in cast and wrought form. However, because there is a significant temperature profile across the ingot cross section during solidification, grain structure and composition can vary from surface to center. For subsequent fabricating operations, it is usually necessary to remove the skin layer by scalping so that the final product has consistent physical properties. The amount of surface to be removed is dependent on shape, surface quality, and the depth of undesirable grain structure and segregation. Scalpings are remelted and reprocessed, which results in additional energy usage and metal losses due to oxidation.

A percentage of DC ingots are rejected for quality reasons. Cracking may occur during or after solidification. Surface defects may form, which affect the acceptability of the ingot for wrought processing. Other specialized standards concern grain structure, segregation, and microstructure. At times, ingots are found to exceed alloy specification limits. The processing energy used to produce the ingot is then lost, and additional energy is required for remelting and reprocessing.

Energy Requirements. More than 2,480,000 metric tons of primary aluminum were cast into ingots in 2005 (Table 10.3). The tacit energy consumed in aluminum processing operations can be divided into three categories:

- Fossil-thermal energy use, which includes furnaces, heating, and heat treatment operations
- Electrical energy required for heating, sawing, and scalping and for motor, pump, and compressor operation
- Other fuel-consuming operations, including transportation

Primary ingot casting has typical metal yields from 88 to 98% and requires approximately 1.01 (1.46^{tf}) kWh/kg of cast ingot product (Appendix E, Table E.2).

The theoretical minimum energy requirements for primary and secondary castings are the same at 0.33 kWh/kg of aluminum (Appendix E, Table E.3). The difference in their actual energy usage results from their respective initial materials. In primary casting, the initial material is molten aluminum in a holding furnace, while the initial material in secondary casting is metal scrap. The scrap must first be melted before it enters a holding furnace, giving secondary casting a higher actual energy use than primary casting. Secondary aluminum was cast into 2,990,000 metric tons of ingots in 2005. Secondary casting has typical yields of 96% and requires approximately 2.50 (2.81^{tf}) kWh/kg of product.

Technological Change in the Next Decade. Ingot casters have strived to improve process yields and to refine practices to provide more consistent internal and surface quality in wrought ingot manufacture. These efforts have resulted in significant progress in surface, sub surface, and metallurgical improvements. Grain refining by heterogeneous nucleation agents has benefited from decades of constant research and development by primary producers in cooperation with master alloy suppliers.

Molten metal fluxing and filtration processes continue to undergo changes, leading to greater efficiencies, higher product quality, reduced environmental impact, and reduced costs.

Numerous research programs are directed at the modeling and prediction of the solidification process for reduced cracking incidence and improved structural uniformity. The evolution of mold designs capable of improved surfaces, reduced scalping, and higher production rates continues in all wrought ingot production.

All of these developments reflect advances in sensors and instrumentation that permit more accurate monitoring and comprehensive control of the casting process.

Rolling

Rolling is the process of reducing ingot thickness by passing it between counterrotating steel rolls. Aluminum-rolled products include plate (typically >0.6 cm, or 0.2 in., thick), sheet (typically 0.02 to 0.6 cm , or 0.008 to 0.24 in., thick), and foil (typically <0.02 cm, or 0.008 in., thick). Figure 10.25 shows the unit operations of a typical rolling mill.

Hot, cold, and continuous rolling operations are used industrially to shape material into the broad category of flat rolled products. Hot rolling is generally performed at various temperatures exceeding the recrystallization temperature for differing rolling alloys. Cold rolling takes place at room temperature, but the heat generated can result in metal temperatures as high as 150 °C (300 °F). Continuous rolling pours molten metal directly into strip used for sheet and foil and is significantly more energy efficient than cold or hot rolling. The United States has more

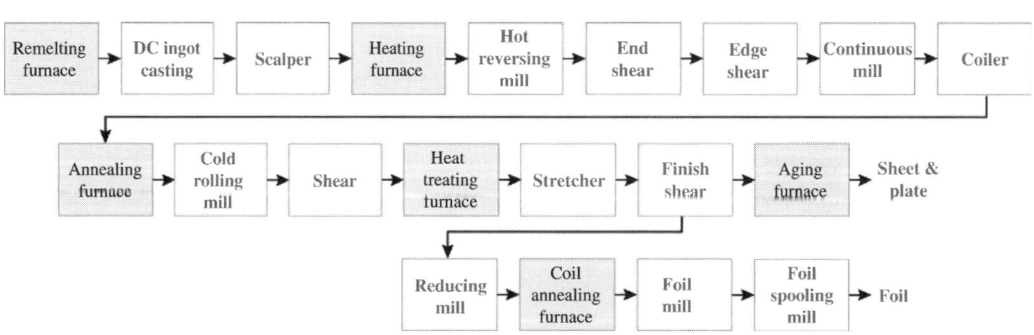

Fig. 10.25 Typical rolling mill processing operations. DC, direct chill

than 50 aluminum rolling mills, 11 of which are continuous casters.

Large, high-strength alloy ingot may require thermal stress relief after casting. Most ingots are homogenized to reduce intergranular segregation and to modify intermetallic particle form. Homogenization may be integral to preheating for hot rolling, or the preheating may take place separately. Although some ingot may be rolled with the cast surface intact, rolling surface faces are normally removed by scalping, and the head and butt of the ingot are cropped to assure finished product uniformity.

Hot rolling begins with repetitive reversing mill reductions on stock preheated to temperatures in the range of 400 to 500 °C (750 to 930 °F). The slab is passed repeatedly back and forth through the rolls until the desired reduction in thickness is achieved. While heat is generated during deformation, the number of passes required at the reversing mills may—because of reducing section thicknesses, time, and coolant application—cause losses in temperature that require reheating before continuing the hot rolling process. Slabs are lifted from the hot line and charged in reheating furnaces until working temperatures are restored. End crops are taken during breakdown rolling as required to maintain squareness.

Edge-trimming knives remove stock from the edges of the sheet before coiling. The amount of edge trim required is determined by the depth of edge cracks and the ragged conditions associated with the unrestrained deformation that occurs during hot rolling. The amount of edge trim required with cropping losses represents a significant percentage of planned scrap that must be returned to the cast shop for remelting. Edge trim losses can be minimized by improved ingot quality and by edge rolling.

The hot rolling process completely changes the microstructure formed during casting and elongates the grain structure in the direction of rolling. Even though deformation temperatures are typically greater than the recrystallization temperature, there is inevitably some degree of equivalent cold work, so that annealing at reroll gage will result in recovery or recrystallization before cold rolling.

Energy Requirements for Rolling Aluminum. Approximately half of U.S. rolled aluminum products are cold rolled. In 2003, 2,421,300 metric tons of cold-rolled products were produced (Appendix E, Table E.4). Cold rolling has typical yields of approximately 84% and requires approximately 0.64 (1.35^{tf}) kWh/kg of rolled product. Hot rolling has typical yields of approximately 82% and requires approximately 0.62 (1.16^{tf}) kWh/kg of rolled product.

Theoretically, it is possible to roll products without the need for heat treatments and with no loss of material due to trimming or slitting. In this case, the minimum theoretical energy to roll a product is composed of only two components:

- The energy required to heat starting stock to the rolling temperature
- The energy required to deform the shape

The hotter the material, the lower the deformation energy required. The heat capacity equations required to calculate the energy requirement for heating pure aluminum are listed in Table 10.9. The energy required for deforming is given by the equation $E = \varepsilon\sigma c$, where ε is the strain or deformation defined as $\varepsilon = \ln (t_i/t_f)$, where t_i represents the initial and t_f the final dimension, σ denotes the yield stress, and c denotes a constant describing the shape of the stress-strain curve. The yield stress value for aluminum can vary by as much as a factor of 10 over the hundreds of alloys that are used by the industry.

The very large variations in alloy properties, particularly the shape and the magnitude of the stress-strain curves, make it possible to calculate theoretical minimum energy requirements for rolling aluminum only for a specific rolling process with a specific alloy. There are a large number of heavy-equipment requirements in rolling mill operations that are not confined to rolling. Roller and tension levelers, slitters, stretchers, roll formers, and paint or coating lines are examples of energy-consuming operations that rely on pumps, motors, and compressors as well as on mechanical design for efficiency. A rough approximation of the rolling sector minimum energy can be made by assuming overall process heating efficiencies and electric/-hydraulic system efficiencies and by looking at the entire rolling sector yield and energy consumption. If the overall sector heating efficiency is 50% and the electric/hydraulic system efficiency is 75%, the estimate of the minimum energy requirement is 0.31 kWh/kg of product for hot rolling and 0.33 kWh/kg of product for cold rolling. These assumptions imply that cold rolling is operating at approximately 52% overall energy efficiency and hot rolling at approximately 50% overall efficiency (Appendix E, Table E.3).

Advanced Rolling Technology. Hot rolling typically requires numerous passes through the rolling mills and is energy-intensive. One approach to improving productivity and reducing heating energy is to continuously cast molten metal into slab or strip (Ref 10.48). Going directly to thin strip, continuous strip and slab casting saves the energy required for homogenization, scalping, preheating, end and side trim, and multiple passes through rolling mills.

Continuous strip casting is in wide current use for some sheet and foil specifications and has demonstrated energy savings of more than 25% relative to conventional ingot rolling. These casters convert molten metal directly into reroll gage sheet at 1 to 12 mm (0.04 to 0.5 in.) thickness. Continuous strip casters employ twin counterrotating water-cooled rolls or belts to accomplish solidification. Strip casting is now restricted to certain low-alloy-content compositions. Technology is being developed for more complex or more highly alloyed compositions. Very high solidification rates and the extrusion component of casting between cylindrical rolls result in high degrees of segregation of solute to the centerline and in cracking tendencies.

While continuous strip casting is recognized as alloy-constrained, an alternative continuous process for coiled reroll, slab casting is highly alloy-tolerant and results in metallurgical structures that closely correspond to those of the ingot and wrought products produced by the DC process. In slab casting, solidification takes place between water-cooled belts or blocks. Slab thickness varies but typically corresponds to continuous hot mill entry gages of 75 to 150 mm (3 to 6 in.). An individual slab-casting line has a production capacity ten times that of the largest strip caster. The cast slab is directly fed into one or more in-line, low-speed, high-torque hot reduction mills that reduces thickness to coilable gage. Slab-casting processes have successfully produced high-strength aluminum alloy sheet as well as challenging products such as beverage can sheet, but subtle differences in product performance—especially in formability and anisotropy—continue to favor the use of conventionally hot-rolled sheet in these applications.

A further advancement in rolling technology is spray rolling. In this process, molten metal droplets are sprayed directly into the nip of twin rolls, and the material is solidified and consolidated directly into sheet material in one step. In concept, this offers the most energy-efficient process, and recent projects have sought to advance this technology.

Extrusion

Extrusion is the process of forcing an aluminum ingot or billet through a steel die to form an elongated shape of consistent cross section. Extruded products include rods, bars, tubes, and specialized products interchangeably called shapes, sections, or profiles. Figure 10.26 shows the unit operations of a typical extrusion plant. After rolling, extrusion is the second most common processing technique for aluminum. Aluminum extrusion is remarkable because the process combines high productivity with an essentially infinite variety of extremely complex shapes, cross sections, or profiles that cannot be economically duplicated in any other process. Furthermore, aluminum can be readily extruded; this process is either extremely difficult or impractical for many other metals. There are 190 extrusion plants in the United States. Many of these plants run multiple extrusion presses.

It is possible to produce almost any cross-sectional shape, within wide limits. Through the use of hollow stock and floating or fixed mandrels, hollow shapes or cross sections with complex enclosed configurations can be produced. The extrusion process is capable of producing a cross section with a weight of a few grams to more than 300 kg/m, a thickness of less than 1 to over 250 mm (0.04 to over 98 in.), circumscribing diameters of 5 to 1000 mm (0.2 to 39 in.), and lengths in excess of 30 m (98 ft). The appropriate choice of alloy and extrusion conditions can result in an optimal combination of properties for a particular

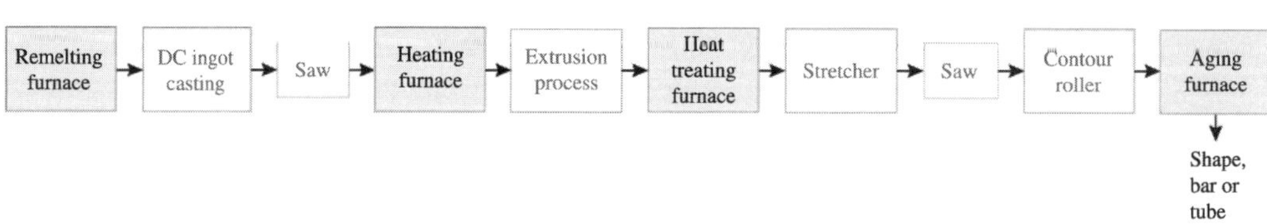

Fig. 10.26 Typical aluminum extrusion processing operations. DC, direct chill

application. Such properties may include tensile strength, toughness, formability, corrosion resistance, and machinability. Countless extrusion products are made in the United States. These include frame and supporting shapes for windows and doors, carpet strips, household bath enclosures, screens, bridge structures, automotive parts, aerospace components, and many other consumer products.

Operations at individual plants vary widely depending on the cross sections and alloys produced. Extrusion presses range in capacity up to 15,000 tons, but the most common are perhaps 2500 tons. Presses may extrude vertically or horizontally, but virtually all modern extrusion presses are horizontal. Typically, billets in diameters up to 275 mm (11 in.) are preheated to temperatures ranging from 450 to 550 °C (840 to 1020 °F) depending on the alloy, product design, and the desired mechanical characteristics. The preheated billet is charged to the extrusion press container and forced by hydraulic pressure through the extrusion die.

There are essentially two processes for extrusion production. In the direct extrusion process, the billet is hydraulically pressed through the die, while in the indirect extrusion process, the die is forced over the billet. In direct extrusion, the billet surface is retained in the extrusion container and contributes to butt loss, which may total 8% of the starting billet weight. Because the billet surface is not extruded to become part of the product, scalping is not required. For indirect extrusion, the billet surface becomes an integral part of the product, and so, scalping before extrusion is essential.

The single largest area for energy improvement in extrusion technology is the reduction of process scrap, including butt losses. Extrusion product specifications include significant surface-quality and chemical-finishing criteria. Defects include torn surface; die pickup, which is often related to the billet microstructure; and nonfill. Variable response to chemical finishing results in appearance and color mismatches that affect product acceptability.

Energy Requirements for Extruding Aluminum. The United States produced over 1,826,000 metric tons of extruded aluminum products in 2003 (Appendix E, Table E.4). Extrusion processes have typical yields of 69% and require approximately 1.30 (1.52^{tf}) kWh/kg of extruded product.

Theoretically, it is possible to extrude products without the need of additional heat treatments and without loss of material. The minimum theoretical energy to extrude a product, in such a case, is composed of only two components:

- The energy required to preheat the billet to extrusion temperature
- The energy required to deform the material through a die

The hotter the material, the lower the deformation energy required. The simpler the die, the lower the extrusion-energy requirement. The heat capacity equations needed to calculate the energy requirements for heating pure aluminum are listed in Appendix G. The energy required to deform the material through the die is highly dependent on the size and shape of the product and the die design. Calculation of the minimum extrusion force is very complex and can only be estimated with theoretical and empirical models. Typical formulae have the following simplified form:

$$F = A_o(\sigma_m/\eta)\ \varepsilon$$

where ε, the strain, corresponds to the reduction area, $\varepsilon = \ln(A_o/A_f)$, σ_m is the mean stress for the strain, A_o is the original cross-sectional area, and η is an efficiency factor.

The very large variations in alloy properties, particularly the infinite numbers of possible shapes, make it impossible to calculate a theoretical minimum energy requirement. This value can only be determined by analyzing a specific process, a specific extruded profile, and a specific alloy. The perimeter of the profile and the radius of intersecting edges have a large influence on the force required for extrusion. A rough approximation of a minimum energy value can be made by examining the entire aluminum extrusion industry yield and energy values and assuming an overall process heating efficiency and electric/hydraulic-system efficiency. This approach provides an estimate of the minimum energy, 0.44 kWh/kg of aluminum, when efficiencies are assumed to be 50% for heating and 75% for the electric/hydraulic system. These assumptions imply that overall extrusion facilities operate at approximately 34% energy efficiency.

Shape Casting

Shape casting, or the casting of engineered designs, enables the production of simple and complex parts that meet a wide variety of needs.

The process produces parts weighing ounces to parts weighing several tons. Over 1000 companies cast aluminum products. Some of these are very small garage-shop operations that supply niche market casters, for example, specialty toymakers. Over 15 plants have capacities to produce >25,000 tons of product per year. Figure 10.27 shows the unit operations of a typical aluminum shape-casting foundry. These operations vary significantly depending on the size of the operation, the processes employed, the complexity of the parts, the alloy compositions, and the type of castings being produced.

The basic casting process consists of melting and alloying aluminum and pouring or injecting molten metal into molds containing single or multiple cavities of the desired shape. The important casting processes for engineered aluminum castings are pressure die, permanent mold, green and dry sand, plaster, and investment casting.

In pressure die or die casting, metal is injected at pressures up to 10,000 psi into water-cooled steel dies. Productivity rates are high, and the process can be highly automated. While surface quality and dimensional accuracy are excellent, the typical die casting contains a degree of internal unsoundness associated with nondirectional solidification, entrapped gases and inclusions resulting from turbulent metal flow, and the presence of air and lubricants in the die cavity.

Permanent mold or gravity die castings are produced by introducing molten metal by gravity or countergravity means into iron or steel molds. Productivity, surface quality and dimensional accuracy are lower than in pressure die casting. Internal soundness, depending on the extent to which sound molten metal treatment for hydrogen elimination and oxide removal are practiced and the principles of directional solidification are employed, can meet the most challenging quality standards. Low-pressure casting is usually considered a variation of the permanent mold process, even though dry sand and plaster cast parts have been produced. In this process, molten metal is forced by the application of pressure to rise through a tube into a mold mounted over the furnace. It has the advantages of a significantly reduced gross-to-net weight ratio and correspondingly lower trimming costs.

Green and dry sand casting can also yield high-integrity parts. In green sand casting, sand, binders such as clays, and moisture are blended to provide the molding medium. Patterns may include loose pieces, wood models, cast matchplates, and molded Styrofoam (Dow Chemical Company). For dry sand molding, air or thermosetting chemicals coat sand particles so that the finished mold after curing offers superior surfaces, dimensional accuracy, and shelf life.

Investment molds are produced by repetitive immersion of plastic, wax, or other low-temperature-melting and volatile pattern material into ceramic slurries. After drying, the hardened casing containing the pattern is heated to a temperature at which the pattern material is eliminated. Typically, the mold is preheated before pouring and may be filled under vacuum. Investment castings offer extremely fine finishes, thin walls, and excellent dimensional accuracy. Plaster molding offers the same advantages and may be used for the production of thicker sections, large parts for which investment is less suited.

Molten aluminum is required as the basis for all foundry production. High-volume casting operations may acquire part or all of their metal requirements as molten metal delivered over-the-road in insulated crucibles. Other foundries melt and process prealloyed ingot, RSI, and internal scrap, including gates and risers. All melting and melt-processing technologies and considerations described in the section "Melting, Alloying, and Melt Treatment" are applicable to foundry operations.

Energy Requirements for Shape Casting Aluminum. Nearly 2,413,000 metric tons of shape-cast products were produced in 2003. Shape casting has typical yields of only 45% and requires approximately 2.56 (2.64^{tf}) kWh/kg of cast product. The energy consumed is almost exclusively related to furnace and heating operations (Appendix E, Table E.4).

The theoretical minimum energy requirement for shape casting can be calculated from the

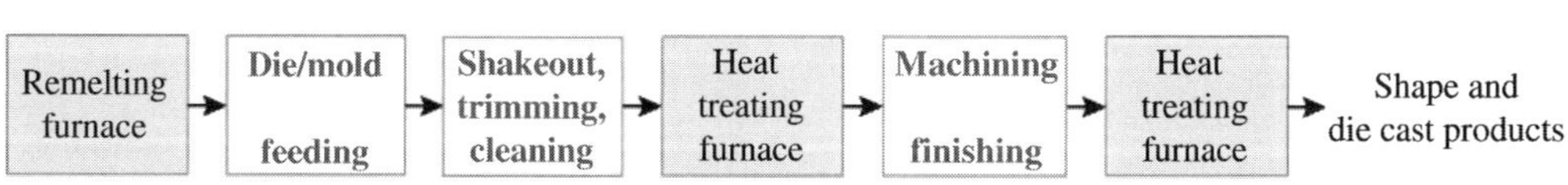

Fig. 10.27 Typical aluminum product shape-casting operations

energy required to go from room temperature to liquid metal plus some superheat value (see the section "Theoretical Energy for Secondary Aluminum" in this chapter). Pure aluminum melts at 660 °C (1220 °F). The minimum energy required to produce liquid aluminum at 660 °C (1220 °F) is approximately 0.3 kWh/kg.

Alloy composition; superheat requirements; mold sprue; gates, runners, and riser systems; and postcasting heat treatments vary by mold design and casting practices. A rough approximation of the aluminum shape cast sector minimum energy requirement can be made by looking at the entire sector yield and energy values (Appendix E, Table E.2) and assuming an overall process heating efficiency and electric/hydraulic-system efficiency. The estimate of the minimum energy requirement using this approach is 0.60 kWh/kg, when a 50% overall heating efficiency and a 75% electric/hydraulic system efficiency are assumed. These assumptions imply that shape casting facilities operate at approximately 23% overall energy efficiency.

Technological Change in the Next Decade. Castings are among the most cost-effective and versatile solutions to part design and performance challenges. The range of available alloys and properties provides combinations of manufacturability and product characteristics for one-of-a-kind, prototype, limited, or high-volume applications. Castings are near-net shape, with the potential for precise integral internal passages and complex shapes. Aluminum is cast in more alloys with a wider range of physical and mechanical properties by more processes than any competing metal system.

The most important trend affecting aluminum casting production is its continuous growth in automotive applications. The advantages of aluminum for many powertrain components, including transmission cases, oil pans, pistons, intake manifolds, cylinder heads, and engine blocks, have been reflected in its wide adoption.

Squeeze casting and semisolid forming have emerged as candidates for a new generation of process capabilities producing heat treatable, high-integrity pressure die castings that use lower-impurity compositions, vacuum, and dry lubrication.

Thermal Treatments

A significant component of energy use in the aluminum industry is the heat treatment of metal and products. The physical and mechanical properties of aluminum alloys in any product form can be controllably altered by thermal treatment. Thermal treatments are used to soften the material and to recrystallize the grain structure. Other aluminum alloys, principally those containing copper, magnesium, silicon, and zinc, can be thermally treated to significantly improve strength through the dissolution and reprecipitation of soluble phases. Aluminum alloys are categorized as heat treatable if thermal treatments have significant hardening benefits or non-heat treatable if the alloy is unresponsive to thermal treatments for hardening purposes. Among the latter are alloys dependent on work hardening for strengthening and those whose properties are essentially fixed after solidification.

All product types, including sheet, plate, foil, wire, rod, bar, extrusions, forgings, and castings, are produced in heat treatable alloys. The majority of extrusions, forgings, and a large percentage of plate and castings are heat treated. Heat treatment facilities are integral to larger operations. Commercial firms also provide contracted heat treatment services.

Annealing is performed at temperatures from 300 to 500 °C (570 to 930 °F) to reduce strength, improve formability and ductility, lower residual stress levels, and improve dimensional stability in cast and wrought products. Because electrical conductivity is adversely affected by elements retained in solution, annealing is also used in electrical and electronic applications. Deep-drawn sheet is normally annealed. Intermediate annealing is the usual practice in rolled product manufacture to permit subsequent cold reductions. The final temper of non-heat treatable rolled products may include annealing, partial annealing, or stabilization treatments. Most annealing is a batch operation.

Heat treatable aluminum alloys contain intermetallic metallurgical phases that can be dissolved at elevated temperature (up to 550 °C, or 1020 °F) and retained in solid solution by rapid quenching. Solution heat treatment is batch or continuous. Sheet and foil can be heat treated continuously through accumulator towers, and plate, castings, and forgings by conveyer furnaces. Coiled sheet, plate, and extrusions are more typically batch treated. In either case, the product must be held at solution temperature long enough for complete solution to occur and for desirable changes in the shape or form of the insoluble intermetallics that are present in the microstructure. The quench medium is generally water but may be glycol where greater control is required at room temperature. The retained metastable solid solution permits precipitation hardening at intermediate temperatures (170 to 300 °C, or 340 to 570 °F) for fully hardened, partially hardened, or overaged conditions. Each offers combinations of strengths, ductilities, toughness, stability, resistance to stress corrosion, and hardness not achievable through solidification or work hardening. Precipitation-hardening furnaces are also batch or continuous.

The high rate of solidification in many casting processes results in a degree of solution retention in heat treatable compositions, permitting age or precipitation hardening to be employed without solution heat treatment. This method is extensively used in extrusions that can be press quenched by forced air or water mist to improve solution retention.

Recommended solution heat treatment practices have been standardized to reflect worst-case conditions. The cycle defines the minimum time at the required temperature for successfully treating the part requiring the longest exposure. A safety factor is usually also applied to assure that reheat treatment will not be necessary. Practices are typically generic, applying to specific alloys and tempers without regard for section thickness or degree of metallurgical refinement. Thinner wall castings, forgings, and extrusions generally respond to heat treatment more rapidly. Finer grain and dendrite cell sizes are reflected in smaller, more dispersed solute particles that can be more rapidly dissolved. Solution heat treatment time can therefore be patterned to specific products and manufacturing processes and the combination of finer metallurgical structures. Decreased variability can result in substantially reduced cycle times and energy costs.

New heating technologies are being studied to reduce energy requirements through more rapid heating to treatment temperature. Fluidized beds and infrared heating can be used to shorten heat-up times but do little to accelerate either the rate of solution or microstructural change when solution temperatures are reached.

Another approach being investigated is the use of sensible heat to reduce energy requirements. While there are metallurgical concerns, extrusions, castings, and forgings can be placed in heat treatment furnaces directly from the mold or die, thereby preserving the latent heat of the casting or final forming operation.

Energy Requirements for Thermal Treatment. The theoretical energy required for all thermal treatments can be calculated from the specific heat or heat capacity of the various aluminum alloys. For example, it requires 0.06 kWh/kg (95 Btu/lb) to heat A356 to its solution heat treatment temperature, 0.05 kWh/kg (80 Btu/lb) to reach annealing temperature for 1100 alloy, and 0.02 kWh/kg (32 Btu/lb) to reach precipitation-hardening temperature for alloy 2024.

Time at temperature for each procedure varies depending on alloy and product form but can exceed 12 h. While most energy is consumed in raising the product and oven to temperature, additional energy is required to maintain the temperature for the duration of the cycle. Sustaining energy requirements is exclusively a function of furnace design and condition. Standard quench temperatures include room temperature: 65, 80, and 100 °C (150, 180, and 212 °F). Large volumes of water must be heated and maintained at temperature by steam or other means for the latter practices.

REFERENCES

10.1. “Report for U.S. Department of Energy,” by W. Choate and J. Green, BCS Inc.; http://www.eere.doe.gov (accessed July 2007)

10.2. *Aluminum Statistical Review for 2005,* The Aluminum Association, Inc., Washington, D.C., 2006, p7 and Table 16. The Aluminum Association’s statistical reviews are issued annually and assemble, in one document, important detailed data from primary production, to markets for finished goods, to the recovery of scrap. Both U.S. and world information are included.

10.3. A.R. Burkin, Ed., *Production of Aluminium and Alumina*, John Wiley & Sons, 1987, p 38, 49, 57, 63. This book, published on behalf of the Society of Chemical Industry, is one of the only texts to provide a comprehensive overview of the state-of-the-art of all science and technologies associated with primary aluminum production in a single volume. It is a detailed, academic volume that covers chemistry, thermochemistry, fluid dynamics, process dynamics, and so on from both a theoretical and practical perspective.

10.4. D. G. Altenpohl, *Aluminum: Technology, Applications, and Environment; A Profile of a Modern Metal,* 6th ed., The Aluminum Association, Inc. and the Minerals, Metals and Materials Society, 1998. This book is described in its foreword as having "global recognition as the definitive educational text and reference book for aluminum industry participants; a broad range of aluminum fabricators and users; students; and the scientific, engineering, and academic community."

10.5. *Aluminum Industry Vision: Sustainable Solutions for a Dynamic World,* The Aluminum Association, Inc., Washington, D.C., Nov 2001, p 3, 22. The *Vision* document identifies key needs of the industry and outlines a comprehensive aluminum research and development agenda.

10.6. R. D. Naranjo, E.-P.H. Fu, and M. Gwyn, Castings Drive Fuel Efficiency, *Mod. Cast.*, Vol 94 (No. 9), Sep 2004, p. 20

10.7. *Life Cycle Inventory Report for the North American Aluminum Industry,* The Aluminum Association, Inc., Washington, D.C., 1998. This report provides information on a life-cycle inventory study of the North American aluminum industry in 1995. The report is the result of extensive surveying and contains the best and most complete industry performance information of any recent study. Chapter 3 is an adapted version and Appendix E is a summary of the report.

10.8. *Energy and Environmental Profile of the U.S. Aluminum Industry,* U.S. Department of Energy, Office of Industrial Technologies Program, Washington, D.C., July 1998. This report reviews the energy and environmental characteristics of the key technologies used in the major processes of the aluminum industry.

10.9. *Aluminum Statistical Review for 2003,* The Aluminum Association, Inc., Washington, D.C., p 7, 10–12, 19–23, 38, 42, 92

10.10. "Bauxite and Alumina Statistics and Information," U.S. Geological Survey, Minerals Information; http://minerals.usgs.gov/minerals/pubs/commodity/bauxite/ (accessed July 2007)

10.11. *Life Cycle Inventory Report for the North American Aluminum Industry,* The Aluminum Association Inc., Washington, D.C., p E-3, E-6, E-8, E-12

10.12. *Technology Roadmap for Bauxite Residue Treatment and Utilization,* The Aluminum Association, Washington, D.C., Feb 2000. This booklet contains a comprehensive discussion of bauxite residue.

10.13. *Energy and Environmental Profile of the U.S. Aluminum Industry,* U.S. Department of Energy, Office of Industrial Technologies Program Washington, D.C., July 1998, p 29

10.14. *Alumina Technology Roadmap,* The Aluminum Association, Inc., Washington, D.C. Nov 2001. This document identifies performance goals and describes 25 research and development areas that are a priority for the global alumina industry.

10.15. P.J. Ellis, "Tutorial: Petroleum Coke Calcining and Uses of Calcined Petroleum Coke," Speech given at AIChE 2000 Spring National Meeting, Third International Conference on Refining Processes, March 2000, p 13

10.16. The Manufacture of Carbon and Graphite, *Industrial Graphite Engineering Handbook,* UCAR Carbon Company, Inc., Parma, OH, 2001, p 1–6

10.17. W. Haupin, History of Electrical Consumption by Hall-Heroult Cells, *Hall-Heroult Centennial,* W. Peterson and R. Miller, Ed., The Metallurgical Society of AIME, New York, 1986, p 106. This article provides a detailed historical-review of energy usage in the Hall-Heroult process, from the first cells to 1986.

10.18. R.P. Pawlek, 75 Years of Development of Aluminum Electrolysis Cells, *Aluminium,* Vol 75 (No. 9) 1999, p 734–743

10.19. B.J. Welch, Aluminum Production Paths in the New Millennium, *J. Met.,* Vol 51(No. 5), Feb 1999, p 24–28

10.20. Appendix A-9: Thermodynamics of Electrochemical Reduction of Aluminum, *Report of the American Society of Mechanical Engineers: Technical Working Group on Inert Anode Technologies,* U.S. Department of Energy, Office of Industrial Technologies, Washington, D.C., July 1999, p 3, 28. This report reviews the literature and patents concerning inert anode technologies. It presents a theoretical discussion of aluminum reduction and has input from over eleven experts in the field. Carbon anode voltage equation $E = 1.898 - 0.0005728 \times T$, where E is in V (direct current), and T is in degrees Kelvin.

10.21. N. Jarrett, W.B. Frank, and R. Keller, Advances in Aluminum Smelting, *Metall. Treatises AIME,* Vol 1 (No. 93), 1981, p 137

10.22. U.S. Environmental Protection Agency, Office of Air and Radiation; www.epa.gov/globalwarming/ (accessed July 2007)

10.23. G.D. Brown, M.P. Taylor, G.J. Hardie, and R.W. Shaw, "TiB_2 Coated Aluminum Reduction Cells: Status and Future Direction of Coated Cells in Comalco," paper presented at the Queenstown Aluminum Smelting Conference, Nov 26, 1998

10.24. M.P. Taylor, G.J. Hardie, F.J. Stevens McFadden, and W. Uru, "Use of Refractory Hard Cathodes to Reduce Energy Consumption in Aluminum Smelting," paper presented at the Second International Conference on Processing Materials for Properties, TMS 2000 Technical Program

10.25. L.Boxall, A.V. Cooke, and W. Hayden, TiB_2 Cathode Material: Application in Conventional VSS Cells, *J. Light Met.,* Vol 36 (No. 11), Nov 1984, p 35–39

10.26. Low Energy Aluminum Reduction Cell with Induced Bath Flow, U.S. Patent 4,602,990

10.27. T.M. Leeuwen, An Aluminum Revolution, *Desk Notes*, Credit Suisse First Boston Corporation, June 22, 2000

10.28. Appendix A-9: Thermodynamics of Electrochemical Reduction of Aluminum, *Report of the American Society of Mechanical Engineers: Technical Working Group on Inert Anode Technologies*, U.S. Department of Energy, Office of Industrial Technologies, July 1999. Inert anode voltage equation, $E = 2.922 - 0.0005712 \times T$, where E is in V (direct current), and T is in degrees Kelvin

10.29. Appendix A-9: Thermodynamics of Electrochemical Reduction of Aluminum, *Report of the American Society of Mechanical Engineers: Technical Working Group on Inert Anode Technologies*, U.S. Department of Energy, Office of Industrial Technologies, July 1999

10.30. *Inert Anode Roadmap: A Framework for Technology Development,* The Aluminum Association, Inc. and the U.S Department of Energy, Office of Industrial Technologies Program, Feb 1998, p 1. This document describes the potential benefits, performance targets, technical barriers, and technology development steps associated with the research and development and production of inert anodes

10.31. D.B. Rao, U.V. Choudry, T.E. Erstfeld, R.J. Williams, and Y.A. Chang, Extraction Processes for the Production of Aluminum, Titanium, Iron, Magnesium, and Oxygen from Nonterrestrial Sources, *Space Resources and Space Settlements,* NASA Ames Research Center Summer Study, Moffett Field, CA, 1977, p 15

10.32. W. Haupin, Reflections on the Hall-Heroult Process, *Molten Salts Bulletin*, 59th ed., Ecole Polytechnique, Universitarire de Marseille

10.33. C. Brown, Next Generation Vertical Electrode Cell, *J. Met.* Vol 53 (No. 5), May 2001, p 39–42

10.34. J.N. Hryn, "Inert Metal Anode," project summary form from the 2004 Annual Aluminum Portfolio Review, Argonne, IL, Oct 19–20, 2004

10.35. V. Garcia-Osorio, B.E. Ydstie, and T. Lindstad, "Dynamic Model for Vapor Recovery in Carbothermic Aluminum Process," Carnegie-Mellon University and SINTEF Materials Technology

10.36. W.T. Choate, A.M. Aziz, and R. Friedman, New Technologies Will Sustain the U.S. Primary Aluminium Industry, *Light Metals 2005*, H. Kvande, Ed. TMS, p 495

10.37. W. Choate and J. Green, Technoeconomic Assessment of the Carbothermic Reduction Process for Aluminum Production, *Light Metals 2006, Vol. 2: Aluminum Reduction Technology,* T.J. Galloway, Ed., TMS, p 94

10.38. K. Johansen, J.A. Aune, M. Bruno, and A. Schei, "Carbothermic Aluminum—Alcoa and Elkem's New Approach Based on Reactor Technology to Meet Process Requirements," paper presented at the Sixth International Congress on Molten Slags, Fluxes, and Salts, June 2000 (Stockholm)

10.39. H. Myklebust and P. Runde, Greenhouse Gas Emissions from Aluminum

Carbothermic Technology Compared to Hall-Heroult Technology, *Light Metals 2005*, H. Kvande, Ed., TMS, p 519

10.40. "Toth Aluminum Corporation's Bipolar Clay-Aluminum Technology and Project," seminar ed., paper presented at TMS annual meeting, Feb 11, 2001 (New Orleans, LA)

10.41. New Technology is a Breakthrough for Automotive Aluminum Recycling, *Alum. Now,* Nov/Dec 2000, p 28

10.42. W.T. Choate and J.A.S. Green, Modeling the Impact of Secondary Recovery (Recycling) on U.S. Aluminum Supply and Nominal Energy Requirements, *Light Metals 2004,* TMS, p 913–918

10.43. J.H.L. Van Linden, *Melt Loss Reduction in Recycling Processes,* Katholieke Universiteit Leuven, May 1988

10.44. E.L. Rooy, Aluminum Dross—Liability into Opportunity, *Light Met. Age,* June 1995

10.45. Hertwich Engineering, "Advanced Remelt Technology at Hydro Aluminum Portalex," Aluminum 2004 (Chicago, IL)

10.46. New Aluminum Melting Process Provides Energy Savings, *JOM,* June 2006, p 6

10.47. *Metal Melting Practices and Procedure for Efficiency and Effectiveness*, California Cast Metals Association, sponsored by the California Energy Commission, Summer 2001, p 20

10.48. *Metal Melting Practices and Procedure for Efficiency and Effectiveness,* California Cast Metals Association, sponsored by The California Energy Commission, Summer 2001. This and its companion booklet, *Foundry Energy Use Study and Conservation Manual,* provide detailed furnace design information and operational survey data on melting and furnace practices for nonferrous metals in the State of California

APPENDIX A

Energy Intensity of Materials Produced in the United States

ALUMINUM, by several measures, is one of the most energy-intensive (Btu/lb) materials produced, ranking at the top among the major products of the United States. Table A.1 shows a comparison of the on-site energy requirements for several major products manufactured in the United States. These values do not include the energy content of fuels used as materials or the generation and transmission losses associated with electricity production.

The data in Table A.2 are the on-site process energy requirements to produce the corresponding products plus the energy content of fuels used as materials and the generation and transmission losses associated with electricity production. Examples of the energy content of fuels used as materials (feedstock energy) are: 22,681 Btu/lb is the feedstock energy for ethylene; wood products are commonly assumed to have no feedstock value because they are renewable

Table A.1 Energy requirements to produce materials in the United States (2002)

Material	Btu/yr	Btu/lb	lb/yr	Data sources
Paper and paper board	2.75×10^{15}	15,590(a)	1.76×10^{11}	Ref A.1
Gasoline	2.41×10^{15}(b)	2659	9.07×10^{11}	Ref A.2, A.3
Iron and steel	1.79×10^{15}(b)	8700	2.06×10^{11}	Ref A.4, A.5
Ethylene	4.22×10^{14}(b)	8107	5.21×10^{10}	Ref A.6, A.7
Aluminum (primary ingot)	2.66×10^{14}(b)	44,711	5.95×10^{9}	Appendix E and Table 10.3 in Chap 10
Distillate	3.63×10^{14}(b)	990	3.67×10^{11}	Ref A.2, A.3
Ammonia	3.53×10^{14}(b)	12,150	2.90×10^{10}	Ref A.6, A.7
Propylene	4.30×10^{13}(b)	1351	3.18×10^{10}	Ref A.6, A.7
Jet fuel	1.46×10^{14}(b)	990	1.47×10^{11}	Ref A.2, A.3
Coal	1.29×10^{14}(b)	60	2.14×10^{12}	Ref A.8
Benzene	2.03×10^{13}(b)	1255	1.62×10^{10}	Ref A.6, A.7

(a) Calculated from Btu/yr value and lb/yr value. (b) Calculated from Btu/lb value and the value for lb/yr

Table A.2 Gross energy requirements to produce materials in the United States (2002)

Material	Btu/yr	Btu/lb	lb/yr	Data sources
Paper and paper board	2.75×10^{15}	15,590(a)	1.76×10^{11}	Ref A.1
Gasoline	2.22×10^{15}(b)	2659	8.34×10^{11}	Ref A.2, A.3
Iron and steel	1.79×10^{15}(b)	8700	2.06×10^{11}	Ref A.4, A.5
Ethylene	1.74×10^{15}	33,470(a)	5.21×10^{10}	Ref A.6, A.7
Propylene	8.90×10^{14}	27,978(a)	3.18×10^{10}	Ref A.6, A.7
Ammonia	6.95×10^{14}	23,917(a)	2.90×10^{10}	Ref A.6, A.7
Benzene	5.43×10^{14}	91,317(a)	5.95×10^{9}	Ref A.6, A.7
Aluminum (primary ingot)	5.38×10^{14}(b)	33,305	1.62×10^{10}	Appendix E and Table 10.3 in Chap 10
Distillate	3.63×10^{14}(b)	990	3.67×10^{11}	Ref A.2, A.3
Coal	1.29×10^{14}(b)	60	2.14×10^{12}	Ref A.8
Jet fuel	1.46×10^{14}(b)	990	1.47×10^{11}	Ref A.2, A.3

(a) Calculated from Btu/yr and lb/yr data. (b) Calculated from Btu/lb and lb/yr data

resources; and petroleum-calcined coke used as a raw material in aluminum production has a feedstock energy of 15,250 Btu/lb.

REFERENCES

A.1. "Manufacturing Energy Consumption Survey," Energy Information Agency, DOE

A.2. H. Brown, *Energy Analysis of 108 Processes*, Drexel University, 1996

A.3. "Petroleum Annual 2003," Energy Information Agency, DOE

A.4. J. Stubbles, *Energy Use in U.S. Steel Industry: A Historical Perspective and Future Opportunities*, Sept 2000, p 23

A.5. *Steel Industry of the Future, FY 2004 Annual Report*, DOE, p 1

A.6. *Energy and Environmental Profile of the U.S. Chemical Industry*, DOE-OIT, May 2000, p 28, 30, 32

A.7. *Guide to the Business of Chemistry*, American Chemical Council, 2003, p 31

A.8. "Coal Industry Annual 2003," Energy Information Agency, DOE

APPENDIX B

Energy Values for Energy Sources and Materials

CALORIFIC ENERGY VALUES are the energy content inherent to the material. Except for pure materials, for example, propane, these values vary depending on the raw materials used and the final product formulations.

Process energy is a measure of the energy required to manufacture the material. Process energy values are also variable and depend on equipment efficiency estimates and system boundaries. These values for crude-oil-derived products are variable and depend on the specific crude processed, refinery configuration, local product specifications, and refinery efficiency. Process energy values from the California Energy Commission include the energy for processing as well as factors for raw material, for example, delivery of crude to refiners and transporting fuels to their point of use.

Tacit or gross energy is the sum of the calorific and process energy values.

Table B.1 Energy values for energy sources and materials

		Calorific energy values (primary energy)			Process energy values (secondary energy)			Tacit energy	
Energy source	Input unit	Btu per common unit	Btu per input unit	Data source	Btu per common unit required to produce	Btu per input unit	Data source	Btu per Appendix E input unit	Input unit
Fuels and fuels used as materials									
Fuel oil, medium	kg	139,000 Btu/gal	43,380	Ref B.1	5000 Btu/gal	1321	Ref B.2	44,700	kg
Fuel oil, light	kg	150,000 Btu/gal	46,813	Ref B.1	5000 Btu/gal	1321	Ref B.2	48,130	kg
Diesel	L	5,670,000 Btu/bbl	35,667	Ref B.3	5200 Btu/gal	1374	Ref B.4	37,040	L
Kerosene	L	5,670,000 Btu/bbl	35,667	Ref B.5	5200 Btu/gal	1374	Ref B.4	37,040	L
Gasoline	L	5,198,000 Btu/bbl	32,698	Ref B.5	12,000 Btu/gal	3170	Ref B.4	35,870	L
Natural gas	m^3	1027 Btu/ft^3	36,268	Ref B.5	30 Btu/ft^3	1059	...	37,330	m^3
Bituminous/subbituminous	kg	11,110 Btu/lb	24,493	Ref B.6	60 Btu/lb	132	Ref B.7	24,630	kg
Calcined coke	kg	15,250 Btu/lb	33,620	Ref B.8	179 Btu/lb	395	Ref B.7	34,010	kg
Pitch	kg	6,065,000 Btu/bbl	38,152	Ref B.7	18 Btu/lb	40	Ref B.7	38,190	kg
Green coke	kg	14,200 Btu/lb	31,305	Ref B.8	Btu/lb	500	Ref B.2	31,810	kg
Propane	L	3,824,000 Btu/bbl	24,055	Ref B.9	Btu/lb	. . .	. . .	24,050	L
Coal	kg	10,240 Btu/lb	22,574	Ref B.5	60 Btu/lb	132	. . .	22,710	kg
Electricity									
Electric	kWh	1 kWh	3412	Ref 10	Hydroelectric utility	0	Ref B.10	3412	kWh
Avg. U.S. electric(a)	kWh	1 kWh	3412	Ref 10	Average U.S. grid	6478	Tacit Btu	9890	kWh
Primary Al electric(b)	kWh	1 kWh	3412	Ref 10	Primary Al electric(b)	4158	Tacit Btu	7570	kWh
Coal-fired electric	kWh	1 kWh	3412	Ref 10	Coal-fired utility	6891	Tacit Btu	10,300	kWh

(a) Tacit Btu accounts for generation and transmission energy losses for average U.S. generation. (b) Tacit Btu for electrolysis and anode manufacture is lower then general electricity because of the large (50.2%) hydro component.

Table B.2 Fuels consumed worldwide (kWh/yr) for U.S. aluminum processing, excluding electricity and coke feedstock energy

	Mining	Refining	Anode with feedstock	Electrolysis	Primary casting	Secondary casting	Hot rolling	Cold rolling	Extrusion	Shape casting	Total
No process (secondary energy) inputs for fuels											
kWh/yr	8.13×10^{8}	1.79×10^{10}	1.23×10^{9}	4.59×10^{8}	1.98×10^{9}	7.13×10^{9}	8.62×10^{8}	6.93×10^{8}	2.20×10^{9}	5.84×10^{9}	3.91×10^{10}
With process (secondary energy) inputs for fuels											
kWh/yr	8.45×10^{8}	1.84×10^{10}	1.27×10^{9}	4.72×10^{8}	2.04×10^{9}	7.37×10^{9}	8.87×10^{8}	7.17×10^{8}	2.26×10^{9}	6.01×10^{9}	4.02×10^{10}
Percent process (secondary energy) fuel contribution											
	4	3	3	3	3	3	3	3	3	3	3

Table B.3 Impact of electric tacit conversion factors on kWh/yr consumed worldwide for U.S. aluminum production

Unit	Mining	Refining	Anode with feedstock	Electrolysis	Primary casting	Secondary casting	Hot rolling	Cold rolling	Extrusion	Shape casting	Total
Average U.S. aluminum processing facilities											
7570 Btu of energy required to produce 1 kWh of on-site electrical power for an average U.S. aluminum smelting, anode, and primary casting facility											
9890 Btu of energy required to produce 1 kWh of on-site electrical power for all other U.S. aluminum facilities											
kWh/yr	8.60×10^{8}	1.99×10^{10}	1.50×10^{10}	1.13×10^{11}	3.55×10^{9}	8.37×10^{9}	2.75×10^{9}	3.17×10^{9}	2.75×10^{9}	6.04×10^{9}	1.76×10^{11}
Percent increase relative to facilities with 100% hydroelectric energy resource											
%	4	8	4	188	42	12	83	106	16	3	86
U.S. aluminum processing facilities with 100% electrical power generated by hydroelectric utilities											
3412 Btu of energy required to produce 1 kWh of on-site electrical power from hydroelectric resources for all U.S. aluminum facilities											
kWh/yr	8.26×10^{8}	1.84×10^{10}	1.44×10^{10}	3.94×10^{10}	2.50×10^{9}	7.47×10^{9}	1.50×10^{9}	1.54×10^{9}	2.37×10^{9}	5.85×10^{9}	9.43×10^{10}
U.S. aluminum processing facilities with 100% electrical power generated by coal-fired utilities											
10,300 Btu of energy required to produce 1 kWh of on-site electrical power from coal resources for all U.S. aluminum facilities											
kWh/yr	8.62×10^{8}	2.01×10^{10}	1.52×10^{10}	1.29×10^{11}	3.77×10^{9}	8.51×10^{9}	3.01×10^{9}	3.51×10^{9}	2.82×10^{9}	6.04×10^{9}	1.93×10^{11}
Percent increase relative to facilities with 100% hydroelectric energy resource											
%	4	10	5	228	51	14	100	128	19	3	105

The process energy (secondary energy) values contribute approximately 3% additional energy to the total carbon-base fuel energy of the U.S. aluminum industry. This includes the energy expended worldwide for the production of aluminum in the United States. The United States does not mine bauxite but refines it to supply 47% of the alumina needed and imports the remaining 53% of alumina required for aluminum production.

Tacit electric energy conversion factors vary significantly depending on the fuel source used to generate electricity. The large variation in electric tacit conversion factors creates the need for careful analysis when comparing different studies. A completely coal-fired electric-based smelting operation requires 2.5 times greater tacit energy than a completely hydroelectric smelting operation. This report uses an average tacit based on the electric fuel sources of aluminum smelting operations. (Appendix C, Table C.1)

REFERENCES

B.1. Cogeneration Technologies, Houston, TX. http://www.cogeneration.net (accessed July 2007)

B.2. Data derived from Drexel University study.

B.3. Energy Information Agency, DOE

B.4. California Energy Commission

B.5. *Annual Energy Outlook 2004*, Energy Information Agency, Jan 2004, p 262

B.6. *Coal Industry Annual 2000*, Energy Information Administration, Jan 2002, p 284. This report uses an average value for bituminous and subbituminous coals

B.7. "Energy and Environmental Profiles, Aluminum and Mining," DOE-ITP, p 52

B.8. Data from Mid-Continent Coal & Coke Co.

B.9. *Emissions of Greenhouse Gases in the United States 1987–1992*, Energy Information Agency, Oct 1994, Appendix A

B.10. International Organization for Standardization

APPENDIX C

Hydroelectric Distribution and Electrical Energy Values

SIGNIFICANT ENERGY is consumed in the generation and transmission of electricity. Tacit electric energy conversion factors include the energy associated with production, processing, and distribution of the primary energy sources used in the production of electricity. Tacit values vary significantly depending on the source of energy. In the United States, 39.4% of aluminum smelting capacity uses hydroelectric power.

The electrical power used in primary aluminum production worldwide is shown in Tables C.2 and C.3.

Table C.1 Electric tacit energy and emission data for fuels used for aluminum production

Electrical energy source	U.S. primary aluminum capacity(a)		Heat rate (2005)(b), Btu/kWh	Carbon emission coefficient(c), Mt/Qbtu
	Metric tons	%		
Hydro	1,633,696	39.4	3412	0
Coal	2,415,826	58.2	10,303	21.25
Oil	8245	0.2	10,090	19.08
Natural gas	31,120	0.8	7937	12.50
Nuclear	60,112	1.4	10,420	0
Total	**4,149,000**	**100.0**		
	Weighted averages based on aluminum capacity		7570	12.51
	Average U.S. grid(d)		9890	13.56

(a) The distribution of electrical sources for U.S. primary aluminum capacity is obtained by subtracting the Canadian capacity of 2.827 million metric tons (2002), which is 100% hydroelectric, from the North American totals presented in Table C.2.
(b) Heat rate values are derived using the primary energy values in Appendix B, Table B.1, and the net generation and consumption values listed in the *Annual Energy Review 2005*, Energy Information Administration (see Table C.4).
(c) Table C.4 of this appendix.
(d) 9890 tacit Btu/kWh is used here for all electricity consumed in U.S. aluminum processing operations. This value overstates the actual aluminum-related tacit values (which is 7570 Btu/kWh), but it is more useful for comparing the aluminum industry to other U.S. manufacturing operations.

Table C.2 Energy sources of electrical power in 2002 (electrical power used in gigawatt hours)

Electric energy source	Africa, GWh	North America		Latin America, GWh	Asia, GWh	Europe, GWh	Oceania, GWh	Total, GWh	Grand total, %
		GWh	%						
Hydro	6444	50,312	64	32,207	3030	29,969	7380	129,342	50
Coal	13,443	27,248	35	0	10,080	15,773	24,120	90,664	35
Oil	0	93	0	0	109	1279	0	1481	1
Natural gas	38	351	0	1343	16,336	6199	0	24,267	9
Nuclear	189	678	1	0	0	12,672	0	13,539	5
Total	**20,114**	**78,682**	**100**	**33,550**	**29,555**	**65,892**	**31,500**	**259,293**	**100**

Source: International Aluminum Institute

Table C.3 Sources of supply of electrical power in 2002 (electrical power used in gigawatt hours)

Electric source of supply	Africa, GWh	North America GWh	North America %	Latin America, GWh	Asia, GWh	Europe, GWh	Oceania, GWh	Total, GWh	Grand total, %
Self-generated	0	27,702	35	4386	28,228	9193	1244	70,753	27
Purchased—grid	20,114	39,015	50	28,863	1327	56,698	23,237	169,254	65
Purchased—other	0	11,965	15	301	0	1	7019	19,286	7
Total	**20,114**	**78,682**	**100**	**33,550**	**29,555**	**65,892**	**31,500**	**259,293**	**100**
Self-generated other purposes	0	351	...	253	2662	1	0	3267	...

Source: International Aluminum Institute

Table C.4 Average U.S. grid connection tacit energy

Source	U.S. electricity net generation in 2005 (Ref C.1) Billion kWh	%	Units consumed (Ref C.2)	Heat content	Energy consumed, 10^9 Btu	Btu/kWh net generation(a)
Coal/	2014.2	50	1051.2×10^6 short tons	20,620,000 Btu/ton (Ref C.3)	20,752,000 (Ref C.4)	10,303
Petroleum	...	...	21.9×10^6 bbl distillate	5,825,000 Btu/gal (Ref C.3)	127,626(b)	...
	...	...	146.8×10^6 bbl residual	6,287,000 Btu/bbl (Ref C.3)	923,126(b)	...
	...	...	3.7×10^6 bbl other liquids	5,670,000 Btu/bbl (Ref C.3)(c)	20,786(b)	...
	...	...	8.5×10^6 short tons coke	14,200 Btu/lb (Ref C.5)	241,684(b)	...
Subtotal	**121.9**	**3**			**1,230,000 (Ref C.4)**	**10,090**
Natural gas	751.5	19	6466×10^9 ft^3	1019 Btu/ft^3 (Ref C.3)	5,965,000 (Ref C.4)	7937
Nuclear	780.5	19	10,000 Btu/kWh (Ref C.6)		8,133,000 (Ref C.4)	10,420
Hydroelectric	265.1	7	10,117 Btu/kWh(d)		2,682,000 (Ref C.4)	10,117
Renewable	92.1	2	957×10^{12} Btu		1,004,000 (Ref C.4)	10,901
Other	3.7	0	27×10^{12} Btu		22,000 (Ref C.4)	5946
Totals	**4029.0**	**100**			**39,851,000 (Ref C.4)**	**9891**

Note: Transmission and distribution losses = 1.31 of 13.78 quad Btu generated. Therefore 13.78/12.58, or 1.1051 times delivered-to equals generation (*Annual Energy Review 2005*, p 223). (a) 10^9 Btu ÷ Billion kWh. (b) Heat content × units consumed. (c) Assumes diesel fuel is the majority of other fuels. (d) Assumes that a substitute for hydropower is equivalent to the average of the other sources of electricity

Significant energy is consumed in the generation and transmission of electricity. Tacit electric energy conversion factors (Btu/kWh) include the energy associated with production, processing, and distribution of the primary energy sources used in the production of electricity. Tacit values vary significantly depending on the source of energy to produce electric power.

REFERENCES

C.1. *Annual Energy Review 2005*, Table 8.2a Electricity net generation, p 228. Note: Net generation includes transmission losses.

C.2. *Annual Energy Review 2005*, Table 8.5a Consumption of combustible fuels for electricity generation. p 242

C.3. *Annual Energy Outlook 2004*, Energy Information Agency, Jan 2004, p 262

C.4. *Annual Energy Review 2005*, Table 2.1f Electric power sector energy consumption, p 43

C.5. Data from Mid-Continent Coal & Coke Co. for green petroleum coke

C.6. Data from Nuclear Energy Institute

APPENDIX D

Emission Data and Calculations

Table D.1 presents the carbon dioxide equivalent emission values for the fuels and materials associated with the production of aluminum metal and aluminum products.

Table D.2 uses the units of energy input from Appendix E, Table E.1, and calculates the CO_{2e} for each energy input based on the CO_{2e} values for fuels presented in Table D.1.

Table D.3 uses the units of energy input from Appendix E, Table E.2, and calculates the carbon dioxide equivalent (CDE) (CO_{2e}) emissions for each energy input base on the CDE values for fuels presented in Table D.1.

Table D.4 provides comparison of the ore-to-metal carbon dioxide equivalent emissions for a modern Hall-Heroult cell to those for improved Hall-Heroult and alternative technologies.

Table D.1 Carbon dioxide equivalent emission coefficients for fuels associated with aluminum production

Source	Appendix E input unit	Btu per input unit	Carbon emission coefficient(a), Mt/Qbtu	Carbon emission coefficient source	Percentcarbon (Ref D.1)	API gravity (Ref D.1)	Density, (Ref D.1)' lb/gal	Million Btu/bbl (Ref D.1)	Kilogram carbon dioxide equivalent per input unit
Fuel									
Fuel oil, heavy (#6)	kg	...	21.49	Ref D.1	85.7	17.0	...	6.287	...
Fuel oil, medium	kg	44,700	20.72	(b)	...	...	...	...	3,40
Fuel oil, light (#2)	kg	48,130	19.95	(c)	86.3	33.9	7.064	5.825	3.52
Diesel	L	37,040	19.95	Ref D.1	86.3	35.5	7.064	5.825	2.71
Kerosene	L	37,040	19.72	Ref D.2	86.1	41.4	...	5.670	2.68
Gasoline	L	35,870	19.34	Ref D.2	86.6	58.6	...	5.253	2.54
Natural gas	m^3	37,330	14.47	Ref D.2	...	...	...	...	1.98
Bituminous/subbituminous	kg	24,630	25.81	Ref D.1	...	N/A	...	...	2.33
Calcined coke	kg	34,010	27.85	(d)	...	...	...	...	3.47
Pitch	kg	38,190	20.62	(e)	85.8	25.6	...	...	2.89
Green coke	kg	31,810	27.85	Ref D.2	92.3	N/A	9.543	6.024	3.25
Propane	L	24,050	17.20	Ref D.1	...	...	...	...	1.52
Coal	kg	22,710	25.76	Ref D.2	...	N/A	...	...	2.15
Electricity									
Hydroelectric	kWh	3412	0.0	Ref D.2	...	...	...	...	0.00
Average U.S. electric	kWh	9891	13.56	Ref D.2	...	...	...	...	4.92×10^{-1}
Primary Al electric	kWh	7620	12.51	(f)	...	...	...	...	3.49×10^{-1}
Coal-fired electric	kWh	10,303	21.25	Ref D.2	...	...	...	...	8.03×10^{-1}

(a) Mt/Qbtu, million metric tons per quadrillion Btu (10^6 metric tons/10^{15} Btu).
(b) Medium fuel oil values are the average of the light and heavy fuel oil values.
(c) Diesel and light fuel oil is assumed to have the same carbon coefficient in Ref D.1.
(d) Green coke and calcined coke are assumed to have the same carbon coefficient.
(e) Pitch is assumed to have the same carbon coefficient as asphalt, reported in Ref D.1.
(f) Primary aluminum carbon dioxide equivalents are based on the industry's mix of fuels (Appendix C, Table C.1).

Table D.2 Carbon dioxide equivalent (CO_{2e}) emissions associated with primary aluminum production

Input/output		Mining		Refining		Anode		Electrolysis		Primary aluminum			
Rich soil	kg	1150		...		...		...		Cumulative CO_2 emissions from manufacturing the raw materials required to produce aluminum		Total CO_2 emissions from manufacturing the raw materials and electrolytic reduction required to produce aluminum	
Bauxite	kg	1000		2640		...		...					
Alumina	kg	...		1000		...		1930					
Calcined coke	kg	...		...		820		...					
Pitch	kg	...		...		231		...					
Green coke	kg	...		...		85		...					
Anode	kg	...		...		1000		446					
Fluorides	kg	...		...		...		18.6					
Aluminum	kg	...		...		...		1000					
Ratio to aluminum		Bauxite 5.10		Alumina 1.93		Anode 0.45		Aluminum 1.00					
Energy inputs per 1000 kg		**Units**	**kg equivalent of CO_2**	**Units**	**kg equivalent of CO_2**	**Units**	**kg equivalent of CO_2**	**Units/kg All**	**kg equivalent of CO_2**	**Units/kg Al**	**kg CO_2/kg Al**	**Units/kg Al**	**kg CO2/kg Al**
Fuel oil, medium	kg	1.160	3.94	93.200	3.17×10^2	3.790	1.29×10	...		187.48	6.37×10^2	187.477	6.37×10^2
Fuel oil, light (#2)	kg	...		...		0.814	2.87	4.310	1.52×10	0.36	1.28	4.673	1.65×10
Diesel	L	4.370	1.18×10	1.670	4.52	0.110	2.98×10^{-1}	1.840	4.99	25.54	6.92×10	27.378	7.42×10
Kerosene	L	...		...		...		...		...	0.00	...	0.00
Gasoline	L	0.274	6.97×10^{-1}	0.024	6.05×10^{-2}	0.046	1.17×10^{-1}	0.285	7.25×10^{-1}	1.46	3.72	1.748	4.45
Natural gas	m^3	...	..	225	4.46×10^2	97.1	1.92×10^2	7.63	1.51×10	477.56	9.46×10^2	485.187	9.61×10^2
Bituminous/ subbituminous	kg	...	8.59	2.00×10	...	...	...	...	16.58	3.86×10	16.579	3.86×10	
Calcined coke	kg	...	...	0.00134	4.65×10^{-3}	820	2.85×10^3	...	...	365.72	1.27×10^3	365.723	1.27×10^3
Pitch kg		...	...	...	...	231	6.67×10^2	...	...	103.03	2.97×10^2	103.026	2.97×10^2
Green coke	kg	...	...	...	...	85	2.76×10^2	...	...	37.91	1.23×10^2	37.910	1.23×10^2
Propane	L	...	...	...	...	0.136	2.06×10^{-1}	2.72	4.13	0.06	9.20×10^{-2}	2.781	4.22
Coal	kg	...	...	...	...	...	...	...	...	...	...	...	...
Electric	kWh	0.4	1.97×10^{-1}	109	5.36×10	266	1.31×10^2	15,400	7.57×10^3	1.57×10^4	1.63×10^2	31,066.00	7.74×10^3
Total per 1000 kg													
Total kg CO_{2e}			1,67 × 10		8.40×10^2		4.13×10^3		7.61×10^3		3.55×10^3		1.56×10^4
Manufacturing portion kg CO_{2e}			1.67×10		8.40×10^2		3.39×10^2		7.61×10^3		1.86×10^3		8.33×10^3
Feedstock portion CO_{2e}			...		...		3.79×10^3		...		1.69×10^3		7.32×10^3
						Volatile organic compound	4.99×10^{-1}						
kg CO_{2e}/kg Al			0.085		1.622		1.84		7.61		3.55		15.65
Total material production in the U.S. for aluminum manufacturing													
metric tons			0	4,419,000	1,128,000		2,530,000		...		...		
Total U.S. production carbon dioxide emissions													
metric tons CO_{2e}			0		3,713,552		4,659,146		19,263,471		9,245,134		28,508,605
Total material production worldwide for use in the U.S. for aluminum manufacturing													
metric tons			12,391,000		4,883,000		1,128,000		2,530,000		...	...	...
Total worldwide production emissions													
Metric tons CO_{2e}			2[illegible]4,938		4,103,479		4,659,146		19,263,471		11,092,844		30,356,315
kg CO_{2e}/ kg AL			...		...		...		...		4.38		12.00

Table D.3 Carbon dioxide equivalent (CO_{2e}) emissions associated with aluminum production

Input/output		Primary ingot casting		Secondary ingot casting		Hot rolling		Cold rolling		Extrusion		Shape casting	
Electrolysis metal	kg	1020		...		...		...		...		...	
Alloy additives	kg	...		21		...		...		...		...	
Scrap	kg	...		676		...		...		...		...	
Primary ingot	kg	...		346		498		474		559		...	
Secondary ingot	kg	...	Yield	...	Yield	723	Yield	722	Yield	885	Yield	2200	Yield
Product	kg	1000	98%	1000	96%	1000	82%	1000	84%	1000	69%	1000	45%
Energy inputs per 1000 kg		**Units**	**kg equivalent of CO_2**	**Units**	**kg equivalent of CO_2**	**Units**	**kg equivalent of CO_2**	**Units**	**kg equivalent of CO_2**	**Units**	**kg equivalent of CO_2**	**Units**	**kg equivalent of CO_2**
Fuel oil, medium	kg	...		...		...		...		...		...	...
Fuel oil, light (#2)	kg	17.4	61	42.3	149	...		...		0.002	0	...	...
Diesel	L	0.184	...	31.464	85	0.040	0.109	0.038	0	0.417	1	...	...
Kerosene	L	...		...		0.007	0	...		0.028	0	...	...
Gasoline	L	0.075	0	12.757	32	0.007	0	2.150	5	0.044	0	...	...
Natural gas	m^3	51.8	103	126.0	250	33.4	66	24.8	49	103.0	204	240.0	475
Bituminous/ subbituminous	kg	...	...	...	...	...	...	...	...	...	...	...	...
Calcined coke	kg	...	...	...	...	...	...	...	...	...	...	...	...
Pitch	kg	...	...	...	...	...	...	...	...	...	...	...	...
Green coke	kg	...	...	...	...	...	...	...	...	...	...	...	...
Propane	L	0.798	1	1.941	3	0.034	0	0.238	0	4.460	7	...	...
Coal	kg	...	...	...	...	...	...	...	...	11.200	24	...	...
Electric	kWh	211	103.77	115.000	56.56	265	130.33	349	171.64	93	45.74	4	1.98
Total per 1000 kg													
	CO_{2e}		2.69×10^2		5.76×10^2		1.97×10^2		2.27×10^2		2.82×10^2		4.77×10^2
Total material production in the U.S. for aluminum manufacturing													
	Metric tons	2,480,400		2,990,000		2,421,300		2,421,300		1,826,000		2,289,000	
Total U.S. production carbon dioxide emissions													
	Metric tons CO_{2e}		667,297		1,721,552		476,218		548,896		514,666		1,092,602
Kilograms of carbon dioxide equivalent per kilogram of product													
	kg/kg Al	12.27		0.58		0.20		0.23		0.28		0.48	

Table D.4 Carbon dioxide equivalent (CO_{2e}) emissions associated with new aluminum production technologies

Mineral material	Typical modern Hall-Heroult cell operating at 95% current efficiency and with an ACD = 4.5 cm (1.8 in.)		Typical modern Hall-Heroult cell operating at 95% current efficiency, retrofitted with a sloped and wetted cathode surface, aluminum sump, and a reduced ACD (ACD = 2.0 cm or 0.8 in.)		Inert anode operating at 95% current efficiency with oxygen polarization differences and ACD = 2.0 cm (0.8 in.) using wetted cathode technology		Carbothermic reduction with a reaction efficiency of 95% and a furnace efficiency of 85%. Electric furnace operating on the average U.S. grid		Chloride reduction of kaolinite clays, electrolysis current efficiency 95%, reaction efficiency 95%, and heating efficiencies 85%. On-site electric furnace and electrolysis cell operating on the average U.S. grid	
	kWh/kg Al	$kgCO_2$/kg Al	kWh/kg Al	$kgCO_2$/kg Al	kWh/kg Al	$kgCO_2$/kg Al	kWh/kg Al	$kgCO_2$/kg Al	$kgCO_2$/kg Al	kWh/kg Al
Tacit energy required	8.21	1.71	8.21	1.71	8.21	1.71	8.21	1.71	8.81	1.83
On-site reaction energy requirements										
Reaction thermal	...	...	...	...	...	...	7.71	3.79	–1.90	–0.93
Furnace losses	...	...	...	...	...	...	1.36	0.67	0.40	0.20
Reaction electrolysis	3.76	1.85	3.76	1.85	6.90	3.39	...	...	6.48	3.19
Cell ohmic	10.67	5.25	7.62	3.75	6.20	3.05	...	...	2.93	1.44
Total reaction energy	14.43	7.10	11.38	5.60	13.11	6.45	9.07	4.46	7.91	3.89
Anode-related emissions										
Anode manufacturing	0.61	0.15	0.61	0.15	0.76	0.27	...	...	...	...
Anode use										
Anode reaction	0.33 kg	1.22	...	1.22	...	...	...	...	...	...
Anode excess (air burning)	0.11 kg	0.30	...	0.30	...	...	...	...	...	...
Total anode	...	1.68	...	1.68	...	...	...	...	...	...
Process-related emissions										
Carbon reactant	...	...	...	...	...	...	0.66kg	2.45	0.89 kg	3.27
Perfluorocarbon	...	2.20	...	0.55	...	0.55	...	0.00	...	0.00
Total CO_2 emissions										
Mineral material	...	1.71	...	1.71	...	1.71	...	1.71	...	1.83
Reactions	...	7.10	...	5.60	...	6.45	...	4.46	...	3.89
Carbon	...	1.68	...	1.68	...	0.00	...	2.45	...	3.27
Process	...	2.20	...	0.55	...	0.55	...	0.00	...	0.00
Total	...	12.68	...	9.53	...	8.70	...	8.61	...	8.99

ACD, anode-cathode distance

REFERENCES

D.1. *Emissions of Greenhouse Gases in the United States 1987–1992,* Energy Information Agency, Oct 1994, Appendix A

D.2. *Emission of Greenhouse Gases in the United States 2000,* Energy Information Agency, Nov 2001, p 140

APPENDIX E

U.S. Energy Use by Aluminum Processing Area

THE ENERGY INPUT UNITS and raw material quantities used in Tables E.1 and E.2 are from the *Life Cycle Inventory Report for the North American Aluminum Industry*, published by the Aluminum Association in November 1998. The report is the result of extensive surveying and contains the best and most complete industry performance information of any recent study. The report is consistent with ISO procedures and was favorably peer-reviewed by groups outside the industry. The Aluminum Association has provided permission to use these data.

The conversion of input units to common energy units are based on the values presented in Appendix B, Table B.1.

Table E.3 lists the theoretical minimum energy requirements to produce raw materials and aluminum products. It also shows the process efficiency for each operation. Mining, refining, anode manufacturing, and electrolysis minimum energy values are based on the net chemical changes that result from these processes. Primary casting, secondary casting, and shape casting minimum energy values are based on the energy required to produce molten pure aluminum at 775 °C (1425 °F). The minimum energy requirements for rolling and extrusion operations are estimated from their yield and on-site energy consumption values in Table E.2, and from an overall assumed electric and hydraulic system efficiencies of 75% and thermal furnace and heating efficiencies of 50%.

Table E.4 lists U.S. production quantities and energy and tacit energy consumption associated with producing aluminum products within the United States. The United States does not consume energy to produce metallurgical bauxite and consumes only 56% of the energy required for alumina. This report distinguishes between the worldwide energy values and the U.S. values in order to measure the impact of market or process changes to the energy demands within the United States. Table E.9 lists the worldwide values. The theoretical magnitude of the potential U.S. energy savings can be measured by subtracting the theoretical energy requirements from the actual energy consumption numbers.

Table E.5 shows the total energy values consumed by the U.S. aluminum industry and the energy savings possible if:

- It was possible to reduce energy consumption to the theoretical limit
- Energy were reduced by 30%
- Electrolysis electrical energy were reduced from 15 to 11 kWh/kg of aluminum

Table E.6 shows 94% reduction in energy consumption associated with secondary metal production. Recycling aluminum essentially recaptures all the energy associated with mining, refining, and smelting.

Table E.7 shows the electric energy consumption of the U.S. aluminum industry and compares it to the total electric energy produced in the United States.

Electrolysis and process heating are the two most energy-intensive process areas of the aluminum industry. Electrolysis accounts for 43% of all on-site and 66% of all tacit energy use. Process heating accounts for 27% of all on-site and 14% of all tacit energy use in the industry.

Table E.9 lists the worldwide production quantities, energy, and tacit energy consumption associated with producing aluminum products

Table E.1 Materials and energy associated with primary metal electrolysis

Input/output		Mining		Refining		Anode		Electrolysis		Primary aluminum			
Rich soil	kg	1150		...		...		...		Cumulative energy requirement for manufacturing the raw materials required to produce aluminum		Total energy requirement for manufacturing the raw materials and electrolytic reduction required to produce aluminum	
Bauxite	kg	1000		2640		...		...					
Alumina	kg	...		1000		...		1930					
Calcined coke	kg	...		...		820		...					
Pitch	kg	...		...		231		...					
Green coke	kg	...		...		85		...					
Anode	kg	...		...		1000		446					
Fluorides	kg	...		...		...		18.6					
Aluminum	kg	...		...		...		1000					
Ratio to aluminum		Bauxite 5.10		Alumina 1.93		Anode 0.45		Aluminum 1.00					
Energy inputs per 1000 kg of material produced													
		Units	Btu	Units	Btu	Units	Btu	Units	Btu	Units/kg Al	Btu/kg Al	Units/kg Al	Btu/kg Al
Fuel oil, medium	kg	1.160	50,321	93.200	4,043,057	3.790	164,412	...	...	187.48	8,132,824	187.477	8,132,824
Fuel oil, light	kg	...	...	...	...	0.814	38,106	4.310	201,766	0.36	16,995	4.673	218,761
Diesel	L	4.370	155,865	1.670	59,564	0.110	3923	1.840	65,627	25.54	910,873	27.378	976,501
Kerosene	L	...	...	...	...	...	...	...	...	...	...	...	...
Gasoline	L	0.274	8959	0.024	778	0.046	1504	0.285	9319	1.46	47,822	1.748	57,141
Natural gas	m^3	...	...	225	8,160,344	97.1	3,521,642	7.63	276,726	477.56	17,320,117	485.187	17,596,843
Bituminous/ subbituminous	kg	...	...	8.59	210,396	...	...	...	...	16.58	406,064	16.579	406,064
Coke	kg	...	...	0.00134	45	...	...	...	...	0.003	87	0.003	87
Propane	L	...	...	...	...	0.136	3271	2.72	65,429	0.06	1,459	2.781	66,888
Coal	kg	...	...	...	...	...	...	...	...	...	...	...	...
Electric	kWh	0.4	1365	109	371,908	266	907,592	15,400	52,544,800	331.04	1,129,522	15,731.04	53,674,322
Tacit electric	kWh	1.2	...	316	...	771	...	44,638	...	1088	...	45,726	...
Total energy inputs (energy per kg of material output)													
Btu/kg			217		12,846		4640		53,164	Note: The on-site or primary electrical energy requirement to produce aluminum from one accounts for 66% of the total energy requirement.			
kWh/kg			0.06		3.76		1.36		15.58				
Tacit kWh/kg			0.07		4.08		1.90		44.82				
Ratio adjusted energy inputs (energy per kg of aluminum)													
Btu/kg			1103		24,793		2070		53,164		27,966		81,129
kWh/kg			0.32		7.27		0.61		15.58		8.20		23.78
Tacit kWh/kg			0.34		7.87		0.85		44.82		9.06		53.89
Feedstock energy per kg Al (energy inherent in fuels used as materials)						**Feedstock ratio adjusted**							
Btu/kg			...		...		17,414		...		...		98,544
kWh/kg			...		...		5.10		...		...		28.88
Tacit kWh/kg			Total tacit (Table 10.9) (process 0.90 + feedstock 5.10)				6.01		...		14.22		59.05

Table E.2 Materials and energy associated with aluminum manufacturing operations

Input/output		Primary ingot casting		Secondary ingot casting		Hot rolling		Cold rolling		Extrusion		Shape casting	
Electrolysis metal	kg	1020		...		...		...		...		...	
Alloy additives	kg	...		21		...		...		...		...	
Scrap	kg	...		676		...		...		...		...	
Primary ingot	kg	...		346		498		474		559		...	
Secondary ingot	kg	...	Yield	...	Yield	723	Yield	722	Yield	885	Yield	2200	Yield
Product	kg	1000	98%	1000	96%	1000	82%	1000	84%	1000	69%	1000	45%
Energy inputs per 1000 kg		**Units**	**Btu**	**Units**	**Btu**	**Units**	**Btu**	**Units**	**Btu**	**Units**	**Btu**	**Units**	**Btu**
Fuel oil, medium	kg	...	...	...	...	...	...	...	...	...	...	...	...
Fuel oil, light	kg	17.4	814,554	42.3	1,981,346	...	...	...	...	0.002	109	...	...
Diesel	L	0.184	6563	31.464	1,122,230	0.040	1441	0.038	1362	0.417	14,873	...	...
Kerosene	L	...	...	...	...	0.007	249	...	...	0.028	988	...	...
Gasoline	L	0.075	2,439	12.757	417,115	0.007	213	2.150	70,301	0.044	1442	...	...
Natural gas	m^3	51.8	1,878,693	126.0	4,569,793	33.4	1,211,358	24.8	899,451	103.000	3,735,624	240.000	8,704,367
Bituminous/ subbituminous	kg	...	...	...	...	...	...	...	...	...	...	...	...
Coke	kg	...	...	...	...	...	...	...	...	...	...	...	...
Propane	L	0.798	19,196	1.941	46,692	0.034	818	0.238	5725	4.460	107,285	...	...
Anthracite	kg	...	...	...	...	...	...	...	...	11.200	252,829	...	...
Electric	kWh	211	719,932	115	392,380	265	904,180	349	1,190,788	93	317,316	4	13,750
Total energy inputs (energy per kg of output)													
	Btu/kg	3,441		8530		2118		2168		4430		8718	
	kWh/kg	1.01		2.50		0.62		0.64		1.30		2.56	
Tacit energy inputs (energy per kg of output)													
	Btu/kg	4887		9548		3870		4462		5145		8999	
	kWh/kg	1.43		2.80		1.13		1.31		1.51		2.64	

Table E.3 Theoretical minimum energy requirements to produce raw materials and aluminum products

		Mining	Refining	Anode with feedstock	Electrolysis	Primary casting	Secondary casting	Hot rolling	Cold rolling	Extrusion	Shape casting
	kWh/kg	...	0.14	12.97	5.99	0.33	0.33	0.31	0.33	0.44	0.33
Ratioed to Al	kWh/kg Al	...	0.27	4.40	5.99	0.33	0.33	0.31	0.33	0.44	0.33
Current practice on-site process efficiency, %		...	4	77	38	33	13	50	52	34	13

Table E.4 U.S. energy use in the production of domestic aluminum

		Mining	Refining	Anode with feedstock	Electrolysis	Primary casting	Secondary casting	Hot rolling	Cold rolling	Extrusion	Shape casting	Total energy
Material production within the Unitd States												
	Metric tons	0	4,419,000	1,128,000	2,530,000	2,480,400	2,990,000	2,421,300(a)	2,421,300(a)	1,826,000(a)	2,289,000	...
On-site energy consumed n production for U.S. aluminum industry												
	kWh/kg	...	3.76	12.80	15.58	1.01	2.50	0.62	0.64	1.30	2.56	
Ratioed to Al(a)	kWh/kg Al	...	7.27	5.71	15.58	1.01	2.50	0.62	0.64	1.30	2.56	
Total	kWh/yr	...	1.66×10^{10}	1.44×10^{10}	3.94×10^{10}	2.50×10^{9}	7.47×10^{9}	1.50×10^{9}	1.54×10^{9}	2.37×10^{9}	5.85×10^{9}	9.17×10^{10}
Proportion of U.S. energy used, %		...	18	16	43	3	8	2	2	3	6	100
						80%	total primary U.S. proportion					
Total tacit energy consumed in production for U.S. aluminum industry												
	kWh/kg	...	4.08	13.34	44.82	1.43	2.80	1.13	1.31	1.51	2.64	...
Ratioed to Al(a)	kWh/kg Al	...	7.87	6.01	44.82	1.43	2.80	1.13	1.31	1.51	2.64	...
Total	kWh/yr	...	1.80×10^{10}	1.50×10^{10}	1.13×10^{11}	3.55×10^{9}	8.37×10^{9}	2.75×10^{9}	3.17×10^{9}	2.75×10^{9}	6.04×10^{9}	1.73×10^{11}
Proportion of U.S. energy used, %		...	10.4	8.7	65.5	2.1	4.8	1.6	1.8	1.6	3.5	100
						86.7%	total primary U.S. proportion					
Theoretical magnitude of opportunities for U.S. energy savings												
	kWh/yr	...	1.60×10^{10}	-1.82×10^{8}	2.43×10^{10}	1.68×10^{9}	6.48×10^{9}	7.56×10^{8}	7.36×10^{8}	1.57×10^{9}	5.09×10^{9}	5.64×10^{10}
Tacit	kWh/yr	...	1.74×10^{10}	4.23×10^{8}	9.83×10^{10}	2.73×10^{9}	7.37×10^{9}	2.00×10^{9}	2.36×10^{9}	1.96×10^{9}	5.28×10^{9}	1.38×10^{11}

(a) Data based on year 2000 values, the last year numbers specific to U.S. production are available. After 2000, reported numbers are based on North American data, not specifically U.S. production. After 2000, U.S. production declined with the general economy. It is estimated that in 2005, production rose to 2000 levels.

within the United States. These values provide a full measure of the energy associated with producing aluminum and products. The actual energy consumed within the United States is lower because no aluminum metallurgical bauxite is mined in the United States, and approximately 95% of the alumina required is refined in the United States.

Table E.10 divides energy uses in aluminum casting and semifabrication manufacturing operations into three areas: electric, heating, and miscellaneous fuels. The following percentages of energy use are derived from the energy values listed in Table E.2. Fuel oils, natural gas, and propane are assumed to be associated with heating (thermal operations). Miscellaneous fuels include all other fuels, and it is assumed that these are fuels used in nonheating operations, such as transportation and collection of materials. (Note: Rounding causes some totals to differ from 100%.)

The total U.S. primary aluminum capacity operates on over 39% hydroelectric power compared to the average U.S. electric grid use of 7% hydroelectric power. It should be noted that, in reality, the industry has lower tacit energy values than reported in the other tables of this appendix. Using the average U.S. grid connection is the best way to compare various industries.

Table E.5 Total U.S. aluminum industry energy consumption and potential savings

U.S. aluminum industry		kWh/yr	Quads/yr	MW	Households	bbl crude per year
Total energy use	On-site	9.17×10^{10}	0.31	10,500	10,500,000	54,000,000
	Tacit	1.73×10^{11}	0.59	19,800	19,800,000	101,900,000
Theoretical requirement	On-site	3.53×10^{10}	0.12	4100	4,100,000	20,800,000
	Tacit	3.53×10^{10}	0.12	4100	4,100,000	20,800,000
Energy efficiency	On-site	39%	...	...	...	...
	Tacit	20%	...	...	...	...
Theoretical opportunity	On-site	5.64×10^{10}	0.19	6400	6,400,000	33,200,000
	Tacit	1.38×10^{11}	0.47	15,700	15,700,000	81,100,000
Practical goal (vision)	On-site	2.75×10^{10}	0.09	3150	3200	16,200,000
	Tacit	5.19×10^{10}	0.18	5940	6000	30,600,000
Electrolysis savings of 4 kWh/kg	On-site	1.01×10^{10}	0.03	1155	1200	6,000,000
	Tacit	2.25×10^{10}	0.08	2563	2600	13,300,000

Table E.6 Energy impact of recycling

Energy saved with recycling
Tacit energy values
60.5 kWh/kg to produce primary metal ingot
2.8 kWh/kg to produce secondary metal ingot
5.0% secondary-to-primary energy
2,990,000 kg of secondary metal produced in 2003
1.72×10^{11} kWh/yr energy saved 2003
0.59 quads saved per year 2003
19,700 MW saved 2003

Table E.7 U.S. electric on-site energy consumption in the aluminum industry

	Refining	Anode with feedstock	Electrolysis	Primary casting	Secondary casting	Hot rolling	Cold rolling	Extrusion	Shape casting	Total electric energy
Metric tons 2000	4,419,000	1,128,000	2,530,000	2,480,400	2,990,000	2,421,300	2,421,300	1,826,000	2,289,000	
kWh/kg	0.109	0.266	15.400	0.211	0.115	0.265	0.349	0.093	0.004	
kWh/yr	4.82×10^{8}	3.00×10^{8}	3.90×10^{10}	5.23×10^{8}	3.44×10^{8}	6.42×10^{8}	8.45×10^{8}	1.70×10^{8}	9.22×10^{6}	4.23×10^{10}
									quad	0.14
Total U.S. Al industry electrical use	4.23×10^{10}	kWh/yr								
On-site	4826	MW								
	4,826,000	households	Net U.S. generation: 3.858×10^{12}							
Tacit	49,900,000	bbl crude	Aluminum % of U.S. net: 1.1%							

Source: Energy Information Agency, DOE, office of Coal, Nuclear, Electric, and Alternate Fuels; http://www.eia.doe.gov/cneaf/electricity (accessed July 2007)

Table E.8 Smelting and heating fractions of total U.S. aluminum industry energy consumed

Smelting fraction of total U.S. aluminum industry energy consumed			Total industry	Percent of industry
Electrolysis				
On-site	kWh/yr	3.94×10^{10}	9.17×10^{10}	43
Tacit	kWh/yr	1.13×10^{11}	1.73×10^{11}	66

Aluminum metal process heating/melting fraction of total U.S. aluminum industry energy consumed

			Rolling						
		Casting	Hot	Cold	Extrusion	Shape casting	Total process heating	Total industry	Percent of industry
	Btu/kg Al	9244	1212	905	4096	8704	2.42×10^{4}		
	kWh/kg Al	2.71	0.36	0.27	1.20	2.55	7.08		
On-site	kWh/yr	1.48×10^{10}	8.60×10^{8}	6.42×10^{8}	2.19×10^{9}	5.84×10^{9}	2.44×10^{10}	9.17×10^{10}	27
Tacit								1.73×10^{11}	14

Note: The tacit increase for process heating is negligible.

Table E.9 Total worldwide production and energy consumption associated with producing aluminum in the United States

	Mining(a)	Refining(a)	Anode with feedstock(a)	Electrolysis(a)	Primary casting	Secondary casting	Hot rolling	Cold rolling	Extrusion	Shape casting	Total energy
Total worldwide production required to produce aluminum in the United States											
Metric tons 2005	12,891,000	4,883,000	1,128,000	2,530,000	2,480,400	2,990,000	2,421,300(b)	2,421,300(b)	1,826,000(b)	2,289,000	...
Total on-site energy use consumed in worldwide production for U.S. aluminum industry											
kWh/kg	0.06	3.76	12.80	15.6	1.01	2.50	0.62	0.64	1.30	2.56	...
Ratioed to Al(b) kWh/kg Al	0.32	7.27	5.71	15.6	1.01	2.50	0.62	0.64	1.30	2.56	...
Total kWh/yr	8.18×10^{8}	1.84×10^{10}	1.44×10^{10}	3.94×10^{10}	2.50×10^{9}	7.47×10^{9}	1.50×10^{9}	1.54×10^{9}	2.37×10^{9}	5.85×10^{9}	9.43×10^{10}
Percent relative to electrolysis(b)	2.1	46.6	36.6	100.0	6.3	19.0	3.8	3.9	6.0	14.8	...
Proportion of energy used,%	0.9	19.5	15.3	41.8	2.7	7.9	1.6	1.6	2.5	6.2	100.0
					80% total primary proportion						
Total tacit energy consumed in worldwide production for U.S. aluminum industry											
kWh/kg	0.07	4.08	13.34	44.8	1.4	2.8	1.1	1.3	1.5	2.6	...
Ratioed to Al(b) kWh/kg Al	0.34	7.87	6.01	44.8	1.4	2.8	1.1	1.3	1.5	2.6	...
Total kWh/yr	8.60×10^{8}	1.99×10^{10}	1.50×10^{10}	1.13×10^{11}	3.55×10^{9}	8.37×10^{9}	2.75×10^{9}	3.17×10^{9}	2.75×10^{9}	6.04×10^{9}	1.76×10^{11}
Percent relative to electrolysis	0.8	17.6	13.3	100.0	3.1	7.4	2.4	2.8	2.4	5.3	...
Proportion of energy used,%	0.5	11.3	8.6	64.5	2.0	4.8	1.6	1.8	1.6	3.4	100.0
					87% total primary proportion						

(a) Ratios to primary aluminum: Mixing, 5.10; refining, 1.93; anode with feedstock, 0.45; electrolysis, 1.00. (b) Data based on year 2000 values the last year numbers specific to U.S. production are available. After 2000, reported numbers are based on North American data, not specifically U.S. production. After 2000, U.S. production declined with the general economy. It is estimated that in 2005, production rose to 2000 levels.

Table E.10 Percent breakup of different types of energy used in aluminum production operations

Energy	Primary casting	Secondary casting	Hot rolling	Cold rolling	Extrusion	Shape casting
On-site electric,%	21	5	43	55	7	0
Heating,%	79	77	57	42	87	100
Miscellaneous fuels,%	0	18	0	3	6	0
Tacit electric,%	36	9	62	72	72	14
Heating,%	63	73	38	26	26	80
Miscellaneous fuels,%	0	17	0	3	3	6

Table E.11 U.S. energy use in the production of domestic aluminum—tacit electricity aluminum primary grid connection

	Mining	Refining	Anode with feedstock	Electrolysis	Primary casting	Secondary casting	Hot rolling	Cold rolling	Extrusion	Shape casting	Total energy
Material production within the United States											
Metric tons	0	4,419,000	1,128,000	2,530,000	2,480,400	2,990,000	2,421,300	2,421,300	1,826,000	2,289,000	...
On-site energy consumed in production for U.S. aluminum industry											
kWh/kg	...	3.76	12.80	15.58	1.01	2.50	0.62	0.64	1.30	2.56	...
Ratioed to Al kWh/kg Al	...	7.27	5.71	15.58	1.01	2.50	0.62	0.64	1.30	2.56	...
Total kWh/yr	...	1.66×10^{10}	1.44×10^{10}	3.94×10^{10}	2.50×10^{9}	7.47×10^{9}	1.50×10^{9}	1.54×10^{9}	2.37×10^{9}	5.85×10^{9}	9.17×10^{10}
Proportion of U.S. energy used, %	...	18	16	43	3	8	2	2	3	6	100
					80% total primary U.S. proportion						
Total average U.S. grid tacit energy consumed in production for U.S. aluminum industry											
kWh/kg	...	4.08	15.19	34.35	1.43	2.80	1.13	1.31	1.51	2.64	...
Ratioed to Al kWh/kg Al	...	7.87	6.77	34.35	1.43	2.80	1.13	1.31	1.51	2.64	...
Total kWh/yr	...	1.80×10^{10}	1.71×10^{10}	8.69×10^{10}	3.55×10^{9}	8.37×10^{9}	2.75×10^{9}	3.17×10^{9}	2.75×10^{9}	6.04×10^{9}	1.49×10^{11}
Proportion of U.S. energy used, %	...	12.1	11.5	58.5	2.4	5.6	1.8	2.1	1.9	4.1	100
					84.5% total primary U.S. proportion						
Theoretical magnitude of opportunities for U.S. energy savings											
kWh/yr	...	1.60×10^{10}	-1.82×10^{8}	2.43×10^{10}	1.68×10^{9}	6.48×10^{9}	7.56×10^{8}	7.36×10^{8}	1.57×10^{9}	5.09×10^{9}	5.64×10^{10}
Tacit kWh/yr	...	1.74×10^{10}	4.82×10^{8}	7.53×10^{10}	2.73×10^{9}	7.37×10^{9}	2.00×10^{9}	2.36×10^{9}	1.96×10^{9}	5.28×10^{9}	1.15×10^{11}

APPENDIX F

Theoretical Energy Data and Calculations

THE THEORETICAL MINIMUM energy requirement for producing any chemical is determined based on the net chemical reaction used to produce the product. It is defined as the energy required to synthesize a substance in its standard state from substances also in their standard states. It can be calculated by summing the reaction energies of the products minus the energies of the reactants.

This report calculates the theoretical minimum energy by assuming the reactants enter and the by-products leave the system at room temperature and that molten aluminum leaves the system at 960°C (1760 °F). This report has chosen 960°C (1233 K) as the molten metal temperature. This value is an approximation of the average operating cell temperatures of industrial cells.

Some studies assume that the gases evolved during reduction leave the system at the molten metal temperature. In these studies, the theoretical minimum is 2.5 to 3% higher than the numbers calculated here. The additional energy for heating of the emissions is shown in the individual tables. Theoretically, it is possible to capture all the energy associated with these gaseous emissions. However, there is currently no available economic means of recovering this energy. In practice, the emission gas stream is diluted with air in the cells to lower the temperature that the cell hoods and ducts are exposed to. The emission gas is collected from hundreds of hooded pots and treated before release to the atmosphere. Only a very small portion of the heat is actually absorbed and returned to the system.

The minimum theoretical energy requirement for aluminum production requires the evaluation of three energy factors: the energy required to drive the reduction reaction forward, the energy required to maintain the system at constant pressure and temperature, and the energy required to change the temperature of the reactant and/or product. The thermodynamics and chemical equilibrium of reactions are described by the equation $\Delta G = \Delta H - T\Delta S$, and the numeric values are given in Table F.6. The energy required to drive the reaction forward is the Gibbs free energy value (ΔG). The energy required to maintain system equilibrium is the difference between the heat of reaction (ΔH) and the Gibbs free energy value (ΔG), which equals the entropy term ($T\Delta S$). Because the Gibbs free energy requirement is less than the heat of reaction for alumina reduction, additional energy must be added to the system to maintain the system temperature. Otherwise, the system would cool as the reaction proceeds. (Reduction cells operate at atmospheric conditions, and no pressure change results during reduction.) The numeric values for G, H, and S are given in Table F.6.

A detailed discussion of the theoretical requirements is made in "Current and Energy Efficiency of Hall-Héroult Cells—Past, Present and Future," by Warren Haupin and William Frank, published in *Light Metal Age*, June 2002.

Table F.1 Theoretical minimum energy for Hall-Heroult carbon anode system

Reactants temp.		Products temp.	Reaction thermodynamics at 298 K					(Products-reactants) cal/g mole Al
			($2 Al_2O_3 + 3 C$) to ($4 Al + 3 CO_2$) = Net					
$2 Al_2O_3$ 25 °C		4 Al 960 °C	ΔG	−756,358	0	0	−282,779	118,395
			ΔH	−801,000	0	0	−282,155	129,711
	Carbon		ΔS	24.3	4.1	27.1	153.2	11,311
	anode		**Process theoretical minimum energy requirements**					kWh/kg
3C 25 °C	cell	$3CO_2$ 25 °C	Electrolytic work requirement (ΔG)					5.11
			Thermal energy for temperature maintenance (ΔH–ΔG)					0.49
			Thermal energy for Al at 960 °C					0.39
			Theoretical minimum					5.99
			Thermal energy for CO_2 at 960 °C					0.17

Note: Thermodynamic values for *G*, *H*, and *S* are from Table F.6. Heat capacity data are from Table F.9 and Appendix G.

Table F.2 Theoretical minimum energy for Hall-Heroult inert anode system

Reactants temp.		Products temp.	Reaction thermodynamics at 298 K				(products-reactants) cal/g mole Al
			(Al_2O_3) to ($2 Al + 1.5 O_2$) = Net				
			ΔG	−378,179	0	0	189,089
		2 Al 960 °C	ΔH	−400,500	0	0	200,250
			ΔS	12	14	74	11,155
	Inert						
Al_2O_3 25 °C			**Process theoretical minimum energy requirements**				kWh/kg
	anode						
	cell		Electrolytic work requirement (ΔG)				8.16
		$1.5O_2$ 25 °C	Thermal energy for temperature maintenance (ΔH –ΔG)				0.48
			Thermal energy for Al at 960 °C				0.39
			Theoretical minimum				9.03
			Thermal energy for O_2 at 960 °C				0.27

Note: Thermodynamic values for *G*, *H*, and *S* are from Table F.6. Heat capacity data are from Table F.9 and Appendix G.

Table F.3 Theoretical minimum energy for carbothermic reduction

Reactants temp.		Products temp.	Reaction thermodynamics at 298 K					(Products-reactants) cal/g mole Al
			($Al_2O_3 + 3 C$) to ($2 Al + 3 CO$) = Net					
Al_2O_3 25 °C		2 Al 960 °C	ΔG	−378,179	0	0	−98,424	139,878
			ΔH	−400,500	0	0	−79,247	160,626
	Carbo-		ΔS	12.2	4.1	13.5	141.9	21,347
	thermic							
			Process theoretical minimum energy requirements					kWh/kg
	reactor		Work requirement (ΔG)					6.03
3C 25 °C		3CO 25 °C	Thermal energy for temperature maintenance (ΔH–ΔG)					0.90
			Thermal energy for Al at 960°C					0.39
			Theoretical minimum					7.32
			Thermal energy for CO at 960°C					0.19

Note: Thermodynamic values for *G*, *H*, and *S* are from Table F.6. Heat capacity data are from Table F.9 and Appendix G.

Table F.4 Theoretical minimum energy for reduction of aluminum chloride

Reactants temp.		Products temp.	Reaction thermodynamics at 298 K				(products-reactants) cal/g mole Al
			($2\ AlCl_3$) to ($2\ Al + 3\ Cl_2$)= Net				
			ΔG	–300,574	0	0	150,287
	Inert	2 Al 960 °C	ΔH	–336,616	0	0	168,308
			ΔS	48.1	13.5	159.9	18,664
$2\ AlCl_3$ 25 °C	anode		**Process theoretical minimum energy requirements**				kWh/kg
	cell		Electrolytic work requirement (ΔG)				6.48
		$3Cl_2$ 25 °C	Thermal energy for temperature maintenance ($\Delta H - \Delta G$)				0.78
			Thermal energy for Al at 960 °C				0.39
			Theoretical minimum				7.65
			Thermal energy for Cl_2 at 960 °C				0.19

Note: Thermodynamic values for *G*, *H*, and *S* are from Table F.6. Heat capacity data are from Table F.9 and Appendix G.

Table F.5 Theoretical minimum energy for kaolinite/aluminum chloride reduction

Reactants temp.		Products temp.	Net aluminum reaction Carbochlorination (960 °C)								(Products-reactants) cal/g mole Al
			($3\ Al_2O_3{\cdot}2SiO_2 + 14\ C + 21\ Cl_2$) to ($6\ AlCl_3 + 6\ SiCl_4 + 7\ CO + 7\ CO_2$) = Net								
			ΔG	–2,374,053	0	0	–901,721	–884,799	–229,655	–659,819	–50,323
			ΔH	–2,531,520	0	0	–1,009,847	–942,161	–184,910	–658,363	–43,960
$7Al_2O_3{\cdot}2SiO_2$ 25 °C	Kaolinite	14 Al 960 °C	ΔS		19	1119	144	474	331	357	
			Chloride reduction (700 °C) ($2\ AlCl_3$) to ($2\ Al + 3\ Cl_2$) = Net								
	to		ΔG	–300,574		0		0			150,287
	aluminium	$14SiO_2$ 25 °C	ΔH	–336,616		0		0			168,308
	chloride		ΔS	48		14		160			18,664
			Process theoretical minimum energy requirements								kWh/kg
	to	$7CO_2$ 25 °C	Thermal work for $AlCl_3$ production								–1.90
			Electrolytic work requirement (ΔG)								6.48
	reduction		Thermal energy for temperature maintenance ($\Delta H–\Delta G$								0.78
			Thermal energy for Al at 960 °C								0.39
	system		Theoretical minimum								5.76
14 °C 25 °C		7CO 25 °C	Thermal energy for CO_2 at 960 °C								0.11
			Thermal energy for CO at 960 °C								0.25

Note: Thermodynamic values for *G*, *H*, and *S* are from Table F.6. Heat capacity data are from Table F.9 and Appendix G.

Table F.6 Thermochemistry data for elements and compounds associated with aluminum production

	Chemical Abstracts Service Registration No. (CAS RN)	Molecular weight	Chemical formula	H(s)		G(s)		S(s)		C_p (s)	
				J/mol	cal/mol	J/mol	cal/mol	J/mol K	cal/mol K	J/mol K	cal/mol K
Aluminum	7429-90-5	26.98	Al	0	0	0	0	28.3	6.764	24.35	5.82
Aluminum chloride	7446-70-0	133.34	$AlCl_3$	−704,200	−168,308	−628,800	−150,287	100.7	24.061	91.84	21.95
Corundum	1334-28-1	101.96	Al_2O_3	−1,675,700	−400,500	−1,582,300	−378,179	50.9	12.165	79	18.88
Gibbsite(a)	. . .	155.96	$Al_2O_3{\cdot}3H_2O$	−1,293,100	−309,058	−1,154,900	−276,028	. . .	. . .	. . .	. . .
Kaolinite(a)	1332-58-7	. . .	$Al_2O_3{\cdot}SiO_2{\cdot}2H_2O$	−4,119,000	−984,465	−3,793,900	−906,764	. . .	. . .	. . .	. . .
Kaolinite, meta	. . .	162.04	$Al_2O_3{\cdot}SiO_2$	. . .	−843,840	. . .	−791,351	. . .	. . .	. . .	. . .
Graphite	7440-44-0	12.01	C	0	0	0	0	5.7	1.361	. . .	. . .
Chlorine	. . .	70.91	Cl_2 (g)	0	0	0	0	222.9	53.286	. . .	. . .
Carbon monoxide	630-08-0	28.01	CO (g)	−110,523	−26,416	−137,268	−32,808	197.9	47.301	. . .	. . .
Carbon dioxide	124-38-9	44.01	CO_2 (g)	−393,513	−94,052	−394,383	−94,260	213.6	51.061	. . .	. . .
Oxygen	7782-44-7	32.00	O_2 (g)	0	0	0	0	205.0	49.003	. . .	. . .
Water	. . .	18.00	H_2O (g)	−241,826	−57,798	−228,582	−54,632	188.8	45.132	33.598	. . .
Silica dioxide	14808-60-7	60.08	SiO_2	−910,700	−217,663	. . .	−220,615	41.46	9.909	. . .	
Silicon tetrachloride	10026-04-7	169.90	$SiCl_4$ (g)	−657,000	−157,027	−617,000	−147,467	330.70	79.039	90.3	21.6

Source: D.R. Lide, Ed., *Handbook of Chemistry and Physics*, 80th ed., CRC.
(a) Source: S. K. Saxena, Ed., *Advances in Physical Geochemistry*, data courtesy of Mr. B. Hemingway at U.S.G.S.

Table F.7 Changes in heat of formation values as a function of temperature

Temperature		Heats of formation in calories per gram mole					
°C	K	Al_2O_3(a)	C(a)	Al(a)	CO_2(a)	CO(a)	O_2
25	298	−400,300	0	0	−94,050	−26,400	0
727	1000	−404,400	2310	18,710	−94,400	−26,750	9249
827	1100	−404,000	3320	21,710	−94,250	−26,900	10,515
927	1200	−403,600	3850	24,740	−94,300	−27,000	11,843
960	1233	−403,500	4030	25,750	−94,300	−27,100	12,295
1027	1300	−403,200	4390	27,790	−94,300	−27,300	13,233
1127	1400	−402,800	4930	30,850	−94,300	−27,350	14,685

(a) Source: Values from *Technical Working Group on Inert Anode Technologies*, Appendix A.9, p 11

Table F.8 Theoretical minimum energy for gibbsite dehydration

Reactants temp.		Products temp.	Gibbsite dehydration reaction				(Products-reactants) cal/g mole Al_2O_3
			$(Al_2O_3{\cdot}3H_2O)$ to	$(Al_2O_3$ +	$3H_2O)$ =	Net	
		Al_2O_3 → 25 °C	ΔH	−618,117	−400,500	−173,393	44,223
$Al_2O_3{\cdot}3H_2O$ → 25 °C	Kiln						
			Process theoretical minimum energy requirements				
		$3\ H_2O$ → 25 °C					kWh/kg alumina
							0.50

Note: Thermodynamic values for *G*, *H*, and *S* are from Table F.6. Heat capacity data are from Table F.9 and Appendix G.

Table F.9 Heat of capacity equations for gases associated with aluminium production

Standard molar heat capacity,	C=a + bT + cT^2		
cal/mol K	a	b×10^3	c×10^7
O_2	6.148	3.102	−9.23
CO	6.420	1.665	−1.96
CO_2	6.214	10.396	−35.45
Cl_2	7.576	2.424	−0.65

Source: W.F. L udar, *A Different Approach to Thermodynamics, 1967*
T in degrees K

Table F.10 Theoretical minimum energy for kaolinite dehydration

Reactants temp.		Products temp.	Reaction				(products-reactants)
		$Al_2O_3{\cdot}2SiO_2$ → 25		$Al_2Si_2O_5\ (OH)_4$	to $(Al_2O_3{\cdot}2SiO_2 + 2\ H_2O)$ =	Net	cal/g mole $Al_2O_3{\cdot}2SiO_2$
			ΔG	−906,764	−791,351		
$Al_2Si_2O_5\ (OH)_4$ → 25	Kiln		ΔH	−984,465	−843,840	−115,596	25,029
				ΔS			
							kWh/kg
		$2\ H_2O$ → 25				Theoretical minimum	0.18

APPENDIX G

Aluminum Heat Capacity and Heat of Fusion Data

Table G.1 Formulas for calculating heat capacity and heat of fusion

Heat capacity (cal/mol · K) $= A + B \cdot t + C \cdot t^2 + D \cdot t^3 + \frac{E}{t^2}$

Standard enthalpy (Kcal/mol) $= A \cdot t + \frac{(B \cdot t)}{2} + \frac{(C \cdot t^3)}{3} + \frac{(D \cdot t^4)}{4} - \frac{E}{t} + F - H$

Standard entropy (cal/mol · K) $= A \cdot \ln(t) + B \cdot t + \frac{(C \cdot t^2)}{2} + \frac{(D \cdot t^3)}{3} - \frac{E}{(2 \cdot t^2)} + G$

where t = K/1000 and A, B, C, D, E, F, G, and H are constants

Source: Standard Reference Data Program, National Institute of Standard, and Technology, http://webbook.nist.gov (accessed July 2007)

Table G.2 Formula constants for aluminum

	A	B	C	D	E	F	G	H
Solid (298 to 933.45 K, 1 atm):	6.71348	−1.29418	2.04599	0.819161	−0.066294	−2.18623	14.7968	0
Liquid (933.45 to 2790.812 K, 1 atm):	7.588681	9.40685×10^{-9}	4.26987×10^{-9}	6.43922×10^{-9}	1.30976×10^{-9}	−0.226024	17.5429	2.524381

Table G.3 Aluminum energy requirements for heating and melting

Temperature °C	Temperature K		Heat capacity at temperature, cal/mol k	Energy for step change, kWh/kg	Cumulative energy to raise from 25 °C, kWh/kg	Energy for step change, Btu/lb	Cumulative energy to raise from 25 °C, Btu/lb
25	298	Solid	5.79	0.00	0.00	...	...
660	933		7.88	0.19	0.19	286	286
660	...	Fusion	94.5 (cal/g)	0.11	...	170	...
660	933	Liquid	7.59	...	0.29	...	456
775	1048		7.59	0.04	0.33	58	515
960	1233		7.59	0.06	0.39	94	608
2000	2273		7.59	0.34	0.73	526	1134
Smelting		25 to 960 °C		Total	0.39	...	608
Furnace melting		25 to 775 °C		Total	0.33	...	515

Note: Heat capacity for solid aluminum varies significantly with temperature, whereas for molten aluminum, it is nearly constant.

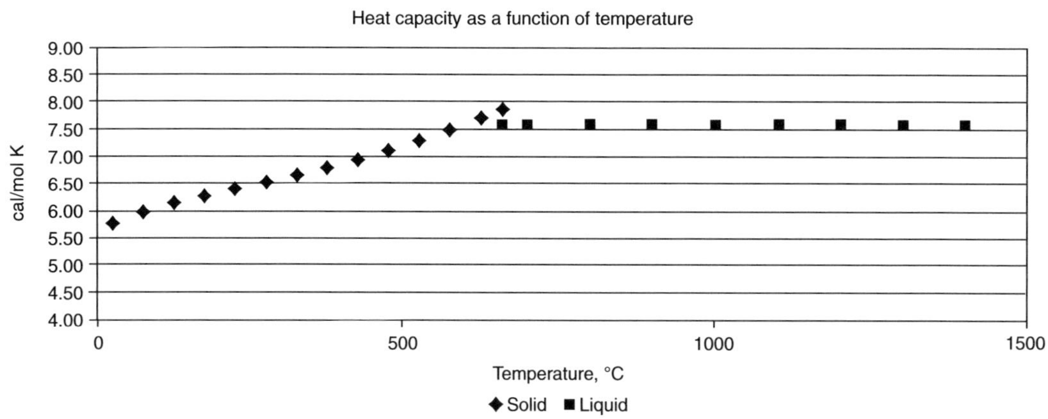

APPENDIX H

Impact of Using Different Technologies on Energy Requirements for Producing Aluminum

WETTED CATHODES allow the anode-cathode distance to be reduced. This results in a lowering of the voltage requirement for amperage to pass through the cryolite bath. The following table shows the effect of lowering the anode-cathode distance on energy consumption.

Table H.1 Impact of wetted cathode technology on primary metal electrolysis

Wetted cathode technology	Typical modern prebaked Hall-Heroult cell operating at 95% current efficiency ACD = 4.5 cm		Typical modern prebaked Hall-Heroult cell operating at 95% current efficiency and retrofitted with a wetted cathode surface and reduced ACD ACD = 3.5 cm		Typical modern prebaked Hall-Heroult cell operating at 95% current efficiency and retrofitted with a wetted cathode surface, aluminum sump, and reduced ACD ACD = 2.5 cm		Typical modern prebaked Hall-Heroult cell operating at 95% current efficiency and retrofitted with a sloped and wetted cathode surface, aluminum sump, and reduced ACD ACD = 2.0 cm	
Energy requirements	**V (dc)**	**kWh/kg Al**	**V (dc)**	**kWh/kg Al**	**V (dc)**	**kWh/kg Al**	**V (dc)**	**kWh/kg Al**
Reaction	1.20	3.76	1.20	3.76	1.20	3.76	1.20	3.76
Additional energy requirements:								
External	0.15	0.47	0.15	0.47	0.15	0.47	0.15	0.47
Anode	0.30	0.94	0.30	0.94	0.30	0.94	0.30	0.94
Anode polarization	0.55	1.73	0.55	1.73	0.55	1.73	0.55	1.73
Cathode polarization	0.05	0.16	0.05	0.16	0.05	0.16	0.05	0.16
Cryolite bath	1.75	5.49	1.36	4.27	0.97	3.05	0.78	2.44
Cathode	0.45	1.41	0.45	1.41	0.45	1.41	0.45	1.41
Other	0.15	0.47	0.15	0.47	0.15	0.47	0.15	0.47
Onsite energy values								
Cell total	4.60	14.43	4.21	13.21	3.82	11.99	3.63	11.38
Energy savings, %		...		8		17		21
Anode manufacturing		0.61		0.61		0.61		0.61
Total on-site cell and anode		15.04		13.82		12.60		11.99
Energy savings, %		...		8		16		20
Tacit energy values								
Cell total	4.60	41.83	4.21	38.29	3.82	34.75	3.63	32.99
Energy savings, %		...		8		17		21
Anode manufacturing		6.01		6.01		6.01		6.01
Total on-site cell and anode		47.83		44.30		40.76		38.99
Energy savings, %		...		7		15		18

ACD, anode-cathode distance

Table H.2 Impact of inert anode and wetted cathode technology on primary metal electrolysis

Inert anode and wetted cathode technology	A Typical modern Hall-Heroult cell operating at 95% current efficiency and ACD of 4.5 cm		B Inert anode direct substitution operating at 95% current efficiency		C Inert anode operating at 95% current efficiency with polarization differences reported for oxygen generation		D Inert anode operating at 95% current efficiency, with oxygen polarization differences and an ACD of 2 cm using wetted cathode	
Energy requirements	**V (dc)**	**kWh/kg Al**	**V (dc)**	**kWh/kg Al**	**V (dc)**	**kWh/kg Al**	**V (dc)**	**kWh/kg Al**
Reaction	1.20	3.76	2.20	6.90	2.20	6.90	2.20	6.90
Additional energy requirements:								
External	0.15	0.47	0.15	0.47	0.15	0.47	0.15	0.47
Anode	0.30	0.94	0.30	0.94	0.30	0.94	0.30	0.94
Anode polarization	0.55	1.73	0.55	1.73	0.10	0.31	0.10	0.31
Cathode polarization	0.05	0.16	0.05	0.16	0.05	0.16	0.05	0.16
Cryolite bath	1.75	5.49	1.75	5.49	1.75	5.49	0.78	2.44
Cathode	0.45	1.41	0.45	1.41	0.45	1.41	0.45	1.41
Other	0.15	0.47	0.15	0.47	0.15	0.47	0.15	0.47
On-site energy values								
Cell total	4.60	14.43	5.60	17.57	5.15	16.15	4.18	13.11
Energy savings, %		...		−22		−12		9
Anode manufacturing		0.61		0.76		0.76		0.76
Total on-site cell and anode		15.04		18.33		16.92		13.87
Energy savings, %		...		−22		−13		8
Tacit energy values								
Cell total	4.60	41.83	5.60	50.92	5.15	46.83	4.18	37.99
Energy savings, %		...		−22		−12		9
Anode manufacturing		6.01		0.76		0.76		0.76
Total on-site cell and anode		47.83		51.68		47.59		38.75
Energy savings, %		...		−8		1		19

ACD, anode-cathode distance. Inert anodes by themselves require additional reaction voltage, as can be seen by comparing columns A and B. However, this additional energy requirement is offset by the elimination of carbon anode manufacturing, along with the elimination of the feedstock energy associated with carbon. The estimate for the energy associated with the manufacture of inert anode is shown in Table H.3. Column C shows the impact that results from the lower anode polarization and the ability to design to release gas more effectively. Wetted cathodes combined with inert anodes can provide additional savings, as shown in column D.

Table H.3 Estimate of energy requirement for manufacturing an inert anode

The energy requirements for manufacturing an inert anode are significantly less than the total manufacturing energy of the many consumable carbon anodes that it replaces. Below is an estimate of the energy required to manufacture an inert anode. Assuming an inert anode has a cell life of two years, the equivalent carbon energy requirements can be calculated. Using 1 cm^2 as a basis, the following can be calculated:

0.85 A/cm^2	is the typical current density for a modern Hall-Heroult cell current density.
2980 Ah/kg Al	is the theoretical amperage required to produce aluminum.
95%	is the typical efficiency of a modern Hall-Heroult carbon anode cell.
0.446 kg	of carbon are required to produce 1 kg of aluminum.

From the above data, the amount of carbon consumed and aluminum produced per cm^2 over a two-year period is calculated to be:

14.82 kg	of carbon are consumed per cm^2 over a two-year operating period.
33.23 kg	of aluminum are produced per cm^2 over a two-year operating period.

From Table 10.9:

6.01 kWh/kg of Al produced	is the tacit energy associated with carbon anode. The total energy associated with the two-year operation of the 1 cm^2 of carbon anode can be calculated.
200 kWh	of tacit energy are consumed for anode manufacture and use.
2.27 g/cm^3	is the density of carbon anode. The height of the 1 cm^2 of carbon can be calculated as below.
6529 cm	is the height of carbon anode with a 1 cm^2 base or anode face for two years of operation.

The materials under consideration for inert anodes have no inherent fuel value, as does the carbon anode. The tacit energy requirement associated with the manufacturing of an inert anode is related to the extraction of materials and the inert anode manufacturing process. Assuming that the inert anode is 5 cm thick per cm^2 of anode face and that it requires five times the total tacit energy of a carbon anode (which includes its fuel value) to manufacture, it can be calculated that:

0.76 tacit kWh/kg of aluminum will be required to produce an inert anode.

Table H.4 Estimate of energy requirement to manufacture aluminum using different technologies

Aluminum production technologies	Typical modern Hall-Heroult prebaked anode cell operating at 95% current efficiency and with ACD of 4.5 cm		Typical modern Hall-Heroult cell operating at 95% current efficiency, retrofitted with a sloped and wetted cathode surface, aluminum sump, and a reduced ACD of 2.0 cm		Inert anode operating at 95% current efficiency with oxygen polarization differences and ACD of 2.0 cm using wetted cathode technology		Carbothermic reduction with a reaction efficiency of 95% and a furnace efficiency of 80%		Chloride reduction of kaolinite clays: electrolysis current efficiency 95%, reaction efficiency 95%, and heating efficiency 85%	
Raw materials	kWh/kg Al		kWh/kg Al		kWh/kg Al		kWh/kg Al		kWh/kg Al	
									(a)	
On-site energy required	Alumina	7.59	Alumina	7.59	Alumina	7.59	Alumina	7.59	Kaolinite	8.14
Tacit energy required	Alumina	8.21	Alumina	8.21	Alumina	8.21	Alumina	8.21	Kaolinite	8.81
On-site reaction energy requirements										
Reaction thermal		. . .		. . .		. . .		7.71		−1.90
Furnace losses		. . .		. . .		. . .		1.36	(b)	0.40
	V(dc)		V(dc)		V(dc)				V(dc)	
Reaction electrolysis	1.20	3.76	1.20	3.76	2.20	6.90		. . .	2.07	6.48
Cell ohmic	3.40	10.67	2.43	7.62	1.98	6.20		. . .	0.93	2.93
Total reaction energy	4.60	14.43	3.63	11.38	4.18	13.11		9.07	3.00	7.91
Anode manufacturing		0.61		0.61		0.76		0		0.76
On-site energy savings, %		. . .	20		8		40		42	
Total on-site energy requirements including raw materials										
Raw ore materials		7.59		7.59		7.59		7.59		8.14
Reactions		14.43		11.38		13.11		9.07		7.91
Carbon/electrodes		0.61		0.61		0.76	(c)	0	(d)	0.76
Total energy savings, %		. . .	13		5		26		26	
Total tacit energy requirements including raw materials										
Raw ore materials		8.21		8.21		8.21		8.21		8.81
Reactions		41.83		32.99		37.99		26.28		24.36
Carbon/electrodes		6.01		6.01		0.76	(c)	8.41	(d)	12.10
Total energy savings, %		. . .	16		16		23		19	

ACD, anode-cathode distance.

(a) Kaolinite contains 27.2% Al by weight. Bauxite contains 45% Al by weight. Both materials contain approximately the same percentage of impurities. For this estimate, it is assumed that the processing and calcination energy of kaolinite is the same as bauxite. However, 66% more kaolinite must be processed to produce a kilogram of aluminum than bauxite. Kaolinic clays contain valuable titanium in addition to their silicon content. These materials will likely be recovered in the processing plant and account for approximately 35% of kaolinic clay content. This table allocates 65% of the total raw material energy requirement to the kaolinite material used for aluminum manufacturing.

(b) The carbochlorination reaction is exothermic (−1.90 kWh/kg Al). However, Toth Aluminum reports that, based on pilot plant experience and the challenges of capturing the offgas energy, a small quantity of energy (0.4 kWh/kg Al) is required to maintain the reaction system temperature.

(c) The carbothermic reaction requires twice the carbon as the Hall-Heroult reaction. The Hall-Heroult reaction, on a theoretical basis, requires 0.33kg C/kg Al. In this estimate, it is assumed that the carbothermic reaction utilized carbon at a 95% efficiency. Unlike a carbon anode, the carbon used does not require the energy associated with carbon anode manufacturing. Hence, only the fuel or secondary energy requirement of the carbon is used.

(d) The carbochlorination reaction requires 0.89 kg of carbon per kilogram of aluminum produced. In this estimate, it is assumed that the carbochlorination reaction utilized carbon at a 95% efficiency. It is also assumed that the energy required and fuel value of the carbon is the same on a weight basis as the carbon utilized for carbothermic reduction of aluminum (see footnote c).

APPENDIX I

Glossary

A

alumina. An oxide of aluminum (Al_2O_3), and the compound from which aluminum metal is commercially obtained.

aluminum. A versatile, silvery-white metal. When exposed to the atmosphere, aluminum rapidly forms an oxide film that prevents it from reacting with air and water. This gives it exceptional corrosion-resistant properties. Aluminum is not found in nature as a free metal, like gold, but is chemically bound to other elements. Aluminum is the most abundant metal in the Earth's crust (8.1%). Atomic number, 13; atomic mass, 26.982; melting point, 993.52 K; boiling point, 2698 K.

anode. A positively charged mass or surface that attracts negatively charged ions (anions). The anode used in the Hall-Heroult process is composed of carbon. The oxygen-containing anions react on the anode surface, releasing oxygen that consumes the carbon to form carbon dioxide.

anode-cathode distance (ACD). The geometric linear distance between the anode and the cathode is a critical measurement in an electrolytic cell. This distance affects the voltage and energy requirement of a cell.

anode effect. An aluminum-industry idiom used to describe a process upset where the anode reaction shifts from producing oxygen to fluorine, and the cell voltage increases. Anode effects are primarily the result of having insufficient alumina dissolved in the bath and available at the anode for reduction.

B

bath. An aluminum-industry idiom referring to the cryolite-based electrolyte pool in the reduction cell.

bauxite. A prime source of alumina, found as a collection of small, reddish-brown nodules in a light-brown, earthy matrix. Commercial bauxite ore contains 30 to 60 wt% alumina.

Bayer process. A process developed by Karl Bayer in 1888 that refines bauxite ore into alumina grains. It is the process currently in use worldwide.

C

calcining. The process of heating a material to a sufficiently high temperature to drive off volatile components or to oxidize the material without fusing it. The aluminum industry uses calcining in the Bayer process to produce alumina and to prepare coke for anodes.

carbon dioxide equivalents (CDE). The preferred unit of measure used to compare the impact of different greenhouse gases. It is calculated by multiplying the quantity of a greenhouse gas emission by the global-warming potential of the gas. The results are commonly expressed in terms of a million metric tons of carbon dioxide equivalent (10^6 TCDE).

carbon equivalents (CE). A unit of measure used to compare the impact of different greenhouse gases. It is calculated by multiplying the carbon dioxide equivalent by 12/44, the mass ratio of carbon to carbon dioxide.

carbothermic reduction. An alternative process to electrolytic reduction. The carbothermic process reduces alumina in a high-temperature furnace with carbon.

castings. Metal objects that are cast into a shape by pouring or injecting molten/liquid metal into a mold. This book divides castings into ingot and shape categories. Ingot castings are produced in molds of very simple cross section, and shape castings are complex structures.

cathode. A negatively charged surface that attracts positively charged ions (cations). The cathode surface in the Hall-Héroult process is the molten aluminum pad, which rests directly on the cell carbon lining. The aluminum-containing cations react on the cathode surface, releasing the aluminum as free metal.

chloride reduction. An alternative process to alumina electrolytic reduction in which aluminum chloride is used as the feed to the reduction cell.

coke. A carbon product of the crude oil refining industry. Green or raw coke contains 8 to 10% moisture and 5 to 15% volatile organic materials. Coke is calcined in thermal kilns to remove moisture and volatile organic materials.

cryolite. Na_3AlF_6, a mineral that, when molten, dissolves alumina to form aluminum and oxide ions. It is the main component used in the electrolyte bath for aluminum production.

D

dross. The material that forms on the surface of molten aluminum as it is held in a furnace. It is composed of impurities that have surfaced as a result of gas fluxing, oxidized aluminum that is the result of molten aluminum exposure to the furnace atmosphere, and aluminum that becomes entrapped in the surface material. Dross is periodically skimmed off the surface of molten aluminum and processed to recover its aluminum content.

dusting. An aluminum-industry idiom used to describe fine carbon anode particles that are lost in the electrolyte bath or atmosphere during electrolytic reduction. Dusting results in a loss of productivity.

E

electrolysis. An electrochemical process in which the charged species in an electrolyte are attracted to electrodes, where they react with the electrons of the electrical current. Positively charged ions migrate to the cathode, and negatively charged ions migrate to the anode.

electrolyte. A nonmetallic electrical conductor in which current is carried by the movement of ions.

extrusion. The process of forcing the metal ingot (or billet) to flow through a die to create a new cross section.

F

feedstock energy. These values represent the energy inherent in a fuel that is used as material. For example, aluminum production uses coke as the raw material in carbon anodes. The energy contribution of a feedstock is expressed in terms of calorific or fuel value plus the tacit/process energy used to produce the feedstock.

G

global-warming potential (GWP). Greenhouse gases differ in their abilities to trap heat. Global-warming potential is used to express the greenhouse effect of different gases in a comparable way. The heat-trapping ability of one metric ton of CO_2 is the common standard, and emissions are expressed in terms of a million metric tons of CO_2 equivalent, or 10^6 TCDE.

greenhouse gases (GHG). Atmospheric gases that contribute to climate change by increasing the ability of the atmosphere to trap heat.

H

Hall-Heroult process. An electrolytic process for reduction of alumina, developed independently by Charles Martin Hall and Paul Lewis Toussaint Héroult in 1886. This process is commonly referred to using both names, the Hall-Heroult process. It is the process used worldwide for commercial aluminum production.

I

ingot. Ingot, as used in this book, describes an aluminum casting of simple shape. It includes billets, pigs, sows, T-bar, and other simple cast semifinished shapes.

K

kilowatt-hour (kWh). A unit of energy.

L

life-cycle assessment (LCA). An internationally recognized analysis model of the impact of a product on energy, environment, economic, and social values. The LCA extends from "cradle-to-grave": from material acquisition and production; through manufacturing, product use, and maintenance; and finally, through the end of the product life in disposal or recycling. The LCA is particularly useful in ensuring that benefits derived in one area do not shift the impact burden to other places within a product life cycle.

O

on-site energy. The energy used within a facility. This is sometimes called primary energy. Electrical on-site energy is the kilowatt hours used and does not include the secondary energy required for generation and transmission of electricity. Fuel on-site energy use is based on the calorific heating value of the fuel and does not include the secondary energy required to produce and transport the fuel.

P

pad. An aluminum-industry idiom used to describe the body of molten aluminum that accumulates within the Hall-Heroult electrolytic cell.

polarization. The nonuniform concentration gradients that form near electrodes during the reduction process. The reactions occurring at the anode and the cathode create localized conditions that are different from the bulk of the bath. The reactions deplete the supply of reactants and increase the quantity of products. Additionally, in aluminum electrolysis, gas is generated at the anode, which lowers the effective bath conductivity. An electric overpotential is required to overcome the effects of polarization.

pot. An aluminum industry idiom used to describe an electrolytic cell. The term was derived from the shape of the first cells.

potline. An aluminum-industry idiom that describes the arrangement of a long row of interconnected electrolytic cells (pots).

potlining. An aluminum-industry idiom that describes the refractory and carbon materials used to line the interior of the cell (pot).

primary aluminum. Refers to aluminum metal produced directly from alumina feedstock by chemical reduction.

Q

quad. A common abbreviation for a quadrillion Btu (1 quad = 10^{15} Btu).

R

red mud. The residue of insoluble materials that results from extracting alumina from bauxite ore. It is also referred to as bauxite residue.

reduction cell. A container holding single or multiple anodes, cathodes, and an electrolytic bath used for reducing a material.

reverberatory furnace. The most commonly used furnace type in the aluminum industry. The furnace is box-shaped and consists of a steel shell with refractory lining. Fuel is fired directly into the box, either from the roof or, more typically, from the sidewall. Heat is transferred to the molten metal with convection and radiation.

rolling. A process that results in the reduction of the cross-sectional area of a metal shape as it is passed through rotating rolls.

S

secondary aluminum. Aluminum metal that is produced from recycled aluminum products and wastes.

T

tacit energy. A term used to describe an energy value that equals the combination of on-site energy (primary energy) consumption, the process energy required to produce and transmit/transport the energy source (secondary energy), and feedstock energy (energy inherent in fuels used as materials). This book uses the superscript "tf" to denote any value that includes the tacit and feedstock energy contributions. The book does not include the energy used to make the equipment or buildings that house the process steps (tertiary energy).

U

urban mining. A term that describes the large source of aluminum available through urban recycling programs as compared to bauxite mining.

V

value chain analysis. A method that captures the energy and material inputs and outputs of each processing step (link) and builds the cumulative value for each product along the chain. A value chain analysis, or "cradle-to-shipping dock" analysis, is an integral part of a life-cycle analysis.

Index

M

N

O

P

Q

R

X